Atomistic and Continuum Fracture Mechanics of Solids

Shaker A. Meguid

Atomistic and Continuum Fracture Mechanics of Solids

An Integrated Treatment

Springer

Shaker A. Meguid
Director, Mechanics and Aerospace Design Laboratory
Department of Mechanical and Industrial Engineering
University of Toronto
Toronto, ON, Canada

ISBN 978-3-031-56084-2 ISBN 978-3-031-56085-9 (eBook)
https://doi.org/10.1007/978-3-031-56085-9

This Springer imprint is published by the registered company Springer Nature Switzerland AG
The registered company address is: Gewerbestrasse 11, 6330 Cham, Switzerland

Paper in this product is recyclable.

Dedication

Affectionately dedicated to my wife Valerie
Her sacrifice was immense, my gratitude is profound!

Preface

Engineering materials contain microdefects in the form of dislocations, cracks, voids, inclusions, and inhomogeneities. The presence of such microdefects effect changes in the mechanical behavior of the host material. In addition, when these microdefects reside in close proximity to one another, they induce coupled elastic fields that determine the structural integrity of their host material. It is therefore crucial to determine the elastic fields that surround the presence of microdefects, their interactions when placed in close proximity, and the material response to these fields at varied length and time scales. This is indeed the motivation behind writing this book.

Existing fracture mechanics books are highly traditional. Typically, they characterize the behavior of microdefects using the traditional continuum mechanics approach. There exists no significant efforts that show the importance of atomistic models and how these atomistic models differ from continuum models, as well as how they can be integrated and combined to generate the bulk response of a continuum under service loads. This book is the first substantial effort that combines both atomistic and continuum in the treatment of fracture mechanics of solids in a single volume.

In this new integrated treatment of fracture mechanics, Meguid examines and establishes the relationship between the structure of materials at the atomic/molecular scales and their macroscopic fracture characteristics. Undoubtedly, this integrated treatment will enhance our understanding of the behavior of microdefects at varied length and time scales.

The book is designed in an orderly manner covering three main components in support of the title: (I) fundamentals, (II) atomistic and continuum fracture mechanics, and (III) multiscale modeling of interacting microdefects. The fundamental component comprises Chapters 1, 2, 3, and 4 covering an introduction to atomistic and continuum mechanics, molecular dynamics, elastostatics of solids, and the finite element method. The atomistic and continuum fracture mechanics component contains five chapters: defects in solids, modes of failure, the Griffith crack, linear elastic fracture mechanics, and atomistic fracture mechanics. The final component is devoted to multiscale modeling, and the analysis of interacting microdefects is covered in Chapter 10.

This book is intended for university and college seniors, postgraduates, postdoctoral, practicing engineers, and researchers working in the fields of mechanical, aerospace, nuclear, civil, material science, chemical, and electrical engineering. To enhance its utility, the book provides the readers with the derivation of the different expressions for greater understanding of the covered topics while at the same time offers a detailed conceptual account of atomistic and continuum modeling and simulations.

I take this opportunity to thank all of my past students, staff, and collaborators for sharing their knowledge, experiences, and efforts with me. Specifically, I wish to acknowledge the efforts of Dr. Ahmed Alian in multiscale modeling and Dr. M. A. N. Dewapriya in atomistic fracture mechanics. I am also indebted to my wife Valerie Meguid for her remarkable efforts with the compilation of the book, affectionate encouragement, and support throughout the different stages of this effort.

Shaker A. Meguid
Toronto, ON, Canada

Contents

Chapter 1
Atomistic and Continuum Treatments of Solids

Summary This introductory chapter highlights the importance of both atomistic and continuum treatments of solids. Additionally, it features a class of fracture mechanics problems involving varied length scales and domain decomposition. The treatment of this class of problems requires multiscale modeling techniques.

1.1 Traditional Continuum Mechanics Framework

Engineering materials contain microdefects/defects in the form of dislocations, microcracks, voids, vacancies, inclusions, and inhomogeneities. The presence of such microdefects/defects effect changes in the mechanical behavior of the host material. In addition, when these microdefects/defects reside in close proximity, they interact with one another and induce coupled elastic fields that determine the structural integrity of their host material. It is therefore crucial to determine the stress fields that surround the presence of microdefects, their interactions when placed in close proximity and the material response to these fields at varied length and time scales.

In this effort, the author provides an integrated treatment of the atomistic-continuum behavior of single and interacting microdefects in solids under elastostatic loading at varied length scales. These microdefects may take the form of nanocracks, microcracks, microvoids, vacancies, inclusions, and inhomogeneities. The author offers a fresh look at molecular and continuum mechanics fundamentals, imperfections in solids, the Griffith crack, modes of failure and diagnostics at varied length scales, linear elastic fracture mechanics, and interacting microdefects. Specifically, "Atomistic and Continuum Fracture Mechanics: An Integrated Treatment" offers a fundamental understanding of the fracture behavior of solids and the techniques adopted in treating these solids at varied length and time scales [1.1, 1.2].

Figure 1.1 shows a turbofan for aviation gas turbine engine (GTE), typically used in wide-bodied commercial aircrafts [1.3]. It utilizes a wide body fan disc as the first stage of the rotor to produce a large thrust. The fan disc is followed by a low-high pressure compressor and high-low pressure turbine stages. The fan, compressor and

S. A. Meguid, *Atomistic and Continuum Fracture Mechanics of Solids*,
https://doi.org/10.1007/978-3-031-56085-9_1

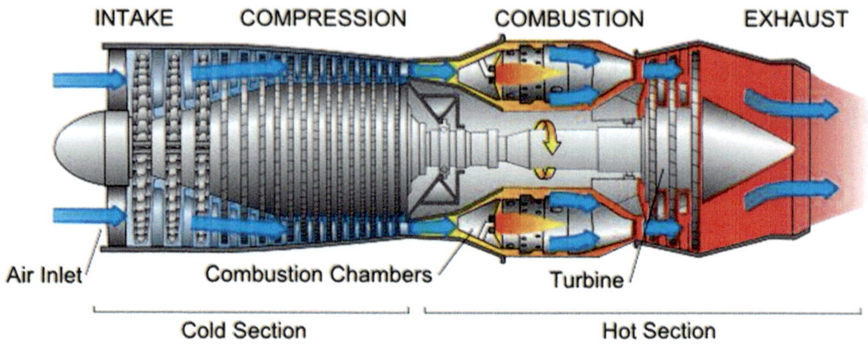

Figure 1.1 A typical aviation gas turbine engine [1.3]

turbine discs are equipped with a number of blades and operate at high rotational speeds (2500–15,000 rev/minute) under extreme operating thermomechanical loads to extract high thrust and increase engine efficiency. The engine is equipped with a containment ring surrounding these rotating discs to safeguard against any potential release due to the failure/fracture of the attached blades within the confines of the fan disc causing maximum damage to the engine.

It has been customary to adopt the traditional continuum framework for decades to describe the fracture environment of deformable solids. This ignores the atomistic nature of these solids. Recent advances in computational mechanics has led to a number of atomistic modeling techniques, with molecular dynamics being the most dominant approach. For example, engine components, subassemblies and the entire assembly of GTEs are mostly designed using a continuum approach that involves theoretical, computational, experimental mechanics using functional prototypes techniques. The most dominant computational techniques adopted are finite element, finite volume method, finite difference and Monte Carlo. These techniques have served us well in the past and will continue to serve us well in the future.

Let us examine two examples to demonstrate what we mean by continuum modeling using the finite element method. For the first example, let us focus our attention to the stress state in the firtree root region in a turbine disc assembly. As can be seen in Figure 1.2, the selected sector relies on the geometrical and loading symmetry of the disc. It is clear that the model makes use of an integrated continuum in assessing the structural integrity of that system. We have also examined the structural integrity of the dovetail region in a compressor disc assembly using finite element modeling. Figure 1.3a–c depict a typical compressor disc without the blades, the finite element model of the dovetail region with an initiated and propagated inclined crack, experimental validation of fretting fatigue crack propagation and the photoelastic stress trajectories known as isochromatics. Further details regarding this work can be obtained from the archived literature.

As a second example, let us consider the hot section of GTEs, where the temperature can reach as high as 1200 °C. Typically, the surfaces of these components (e.g., combustion chambers and high pressure turbine blades) are protected by

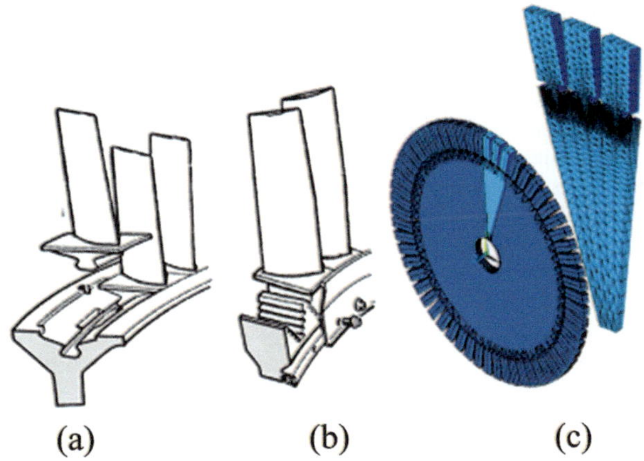

Figure 1.2 Blade attachments in gas turbine engine: (**a**) dovetail attachment in compressor disc assembly, (**b**) firtree attachment in turbine disc assembly, and (**c**) finite element modeling of a sector of a turbine disc assembly

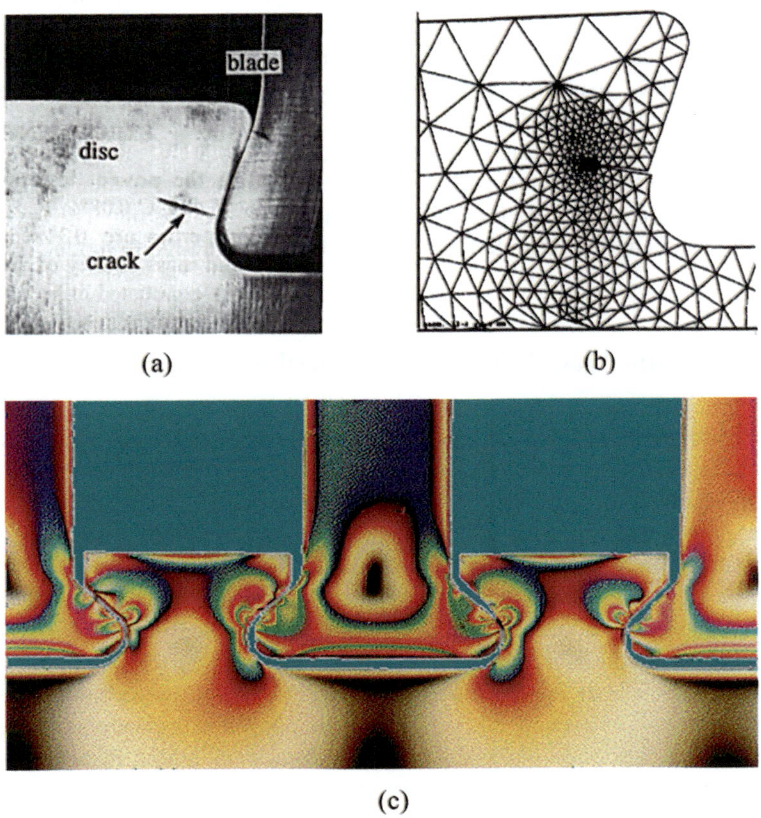

Figure 1.3 Continuum FE modeling of dovetail region in compressor disc and effect of fretting on crack initiation and growth: (**a**) Geometry, (**b**) FE mesh containing a crack, and (**c**) Photoelastic image of shear stress contours known as "isochromatics"

Figure 1.4 Constituents of thermal barrier coating used to protect turbine blades and combustors in gas turbine engines against very high temperatures (1200 °C). Interfaces are assumed straight lines in this figure

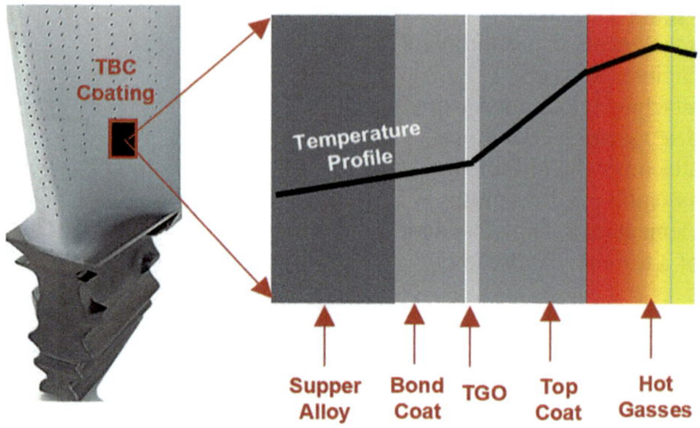

thermal barrier coatings (TBCs) and the cooling strategies of the turbine blades and disc. Failures in thermal barrier coatings (TBCs) typically result from phase transformation, the growth of a thermally grown oxide layer (TGO) and the build-up of thermomechanical residual stresses due to thermal cycling. We modeled these highly coupled processes using transient thermomechanical finite element simulations using ANSYS nonlinear platform; details can be viewed in [1.4]. The temperature profile of the deposited coating is shown in Figure 1.4, while Figure 1.5a–c show the examined geometry and its discretized finite element mesh. Our interest here is to demonstrate continuum modeling of a functionally graded coating. The two examples discussed here demonstrate the appropriate use of continuum modeling.

1.2 Atomistic Modeling Using Molecular Dynamics

Unlike continuum modeling, where bulk modeling is conducted of a system, in atomistic modeling the focus is on the discrete atomistic/molecular behavior of the atoms/molecules contained within a representative volume of that solid or fluid. Modeling the motion of a complex molecule by solving the wave functions of the various subatomic particles could lead to accurate results. The fundamental equation used in molecular dynamics (MD) formulations is Newton's second law of motion. However, there are certain assumptions that should be imposed on the system to facilitate the application of Newton's equation of motion. The first is concerned with the Born-Oppenheimer approximation, which does not allow transitions between electronic states. The second is concerned with approximating the potential energy surface, which describes the energy of a collection of atoms in terms of their positions either from quantum chemistry or from the associated force field(s). The third is concerned with applying Newtonian mechanics to atoms, where the nuclei are assumed to be quantum particles. The fourth with accurate and cost effective computing of a system with a very large degrees of freedom representing the atoms [1.5–1.7].

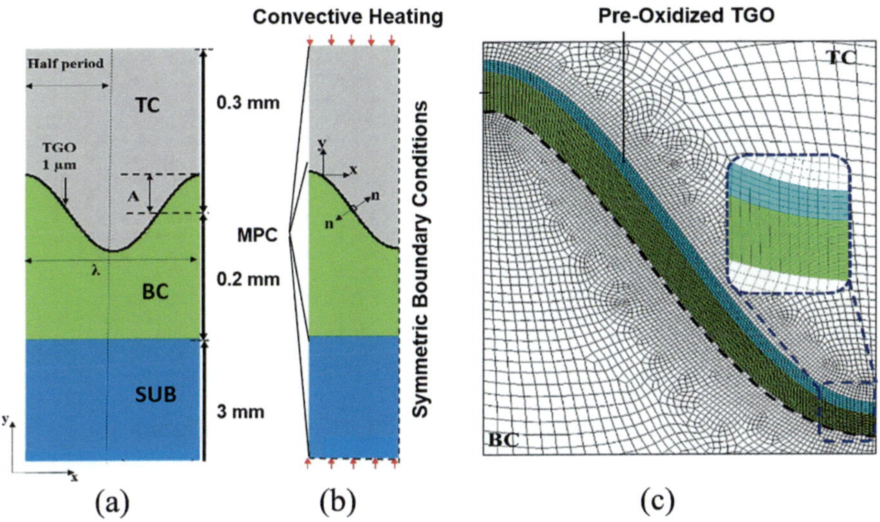

Figure 1.5 Thermal barrier coating: (**a**) A continuum description of thermal barrier coating designed to protect turbine blades from high temperatures: and (**b**) Detailed continuum geometry of coating examined, and (**c**) Discretized FE mesh. In these simulations the top coat-bond coat interface known as thermally grown oxide (TGO) is assumed to follow a sine wave not a line interface as depicted earlier in Figure 1.4 [1.4]

In continuum fracture mechanics, it has been well established that a main crack-microcrack interaction as well as defense hole systems can be effectively used to control the stress state at or ahead a crack. However, we lack understanding of the atomistic behavior of an atomistic crack interacting with a void or a cavity in close proximity to that crack.

To demonstrate atomistic modeling, let us consider the interaction between an atomistic central crack and a symmetrically located system of vacancies in the neighborhood of that crack in a graphene sheet. This should provide an insight into atomistic modeling of one-dimensional materials using MD. Linear elastic fracture mechanics establishes the conditions necessary for crack initiation, propagation and arrest due to the presence of a stress concentration feature such as a hole in close proximity to that crack. The high stress concentration of the stress concentration resulting from the hole could lead to crack initiating in the proximity of that hole. Additional loads would ultimately lead to crack propagation. If on the other hand, the hole is placed in the path of the crack tip, we may be able to limit the crack growth due to an immediate reduction in the crack tip stress field. However, it is unclear how nanoscale crack-vacancy interaction manifests itself at the atomistic level. We conducted numerous interatomic studies using molecular dynamics simulations to study the nanoscale crack vacancy behavior in a graphene sheet. Figure 1.6 shows the molecular dynamics discrete model of half the system, while Figure 1.7 depicts the outcome of the MD simulations of a graphene sheet with a

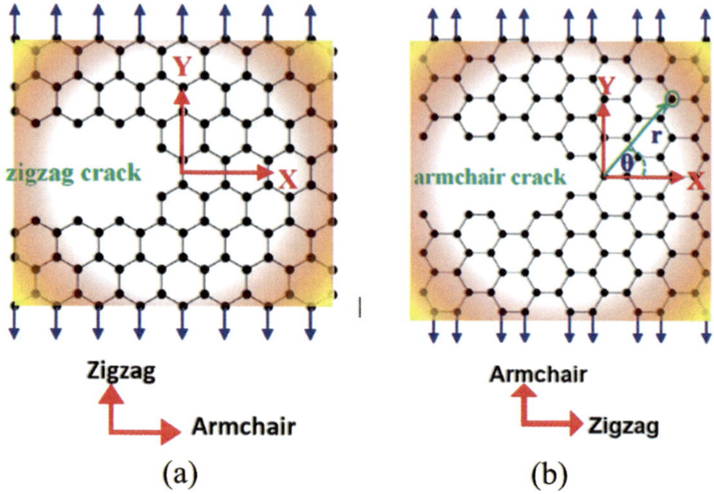

Figure 1.6 Half-MD model of a central crack interacting with atomistic vacancies in a graphene sheet using (**a**) zigzag and (**b**) a reduced armchair arrangements

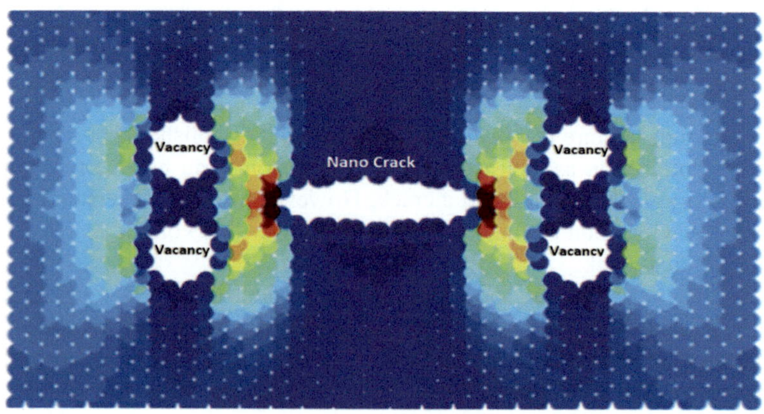

Figure 1.7 MD results of a central crack interacting with two atomistic vacancies located near its tips using zigzag arrangement

central crack interacting with two symmetrically located vacancies. The focus of that work was to determine the effect of the location and orientation of the nanoscale vacancies relative to the central crack upon the crack-tip stress field and the ability of the current system to arrest propagating cracks by the strategic placement of the nanoscale vacancies. Details of a number of case studies are discussed in details in the latter chapters. Here, we are merely interested in showing the importance of atomistic and continuum modeling and the need to link the two approaches via multiscale simulations. In view of its importance to atomistic mechanics, we provide an entire chapter covering the fundamentals of molecular dynamics.

1.3 Micromechanics of Fracture

Micromechanics of fracture is a technique adopted for treating interacting microdefects. The treatment is predominantly at the micron level. It has been advanced in recent years to a level that facilitates the description of the effective stress intensity factor of a system containing interacting microdefects. An excellent example of this approach is demonstrated in the treatment of main crack-microcrack interaction or interacting microcracks; as discussed below.

It is important to recognize the existence of multiple flaw populations as typical failure sources in solids, and the necessity to understand the failure processes associated with each significant population that exists in a specific material. This comprehension is founded on the development of a direct link between the varied micro and macro-defects. As an example, let us consider the following case of a main crack interacting with a microcrack in close proximity, as depicted in Figure 1.8.

In cases involving microcrack interactions, we adopt the complex potentials of Muskhelishvili [1.8] to describe the behavior of the system. Assuming that (2c/a) is very small and making use of the superposition principle together with an appropriate series description of the field variable, we would be able to obtain the stress intensity factor at the main crack as a result of the presence of a microcrack in close proximity. A detailed description of the concepts adopted in the treatment of interacting microdefects is given in a latter chapter.

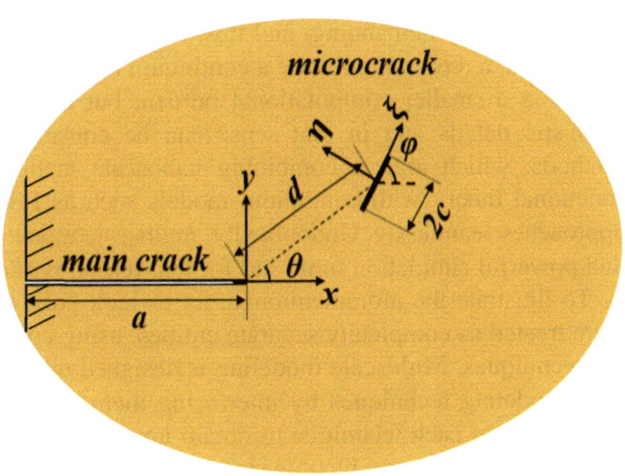

Figure 1.8 Main crack interacting with a microcrack

1.4 Multiscale Modeling

The development of different multiscale analysis methods in different fields began many decades ago and has been accelerated in the past two decades. As a result of this expansion, several review articles have been published. These articles classify the different multiscale methods not only by scope and the underlying assumptions, but also in their approach to broader questions. These broader questions consider hierarchical and concurrent multiscale approaches that seek seamless coupling at the varied length scales; see Refs [1.5–1.7, 1.9].

The recent focus of the scientific community on multiscale computer modeling techniques of nano-engineered materials stems from the desire to develop more realistic methodologies that are capable of accurately describing the varied length and time scales associated with this class of materials. Of importance is the ability to model the atomistic region using the appropriate techniques; such as quantum mechanics/molecular dynamics, and the continuum region using homogenized properties. The continuity of atomistic and continuum regions in a solid necessitates a seamless coupling between these two regions [1.7, 1.9]. This is ensured by using a transition region. In view of the large discrepancy between length and time scales in atomistic and continuum regions, the development of the transition region has been the main concern of the research community.

Figure 1.9 shows a simplified description of the different stages of multiscale modeling; starting from the atomistic structure, followed by a unit cell, then reaching to the component level and finally the assembly. Furthermore, we can easily observe from Sections 1.1 and 1.2 that MD provides an atomistic description of a physical phenomenon, but as can be seen from the degrees of freedom of the problem, MD is computationally demanding, and this, to some extent, limits its usefulness. On the other hand, a representation by a continuum model using the finite element method results in a smaller computational burden, but is incapable of characterizing the atomistic details and in that sense can be considered less accurate. Multiscale methods, which aim at combining nanoscale methods such as MD or density functional theory with continuum models such as FE or others, can join the two approaches seamlessly. Undoubtedly, multiscale modeling techniques are important and powerful simulation tools and for that reason is addressed in Chapter 10.

To illustrate the aforementioned, let us look back at Sections 1.1 and 1.2; they were treated as completely separate entities, using continuum and atomistic modeling techniques. Multiscale modeling is designed to combine atomistic with continuum modeling techniques by interfacing them. In this case, we make use of the advantages of each technique to obtain not only a greater accuracy, but also more economical, robust and flexible simulations. In most cases, a representative volume element (RVE) is used to simulate the local phenomenon. The RVE is then used in a homogenization technique to obtain an effective fiber for dispersion in a polymeric matrix, as depicted in Figure 1.9. Details of different multiscale modeling techniques will be discussed in Chapter 10.

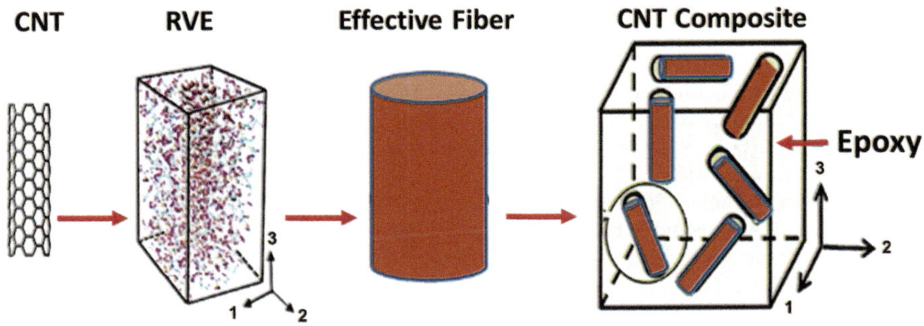

Figure 1.9 A schematic of multiscale modeling approach of CNT-reinforced nanocomposite (Not to scale)

1.5 Applicability of Multiscale Modeling

It is clear from the above that multiscale models that make use of both atomistic and continuum descriptions are essential for treating solids containing interacting microdefects. The fundamental questions that arise in the case of atomistic modeling are typically concerned with the reliability and accuracy of the atomistic description of the varied fracture phenomena experienced. Specifically, in describing a fracture phenomenon, one must seek the most appropriate interatomic potential (force field), effective range of atomic interactions, the appropriate strain rate, the thermodynamics canonical, the correct boundary conditions and time step to limit artificial effects that are motivated by computational convenience.

Atomistic modeling of fracture of solids involves numerous challenges; including the correct description of both the long-range elastic interactions and the short-range chemical interactions, supercomputers to deal with the large degrees of freedom that develop in atomistic modeling, the proper selection of the force field, and the thermodynamic environment of fracture. This means that one has to use a large atomistic model or embed the atomistic region into a flexible surrounding. Thus, one should seek a reasonable compromise between the accuracy of the treatment of atomic interactions and the time used for the solution; details of these challenges and the way to overcome them are discussed in [1.5].

One thing that can be safely stated is that atomistic modeling will play an increasingly important role in computational fracture mechanics. Atomistic fracture mechanics modeling will, in many cases, represent the starting point of multiscale modeling methods. Though classical continuum mechanics techniques may be inadequate for describing the atomistic aspects of fracture, it offers reliable solutions to real world problems with reduced complexity, effort and computational demands. They cannot be discarded in any system analysis. They served us well in the past and will continue to serve us well in the future.

References

1.1 S.A. Meguid (ed.), *Advances in Nanocomposites: Modeling, Characterization and Applications* (Springer, Cham, 2016)

1.2 S.A. Meguid, G.J. Weng (eds.), *Micromechanics and Nanomechanics of Composite Solids* (Springer, Cham, 2018)

1.3 Rolls Royce Technical Publications, *The Jet Engine*, vol 12 (Rolls Royce, Derby, 2015)

1.4 L.Y. Lim, S.A. Meguid, Thermomechanical simulations of the transient coupled effect of thermal cycling and oxidation on residual stresses in thermal barrier coatings. Ceram. Int. **48**, 3133–3147 (2022)

1.5 A.R. Alian, S.A. Meguid, A critical study of the parameters governing molecular dynamics simulations of nanostructured materials. Comput. Mater. Sci. **153**, 183–199 (2018)

1.6 A.R. Alian, S.A. Meguid, Multiscale modeling of the coupled electromechanical behavior of multifunctional nanocomposites. Compos. Struct. **208**, 826–835 (2019)

1.7 A.R. Alian, S.A. Meguid, Multiscale modeling of nanoreinforced composites, in *Advances in Nanocomposites: Modeling, Characterization and Applications*, ed. by S.A. Meguid, (Springer, Cham, 2016), pp. 1–39

1.8 N.I. Muskhelishvili, *Some Basic Problems of the Mathematical Theory of Elasticity* (Noordhoff, Leiden, 1953)

1.9 T. Belytschko, S.P. Xiao, Coupling methods for continuum model with molecular model. Int. J. Multiscale Comput. Eng. **1**, 115–126 (2003)

Chapter 2
Fundamentals of Molecular Dynamics

Summary Molecular dynamics (MD) is a powerful computer simulation tool aimed at studying the physical movement of many-particle systems such as atoms, molecules, clusters, and even macroscopic systems such as gases, liquids and solids at different times and thermodynamic states. The interacting forces between atoms and molecules are computed from the negative gradient of an approximate expression governing their interatomic potential energy. The displacement and velocity trajectories of these atoms and molecules are assumed to follow classical mechanics and are obtained by solving Newton's equation of motion numerically for a system involving many interacting atoms or molecules. MD is used extensively in mechanical engineering, material science, chemical physics, among many other disciplines.

2.1 Schrödinger's Wave Equation

In Newtonian mechanics, the dynamic state of a physical system, such as a traveling tennis ball, can be fully described in terms of its kinetics and kinematics. For example, if we are given a force F acting on a tennis ball of mass m, we will be able to calculate its acceleration and consequently the velocity and displacement vectors of that ball using Newton's second law of motion; viz.: $\vec{F} = \mathrm{m}\ \vec{a}$. The challenge is that atomistic entities do not behave like the tennis ball. In fact, at the quantum mechanics scale, atomistic entities can take one of two forms: a wave or a particle. Take light for example, according to Newton, light can be described in terms of particles, but following the extensive work of many scientists, light was deemed correct to describe it as a wave. This complicates matters because an atomic entity can simultaneously appear as a wave or as a set of particles. Schrödinger [2.1] developed what is known as the wave equation to describe the behavior of atomic entities; as follows:

S. A. Meguid, *Atomistic and Continuum Fracture Mechanics of Solids*,
https://doi.org/10.1007/978-3-031-56085-9_2

$$-\frac{\overline{h}^2}{2m}\nabla^2\psi\left(\vec{r},t\right)+V\left(\vec{r},t\right)\psi\left(\vec{r},t\right)=i\overline{h}\frac{\partial\Psi\left(\vec{r},t\right)}{\partial t} \tag{2.1}$$

Where $\overline{h}$ is the reduced Planck's constant, m an equivalent mass, V entity potential energy, $\vec{r}$ the position vector, t time and ψ is the wave function. The main interest here is to determine ψ. It must be noted that the wave function ψ contains all measurable information about a particle and describe the probability of finding that particle at a certain location, possessing a certain momentum, energy, stiffness and thermal conductivity.

Modeling the kinematics of a complex molecule by solving the wave functions of the various subatomic particles could lead to an accurate outcome. However, it would also be very challenging to program and implement, as it would require enormous computing power. As an alternative to using quantum mechanics, we are encouraged to use classical Newtonian mechanics to model our atomistic system. Clearly, this approximates what is actually happening and would therefore lead to an approximate outcome. To alleviate this problem, we use numbers derived from quantum mechanics for the coefficients in our classical formulations.

2.2 Deformation Trajectory of Atoms

The fundamental equation used in MD is Newton's second law of motion. However, as stated earlier in Chapter 1, there are certain assumptions that need to be imposed on the system to facilitate the application of the equation of motion of Newton. The first is concerned with the Born-Oppenheimer approximation, which does not allow transitions between electronic states. The second is concerned with approximating the potential energy surface, which describes the energy of a collection of atoms in terms of their positions either from quantum chemistry or from an assumed force field. The third is concerned with applying Newtonian mechanics to atoms where the nuclei are represented as quantum particles.

To obtain the trajectory of the different atoms, we must first identify their equilibrated initial positions as depicted in Figure 2.1. It is also important to know the initial velocity of each atom. The initial velocities of the atoms will be governed by the temperature of the medium; with higher initial temperature generating a greater initial atom velocity. Another key ingredient is the interatomic energy or interatomic potential energy, which describes the energy between the different atoms. Effectively, every atom will exert a force on the neighboring atoms.

In Figure 2.1, atoms are treated as spherical particles with mass m. If we to assume that the number of atoms is n, then the initial locations of these atoms will be defined by their position vectors $\vec{r}_1$, $\vec{r}_2$, $\vec{r}_3$,... $\vec{r}_n$ with $\vec{r}_{ij}$ representing the position vector connecting atoms i and j. As Figure 2.1 shows, atom 1 will exert a force $\vec{F}_{1\rightarrow 2}$ on atom 2 and vice versa. In addition, all the other atoms will exert their

Figure 2.1 Position vectors and interacting forces between atoms

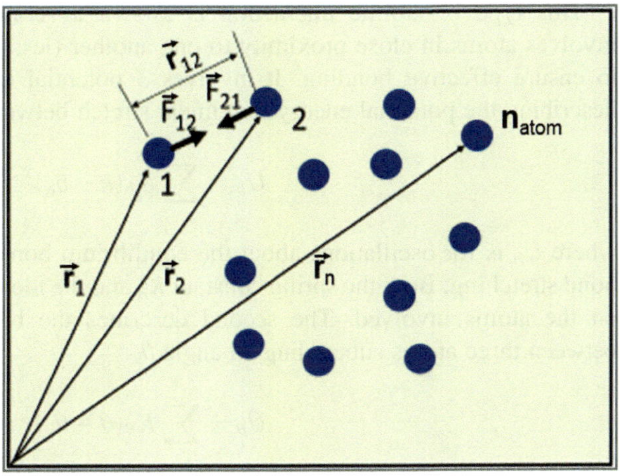

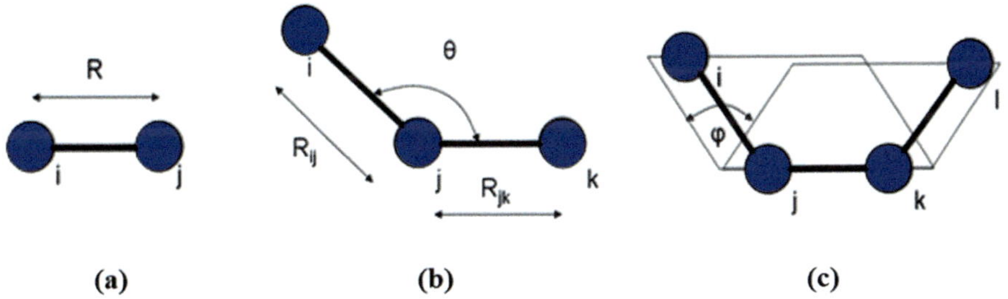

Figure 2.2 A schematic of bonded interactions between the atoms of a small molecule: (**a**) linear stretch, (**b**) bending oscillations, and (**c**) torsional rotations

own forces on atom 2, $\vec{F}_{i\to j}$ and so on. Therefore, in order to calculate the interaction force between atoms 1 and 2, we would need to know the interaction energy between these two atoms. Clearly, this interaction energy will depend upon the interatomic distance.

2.3 Force Field of Bonded Atomic Interactions

A force field is a simple equation that relates the potential energy of a system of atoms with its internal characteristics, such as atomic bond length, bond angle and dihedral angle, as depicted in Figure 2.2. The selected interatomic potential or force field for a system under investigation must be accurate for the quantum mechanical process to yield reliable results. These interatomic potentials and force fields have been developed by several researchers based on quantum mechanics calculations, and then validated by comparing their results with experimental tests (see; e.g., Brenner et al. [2.2] and LeSar [2.3]).

This type of atomic interaction is known as bonded interaction, because it involves atoms in close proximity to one another (less than 3 interatomic spacing) to ensure effective bonding. It involves 3 potential energy terms, with the first describing the potential energy of a linear stretch between two atoms

$$U_b = \sum K_b (b - b_o)^2 \tag{2.2}$$

where U_b is for oscillations about the equilibrium bond length and is the result of bond stretching. Both the spring constant K_b and the ideal bond length are dependent on the atoms involved. The second describes the bond angle potential energy between three atoms subtending an angle θ,

$$U_\theta = \sum K_\theta (\theta - \theta_o)^2 \tag{2.3}$$

where U_θ results from the bending oscillations of 3 atoms about an equilibrium bond angle. The spring constant K_θ and the ideal angle are also dependent on the chemical type of the atoms. The third potential describes the dihedral bond potential energy of 4 atoms subtending an angle φ.

$$U_\varphi = \sum K_\varphi (1 - \cos n\varphi) \tag{2.4}$$

U_φ accounts for the torsional rotation of 4 atoms about a central bond angle φ known as dihedral torsion angle. The sum of the above 3 expressions leads to the following total force field description of bonded interaction; viz.,

$$U_{bonded} = \sum U_b + \sum U_\theta + \sum U_\varphi \tag{2.5}$$

In general, the expression for the bonded atomistic potential energy of a system of atoms can be written as a many body expansion that depends on the position of one, two, three or more atoms at a time [2.3]; viz.:

$$\begin{aligned} U\left(\vec{r}_1, \vec{r}_2, \ldots, \vec{r}_N\right) &= \sum_i^N U_1\left(\vec{r}_i\right) + \sum_{i,j}^N U_2\left(\vec{r}_i, \vec{r}_j\right) \\ &\quad + \sum_{i,j,k}^N U_3\left(\vec{r}_i, \vec{r}_j, \vec{r}_{ik}\right) + \ldots \end{aligned} \tag{2.6}$$

Noting that U_1 is the one-body term (energy of the isolated atom i due to an external force field, such as the electrostatic force). U_2 is the two-body term (pair-wise interactions of the atoms i and j, such as Lennard-Jones potential. U_3 is the three-body term that describes three-body interactions (usually called many body interactions such as Tersoff and Brenner potentials). N is the number of atoms in the system, and $\vec{r}_i$ is the position vector of atom i (Brenner et al. [2.2] and Tersoff [2.4]).

However, most of the polymeric systems need a more generalized interatomic potential, which is mainly defined based on geometrical parameters, such as bond length, angles, and rotations. To treat this class of problems, many force fields have been developed; see Ref. [2.3] for details.

2.4 Force Field of Unbonded Atomic Interaction

Let us now examine the three sources of energy against the atomic spacing r_{12} for two atoms. We begin with the Pauli repulsive energy, which approaches infinity as the two atoms get very close to each other, and vdW attraction energy decreases very rapidly; as the atoms move farther apart it approaches zero, as indicated in Figure 2.3. The columbic energy could be positive or negative depending upon the charges of the atoms; being the same or opposed charges. However, vdW is always attractive. Analogous to the repulsive energy, they decrease with increased distance, but at a much slower rate, as depicted in Figure 2.3. Clearly, at smaller r, the repulsive energy dominates and at larger r, the columbic energy and vdW attractive energy dominate. In between, the balance amongst the repulsive and attractive energies is in evidence and we end up with the resultant variation of the energy between two atoms giving the general shape of the energy profile depicted in Figure 2.3. In terms of molecular dynamics terminology, the aforementioned three types of energy are known as unbonded interaction. This is opposed to bonded interaction such as metallic bonds and covalent bonding.

Figure 2.4 shows that as r approaches zero, the energy required approaches infinity, while as r increases, the interaction between the atoms diminishes and the energy approaches zero. The minimum energy position defines the equilibrium position between the atoms where the force between the atoms is zero (0 K).

The question now is how do we calculate the interaction force from the knowledge of the interatomic potential energy described above? Figure 2.4 shows two

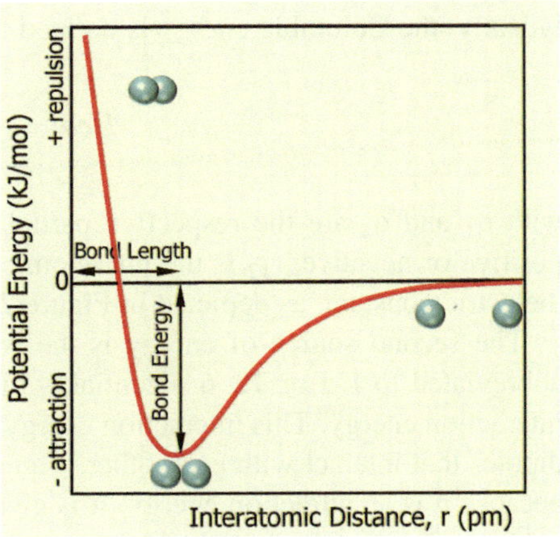

Figure 2.3 L-J potential of a pair of atoms

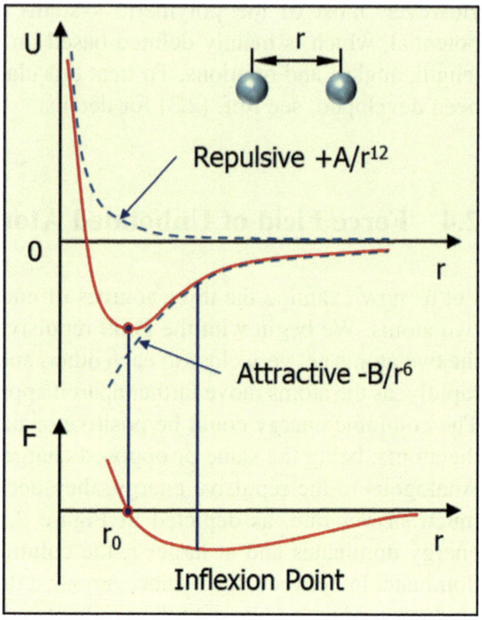

Figure 2.4 L-J Force variation with interatomic spacing between two atoms

types of energies: repulsive and attractive. The repulsive relates to a repulsive force as represented by the negative of the slope of the curve shown in Figure 2.4. In this case, the repulsive force is attempting to reach the equilibrium position that corresponds to the local minimum potential energy well. Similarly, the attraction forces are attempting to equilibrate.

Let us now focus our attention on interatomic potentials of a pair of atoms not in close proximity (>4 interatomic spacing). Atoms are essentially neutral, but when their interatomic-distances become too small or too large, changes in their potential energy takes place via electron orbitals. There exist three interatomic potentials between two interacting atoms. The first interaction is the Columbic energy. Analytically, the Columbic energy is defined as,

$$Ucol = \frac{1}{4\pi\varepsilon_o} \cdot \frac{q_1 q_2}{r_{12}} \tag{2.7}$$

with q_1 and q_2 are the respective partial atomic charges (dipoles), which can be positive or negative, r_{12} is the interatomic spacing between the atoms and ε_o is the dielectric constant, as depicted in Figure 2.5.

The second source of energy is the well-known Lennard-Jones potential. It is abbreviated to L-J or 12–6 potential, which accounts for van der Waal unbonded interaction energy. This interaction energy is caused by the presence or formation of dipoles that interact with each other. Van der Waal energy U_{vdw} is always negative because it is an attraction energy. It is given by the following expression:

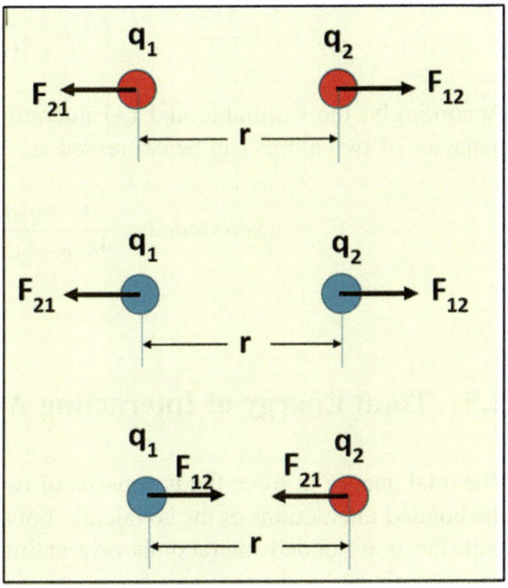

Figure 2.5 A schematic of repulsive and attractive forces associated with Columbic energy

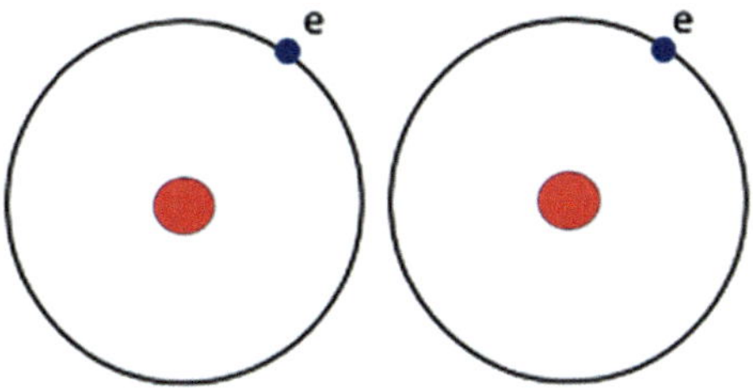

Figure 2.6 A schematic of the Pauli Exclusion Principle

$$Uvdw = -\frac{B}{(r_{12})^6} \tag{2.8}$$

with B being a constant. The third source of atomic interaction energy is the electronic repulsion and develops when the distance between the two atoms becomes too small as to facilitate coincident electrons to share the same orbitals. This would violate the *Pauli Exclusion Principle*. The Pauli Exclusion Principle states that, in an atom or a molecule, no electron orbital overlap allowed, see Figure 2.6. Therefore, one would require considerable amount of energy to bring two atoms close together reaching infinity as the atoms overlap. There is no clear mathematical representation to describe this type of repulsive energy except the energy changes quite rapidly. Leonard-Jones (LJ) expressed the repulsion energy phenomenologically as being U_{rep} as a rapidly decaying function such that,

$$Urep = \frac{A}{(r_{12})^{12}} \tag{2.9}$$

Accordingly, the Columbic and L-J interatomic potential energies that govern the behavior of two atoms can be expressed as:

$$Uunbonded = \frac{1}{4\pi\varepsilon_o} \cdot \frac{q_1 q_2}{r_{12}} + \frac{A}{(r_{12})^{12}} - \frac{B}{(r_{12})^{6}} \tag{2.10}$$

2.5 Total Energy of Interacting Atoms

The total energy in force fields consists of two parts: the first one is concerned with the bonded interactions of the covalently bonded atoms, and the second is concerned with the non-bonded interactions originating from the relatively weak long-range electrostatic and van der Waals forces; as given below:

$$U = \underbrace{\sum U_b + \sum U_\theta + \sum U_\varphi}_{U_{bonded}} + \underbrace{\sum U_{Vdw} + \sum U_{col}}_{U_{unbonded}} \tag{2.11}$$

U_{Vdw} term is for a non-bonded van der Waals (VdW) interactions, and U_{col} term is for a non-bonded electrostatic interactions (Li and Chou [2.5]).

All MD simulations must be conducted under specified ensemble conditions. These ensembles are characterized by fixed values of the following thermodynamic variables: potential energy, temperature, pressure, volume, and total number of particles. The most used ensembles in MD simulations are:

- micro-canonical ensemble: constant number of atoms, volume, and energy (N,V,E)
- isothermal-isobaric: constant number of atoms, temperature, and pressure (N,T,P)
- Canonical ensemble: constant number of atoms, temperature, and volume (N,V,T)

2.6 Computation of Force Field

The force field of interacting particles is given by the negative gradient of the potential energy, viz.,

$$\vec{F} = -\nabla U\left(\vec{r}\right) \tag{2.12}$$

Knowing the force will allow us to determine the accelerations, since,

$$\vec{F} = \mathrm{m}\,\vec{a} = m_i \frac{d^2\vec{r}}{dt^2} \tag{2.13}$$

which in turn will facilitate the computation of the velocities and positions (trajectories) of the atoms and the molecules numerically from the knowledge of their initial conditions for a properly selected time step *dt*. The selection of *dt* is very important because this would be reflected in the imposed strain rate used in the solution of the problem. It is common in the literature to adopt very large time steps to speed the computations. Unfortunately, this leads to misleading and erroneous results, because of the large strain rates imparted to the system. This issue, together with other governing parameters, are discussed near the end of this chapter.

2.7 Numerical Integration of Equations of Motion

The main purpose of MD simulations is to conduct time dependent behavior of a system by computing the current and the future positions and velocities of each atom using Newton's second law of motion. This information would be used to compute the averaged mechanical properties of the considered system.

The initial position and velocity of each atom of the system must be known at the beginning of any MD simulation. The initial velocities are randomly generated based on the provided average temperature of that system. Then, the trajectories of the atoms are determined by solving Newton's equations of motion of the interacting atoms; viz.,

$$\vec{F}_i = m_i\,\vec{a}_i \tag{2.14}$$

where $\vec{F}_i$, m_i, and $\vec{a}_i$ are the respective acting force, mass, and acceleration of atom *i*. The interatomic forces are the gradient of the total potential energy *U* of the system and are given by:

$$\vec{F}_i = -\nabla U\left(\vec{r}\right) \tag{2.15}$$

The respective velocity $\vec{v}_i$ and the acceleration $\vec{a}_i$ of each atom are the respective first and second derivatives of the displacement vector $\vec{r}_i$,

$$\vec{v}_i = \frac{d\vec{r}_i}{dt} \tag{2.16}$$

$$\vec{a}_i = \frac{d\vec{v}_i}{dt} \tag{2.17}$$

The above leads to the following differential equation of motion,

$$-\nabla U\left(\vec{r}\right) = m_i \frac{d^2\vec{r}_i}{dt^2} \tag{2.18}$$

The most popular algorithm to integrate the resulting equations of motion of the system is the Verlet algorithm (Verlet [2.6]). In this algorithm, Newton's equations of motion are approximated by a Taylor series expansion in terms of time; as follow:

$$\begin{aligned} \mathrm{r}(\mathrm{t}+\delta\mathrm{t}) = \mathrm{r}(\mathrm{t}) + \mathrm{v}(\mathrm{t})\,\delta\mathrm{t} + \frac{1}{2}\mathrm{a}(\mathrm{t})\,\delta\mathrm{t}^2 \\ + \frac{1}{6}\frac{d^3r(t)}{\mathrm{dt}^3}\,\delta\mathrm{t}^3 + \mathrm{O}\left(\delta\mathrm{t}^4\right) \end{aligned} \tag{2.19}$$

$$\begin{aligned} \mathrm{r}(\mathrm{t}-\delta\mathrm{t}) = \mathrm{r}(\mathrm{t}) - \mathrm{v}(\mathrm{t})\,\delta\mathrm{t} + \frac{1}{2}\mathrm{a}(\mathrm{t})\,\delta\mathrm{t}^2 \\ - \frac{1}{6}\frac{d^3r(t)}{\mathrm{dt}^3}\,\delta\mathrm{t}^3 + \mathrm{O}\left(\delta\mathrm{t}^4\right) \end{aligned} \tag{2.20}$$

Adding Equations (2.19) and (2.20), and moving the r(t − δt) term to the right hand side leads to:

$$r(t+\delta t) = 2r(t) - \mathrm{r}(\mathrm{t}-\delta\mathrm{t}) + a(t)\,\delta t^2 + O\left(\delta t^4\right) \tag{2.21}$$

This is the general form of the Verlet algorithm for MD, where δt is the time step. The accuracy of this approach increases significantly with the decrease in the time step, because it is a function of the fourth order of δt. The value of a(t) is determined from Equation (2.17), which depends on the location of the atom. Here, we use the positions from the previous and current time steps and acceleration of the current step to predict the trajectory of the atom. The instantaneous velocity $v(t)$ of each atom can be later calculated using the following Taylor series expansion:

$$v(t) = \frac{\mathrm{r}(\mathrm{t}+\delta\mathrm{t}) - \mathrm{r}(\mathrm{t}-\delta\mathrm{t})}{2\delta\mathrm{t}} + O\left(\delta t^3\right) \tag{2.22}$$

The accuracy of the velocity as a function of δt^3 implies that it has lower accuracy than the position, which is a function of δt^4. The kinetic energy K(t) and the averaged instantaneous temperature T(t) of the system, based on the equipartition theory, can be calculated using the obtained velocities; as follows:

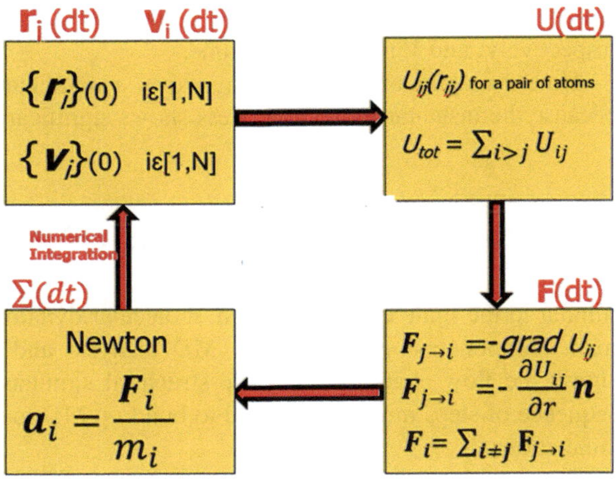

Figure 2.7 The iterative molecular dynamics cycle

$$K(t) = \frac{1}{2}\sum_{i} m_i \,(v_i(t))^2 \tag{2.23}$$

$$T(t) = \frac{2}{3}\frac{K(t)}{N\,K_B} \tag{2.24}$$

where K_B is Boltzmann constant.

Figure 2.7 outlines the input, output and the iterative steps taken to obtain the accelerations, velocities, and displacement from a given force field in MD simulations. It is important to note that MD simulations are chaotic and small differences in the initial conditions could quickly lead to significant errors in the predicted trajectories of the system of atoms under consideration.

2.8 MD Virial Stresses

Stresses in MD simulations can be interpreted using either Cauchy stress or virial stress. The latter is more appropriate in calculating the stress of the individual atoms. The averaged virial stress tensor σ_{ij} is defined in [2.7] as follows:

$$\sigma_{ij} = \frac{1}{V}\sum_{\alpha}\left[\sum_{\beta=1}^{N}\left(R_i^{\beta} - R_i^{\alpha}\right)F_j^{\alpha\beta} - m^{\alpha}v_i^{\alpha}v_j^{\alpha}\right] \tag{2.25}$$

where i and j are the directional indices (i.e. x, y, and z); α is a number assigned to an atom and β is a number assigned to neighboring atoms of α, which varies from 1 to N; R_i^{β} is the position of atom β along the direction i; $F_j^{\alpha\beta}$ is the force on atom α due to atom β along the direction j; m^{α} & v^{α} are the mass and the velocity of atom α, respectively; and V is the total volume.

Time averaging of the atomic stress over a significantly long period is necessary, because the instantaneous virial stress shows significant fluctuations [2.7].

2.9 Implementation of MD using LAMMPS

Similar to the finite element method, molecular dynamics can be divided into three main components: preprocessing, MD analysis and postprocessing. Figure 2.8 shows the flow chart for atomistic structural simulations using MD. A common sequence of steps must be followed to build up MD models and perform successful simulations.

The first step is to build the initial structure of the system using the appropriate preprocessing software such as Nanoengineer, Materials Studio, Packmol, among others. This step consumes a significant amount of time, as it usually requires the building of small units of the system and joining them in a large MD unit cell and ensuring the attainment of its physical properties such as density, thermal conductivity, and others.

The second step is to minimize the structure by changing the initial location of the atoms to reduce the total potential energy of the system and to relief any existing residual stresses. In the third step, knowing the initials atomic positions, the developer should assign an initial velocity to each atom based on the targeted average temperature of the system.

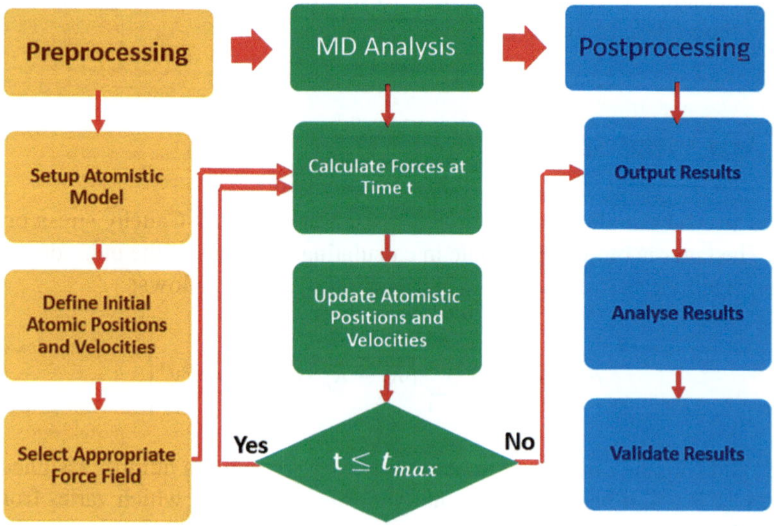

Figure 2.8 Flow chart describing the implementation of MD simulations in structural analysis

The fourth step is concerned with equilibrating the structure to obtain the system at a targeted initial thermodynamic condition(s) (pressure, volume, temperature, energy) for a given number of atoms.

Finally, conducting the required analysis and measuring the system properties of interest. It is essential that the developer must conduct a number of validations of the MD output.

2.10 Periodic Boundary Conditions

It is often the case that large MD simulations (10^{25} atoms) cannot be solved with current supercomputers. Luckily, these systems typically contain periodicity (repeated or cyclic cells), as depicted in Figure 2.9. To reduce the degrees of freedom of these large systems, we make use of these cyclic or repeated cells, known as unit cell (10^6 atoms) in the setup of the molecular dynamics model. It is unfortunate that the reduction of the real system from (10^{25} atoms) to a reduced system (10^6 atoms) results in artificial artifacts resulting from the use of finite size. Effectively, periodic boundary conditions are a set of approximating boundary conditions chosen to limit the degrees of freedom of the system and obtain solutions in a timely fashion. It is also worth noting that in the case of a reduced system size, a large percentage of the surface atoms interact with the surroundings, while only a very small percentage of surface atoms interact with the surroundings. The majority of the atoms in the real system interact with each other. It is also important to note that the number of atoms is conserved all the time in all the cells, including the replica cells.

Figure 2.9 shows a simplified 2D system with a clear periodicity. In this case, we select the unit cell highlighted in yellow, noting that the unit cell experiences the

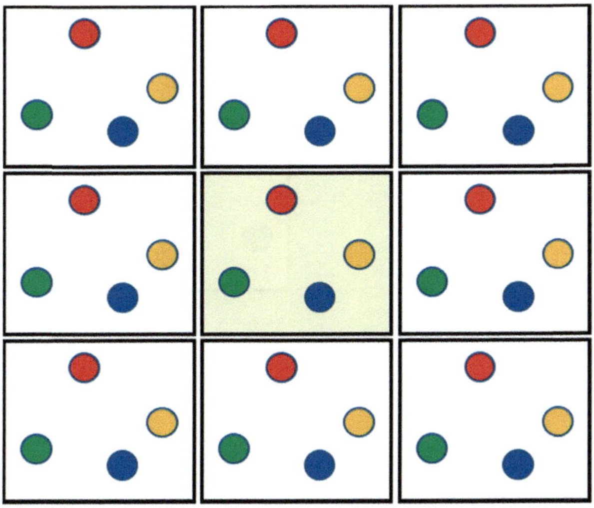

Figure 2.9 Shows a schematic of a system containing periodicity with primary unit cell distinguished by a yellow background

same conditions as every other cell in the system. For example, if the molecule marked in blue moves vertically upward, all blue molecules in the other cells will move the same amount in the same direction.

2.11 Cutoff Interaction Radius

In real molecular dynamics systems, each particle/molecule interacts with all particles/molecules in the system. The major problem here is that long-range interactions would involve an infinite summation. To avoid the computational challenges associated with an infinite summation, the concept of long-range and short-range interactions is introduced. Typically, a cutoff radius is drawn to limit the interactions to short-range interactions, as depicted in Figures 2.4 and 2.10. This effectively means that L-J potential is equal to zero at the cutoff radius. This is ensured in the MD simulations via the use of a switching function. Clearly, this approximation will introduce errors in the simulations. In this case, we compute the interactions (potential energy) of the atoms being confined to within an appropriately selected cutoff radius.

Figure 2.10 Cut-off radius of interacting atoms

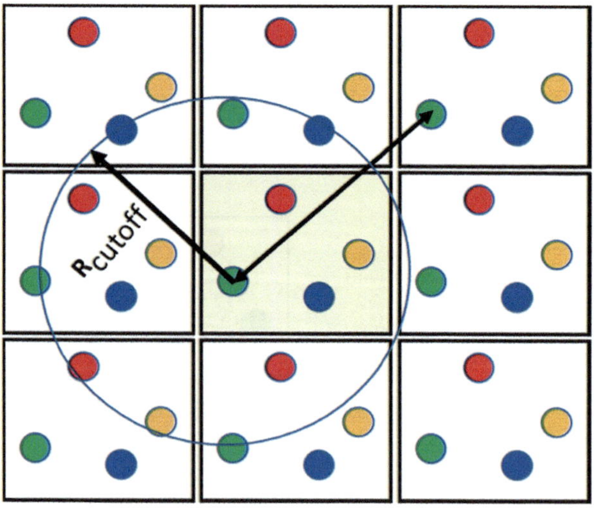

2.12 Parameters Influencing Accuracy of MD Simulations

However, several parameters influence the accuracy of the outcome. These parameters include: selected force field (interatomic potential), cut-off function radii, time step, strain rate, domain size and thermodynamic state. A more detailed account of this section can be viewed in Alian and Meguid in [2.8].

For example, different interatomic potentials have been used to study the reinforcement effect of CNTs and graphene in their nanocomposite host polymer. The selection of inappropriate interatomic potentials would lead to erroneous results. This is analogous to the selection of incorrect constitutive law in finite element analysis. Furthermore, all force fields that use harmonic function to model atomic bond, such as CVFF force field, do not permit bond formation and breakage, thus limiting the simulation to only small deformation [2.9–2.12]. On the other hand, other reactive interatomic potentials such as AIREBO and ReaxFF potentials can be used to model bond breakage and hence allow the study of large deformation and fracture mechanics.

Another source of the contradictions in the reported mechanical properties in the current literature is the selection of the cut-off function parameters. Take for example Tersoff and AIREBO potentials. The original minimum and maximum radii in the cut-off function for the covalent bonds in AIREBO potential are 1.7 Å and 2.0 Å, respectively. However, these values introduce unphysical strain hardening in the structure at high strains. To address this problem, researchers varied the cut-off function parameters to achieve properties that are in good agreement with those obtained experimentally and numerically using DFT techniques.

Selecting the appropriate time step in MD simulations is very critical for obtaining reliable results with a reasonable computational cost, achieving energy conservation, and preventing the MD simulations from exploding [2.13]. Different time steps ranging from 0.1 to 15 fs have been used in previous MD simulation of CNTs, graphene, and BN structures [2.14–2.17]. Budarapu et al. [2.18] have studied the effect of time step on crack initiation and growth mechanisms in graphene using MD simulations. Their results showed a significant dependence of post yielding behavior of graphene on time step, and a time step of 0.1 fs was required to accurately monitor crack initiation and propagation in the fracturing sheet. However, such a small time step will require an enormous computational time/cost to model large systems requiring extended periods of integration.

Conducting MD simulations at very high strain rates means that the molecular structure does not have adequate time to respond (equilibrate) to the applied load [2.19]. The effect of strain rate in MD simulation on the predicted mechanical properties were investigated in the literature to determine a suitable range of loading rates that gives reliable results. However, the reported results of these investigations were contradictory with some papers indicating a significant effect of the strain rate on all mechanical properties, while others showed a slight to no effect [2.20].

The reported results of numerous atomistic studies are contradictory, with the following references showing a growing weakening effect with increasing size [2.21, 2.22], while the work conducted in [2.23] concludes the exact opposite. The influence of the atomistic size originates from the presence of free edges and the associated compressive forces due to dangling bonds in edge atoms [2.24].

All MD simulations should be conducted under specified ensemble conditions. These ensembles are characterized by fixed values of potential energy, temperature, pressure, volume, and total number of particles. The most used ensembles in MD simulations are:

- micro-canonical ensemble: constant number of atoms, volume, and energy (N,V,E)
- isothermal-isobaric: constant number of atoms, temperature, and pressure (N,T,P)
- Canonical ensemble: constant number of atoms, temperature, and volume (N,V,T)

References

2.1 E. Schrödinger, An undulatory theory of the mechanics of atoms and molecules. Phys. Rev. **28**, 1049–1070 (1926)
2.2 D.W. Brenner, O.A. Shenderova, J.A. Harrison, S.J. Stuart, B. Ni, S.B. Sinnott, A second-generation reactive empirical bond order (REBO) potential energy expression for hydrocarbons. J. Phys. Condens. Matter **14**, 783–792 (2002)
2.3 R. LeSar, *Introduction to Computational Materials Science: Fundamentals to Applications* (Cambridge University Press, Cambridge, 2013)
2.4 J. Tersoff, New empirical approach for the structure and energy of covalent systems. Phys. Rev. B **37**, 6991–6999 (1988). https://doi.org/10.1103/PhysRevB.37.6991
2.5 C. Li, T.W. Chou, A structural mechanics approach for the analysis of carbon nanotubes. Int. J. Solids Struct. **40**, 2487–2499 (2003). https://doi.org/10.1016/S0020-7683(03)00056-8
2.6 L. Verlet, Computer "experiments" on classical fluids. I. Thermodynamical properties of Lennard-Jones molecules. Phys. Rev. **159**, 98–103 (1967). https://doi.org/10.1103/PhysRev.159.98
2.7 D.H. Tsai, The virial theorem and stress calculation in molecular dynamics. J. Chem. Phys. **70**, 1375–1382 (1979)
2.8 A.R. Alian, S.A. Meguid, A critical study of the parameters governing molecular dynamics simulations of nanostructured materials. Comput. Mater. Sci. **153**, 183–199 (2018)
2.9 S.A. Meguid (ed.), *Advances in Nanocomposites, Modeling, Characterisation and Applications* (Springer, Cham, 2016)
2.10 A.R. Alian, S.A. Meguid, Large-scale atomistic simulations of CNT-reinforced thermoplastic polymers. Compos. Struct. **191**, 221–230 (2018). https://doi.org/10.1016/j.compstruct.2018.02.056
2.11 J. Zhang, X. He, L. Yang, G. Wu, J. Sha, C. Hou, C. Yin, A. Pan, Z. Li, Y. Liu, Effect of tensile strain on thermal conductivity in monolayer graphene nanoribbons: A molecular dynamics study. Sensors **13**, 9388–9395 (2013). https://doi.org/10.3390/s130709388
2.12 A.R. Alian, S.I. Kundalwal, S.A. Meguid, Multiscale modeling of carbon nanotube epoxy composites. Polymer **70**, 149–160 (2015). https://doi.org/10.1016/j.polymer.2015.06.004

2.13 A.R. Alian, S.I. Kundalwal, S.A. Meguid, Interfacial and mechanical properties of epoxy nanocomposites using different multiscale modeling schemes. Compos. Struct. **131**, 545–555 (2015). https://doi.org/10.1016/j.compstruct.2015.06.014
2.14 B.D. Jensen, K.E. Wise, G.M. Odegard, The effect of time step, thermostat, and strain rate on ReaxFF simulations of mechanical failure in diamond, graphene, and carbon nanotube. J. Comput. Chem. **36**, 1587–1596 (2015). https://doi.org/10.1002/jcc.23970
2.15 K. Mylvaganam, L. Zhang, Important issues in a molecular dynamics simulation for characterising the mechanical properties of carbon nanotubes. Carbon **42**, 2025–2032 (2004). https://doi.org/10.1016/j.carbon.2004.04.004
2.16 Z.Y. Ong, E. Pop, Molecular dynamics simulation of thermal boundary conductance between carbon nanotubes and SiO 2. Phys. Rev. B **81**, 155408 (2010). https://doi.org/10.1103/PhysRevB.81.155408
2.17 H. Bu, Y. Chen, M. Zou, H. Yi, K. Bi, Z. Ni, Atomistic simulations of mechanical properties of graphene nanoribbons. Phys. Lett. A **373**, 3359–3362 (2009). https://doi.org/10.1016/j.physleta.2009.07.048
2.18 P.R. Budarapu, B. Javvaji, V.K. Sutrakar, D. Roy Mahapatra, G. Zi, T. Rabczuk, Crack propagation in graphene. J. Appl. Phys. **118**, 064307 (2015). https://doi.org/10.1063/1.4928316
2.19 M.Q. Chen, S.S. Quek, Z.D. Sha, C.H. Chiu, Q.X. Pei, Y.W. Zhang, Effects of grain size, temperature and strain rate on the mechanical properties of polycrystalline graphene—A molecular dynamics study. Carbon **85**, 135–146 (2015). https://doi.org/10.1016/j.carbon.2014.12.092
2.20 L. Yi, Z. Yin, Y. Zhang, T. Chang, A theoretical evaluation of the temperature and strain-rate dependent fracture strength of tilt grain boundaries in graphene. Carbon **51**, 373–380 (2013). https://doi.org/10.1016/j.carbon.2012.08.069
2.21 X. Liu, Q.S. Yang, Plastic deformation and failure mechanisms of collapsed-carbon nanotube fibers by coarse-grained molecular dynamic simulations. Int. J. Plast. **64**, 104–112 (2015). https://doi.org/10.1016/j.ijplas.2014.08.004
2.22 H. Li, H. Zhang, X. Cheng, The effect of temperature, defect and strain rate on the mechanical property of multi-layer graphene: Coarse-grained molecular dynamics study. Phys. E Low-Dimens. Syst. Nanostruct. **85**, 97–102 (2017). https://doi.org/10.1016/j.physe.2016.07.003
2.23 C. Wei, K. Cho, D. Srivastava, Tensile strength of carbon nanotubes under realistic temperature and strain rate. Phys. Rev. B **67**, 115407 (2003). https://doi.org/10.1103/PhysRevB.67.115407
2.24 Y.Y. Zhang, Q.X. Pei, Y.W. Mai, Y.T. Gu, Temperature and strain-rate dependent fracture strength of graphene. J. Phys. Appl. Phys. **47**, 425301 (2014). https://doi.org/10.1088/0022-3727/47/42/425301

Chapter 3
Fundamentals of the Elastic Behavior of a Continuum

Summary In this chapter, we briefly discuss some of the essential topics of the theories of elasticity, which are pertinent to the subject of fracture mechanics. Every attempt has been made to provide the reader with the necessary background to fully and effectively comprehend the subject matter.

3.1 State of Stress at a Point

In this text, normal stresses are denoted σ_{xx}, σ_{yy} and σ_{zz} and shear stresses are denoted τ_{xy}, τ_{yz} and τ_{zx}. Figure 3.1 shows a three-dimensional element with the six components of the stresses acting on three perpendicular planes. The stresses shown are all positive, and this figure establishes the sign conventions used throughout this textbook. The subscripts define the orientation of these stresses. For example, the shearing stress τ_{xy} denotes a stress acting on the face of the element which is perpendicular to the x-axis with the shearing stress acting along the y-axis.

One cannot ask for the stresses at a given point in a body, since there is an infinite number of stresses in existence at any one point; one resultant stress corresponding to each distinctly inclined plane passing through that point. The request must be made for either the stresses at the point on a given plane passing through it, or more generally, the stresses acting on all planes passing through that point.

Typically, as shown in Figure 3.1, the state of stress at a point can be either be given in the form of a tensor in terms of its Cartesian components,

$$\begin{bmatrix} \sigma_{xx} & \tau_{xy} & \tau_{xz} \\ \tau_{yx} & \sigma_{yy} & \tau_{yz} \\ \tau_{zx} & \tau_{zy} & \sigma_{zz} \end{bmatrix} \tag{3.1}$$

or, equivalently, it may be given in terms of its cylindrical polar components,

S. A. Meguid, *Atomistic and Continuum Fracture Mechanics of Solids*,
https://doi.org/10.1007/978-3-031-56085-9_3

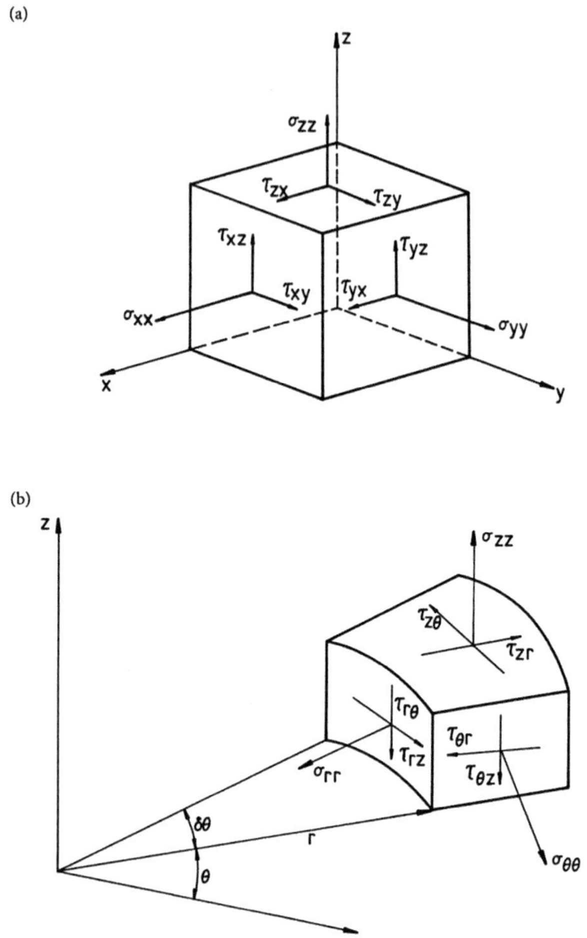

Figure 3.1 Sign conventions: (**a**) stress components referred to Cartesian coordinates; (**b**) stress components referred to Cylindrical Polar coordinates

$$\begin{bmatrix} \sigma_{rr} & \tau_{r\theta} & \tau_{rz} \\ \tau_{\theta r} & \sigma_{\theta\theta} & \tau_{\theta z} \\ \tau_{zr} & \tau_{z\theta} & \sigma_{zz} \end{bmatrix} \tag{3.2}$$

It can also be described in terms of its spherical components; or simply by the stress tensor σ_{ij}.

Actually, not all the nine components of the stresses are independent, since rotational equilibrium enforces that,

$$\left.\begin{aligned}\tau_{ij} &= \tau_{ji} \\ \tau_{xy} = \tau_{yx}, \tau_{yz} &= \tau_{zy} \text{ and } \tau_{zx} = \tau_{xz}\end{aligned}\right\} \quad (3.3)$$

or equivalently,

$$\tau_{r\theta} = \tau_{\theta r}, \tau_{\theta z} = \tau_{z\theta} \text{ and } \tau_{zr} = \tau_{rz} \quad (3.4)$$

Thus, indicating that there are only three independent shear stresses. Consequently, the complete three-dimensional uniform stress field at a point can be determined by six stresses. Since these are tensor quantities, they must be measured in terms of a defined set of coordinate axes, as indicated above by Equations (3.1) and (3.2).

3.1.1 *Stress Boundary Conditions*

The prescribed force intensities acting on the surface of a solid tetrahedron *S* are denoted by the quantities S_{xx}, S_{yy}, and S_{zz}. Force equilibrium at the boundary of the tetrahedron, shown in Figure 3.2, requires that,

$$\begin{bmatrix} S_{xx} \\ S_{yy} \\ S_{zz} \end{bmatrix} = \begin{bmatrix} \sigma_{xx} & \tau_{xy} & \tau_{xz} \\ \tau_{yx} & \sigma_{yy} & \tau_{yz} \\ \tau_{zx} & \tau_{zy} & \sigma_{zz} \end{bmatrix} \begin{bmatrix} l \\ m \\ n \end{bmatrix} \quad (3.5)$$

where *l*, *m* and *n* are the direction cosines of the normal ***n*** and are given by,

$$l = \cos(n, x), m = \cos(n, y), n = \cos(n, z) \quad (3.6)$$

3.1.2 *Eigenvalues and Eigenvectors of Stress Tensor*

The characteristic equation needed for determining the eigenvalues and eigenvector for the stress tensor σ_{ij} is given by:

$$\left[\sigma_{ij} - \sigma\delta_{ij}\right]\left[l_j\right] = 0 \quad (3.7)$$

with

$$\left[l_j\right]\left[l_j\right]^T = 1 \quad (3.8)$$

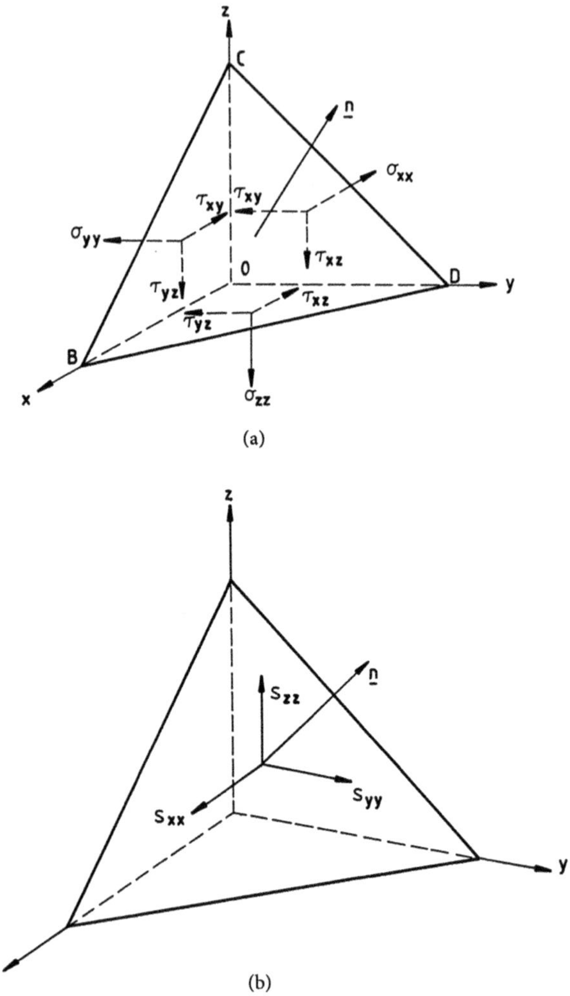

Figure 3.2 Forces on an infinitesimal tetrahedron: (**a**) stress state at a point; (**b**) components of resultant stress vector

or in a long form by

$$\begin{bmatrix} \sigma_{xx} - \sigma & \tau_{xy} & \tau_{xz} \\ \tau_{yx} & \sigma_{yy} - \sigma & \tau_{yz} \\ \tau_{zx} & \tau_{zy} & \sigma_{zz} - \sigma \end{bmatrix} \begin{bmatrix} l \\ m \\ n \end{bmatrix} = 0 \tag{3.9}$$

with

$$\begin{bmatrix} l & m & n \end{bmatrix} \begin{bmatrix} l \\ m \\ n \end{bmatrix} = 1 \tag{3.10}$$

Equation (3.9) are homogeneous in the direction cosines (l, m, n) and can admit a non-trivial solution if, and only if, the determinant,

$$\left|\sigma_{ij} - \sigma\delta_{ij}\right| = 0 \tag{3.11}$$

i.e.,

$$\begin{vmatrix} \sigma_{xx} - \sigma & \tau_{xy} & \tau_{xz} \\ \tau_{yx} & \sigma_{yy} - \sigma & \tau_{yz} \\ \tau_{zx} & \tau_{zy} & \sigma_{zz} - \sigma \end{vmatrix} = 0 \tag{3.12}$$

or

$$\sigma^3 - I_1\sigma^2 + I_2\sigma - I_3 = 0$$

where

$$I_1 = \sigma_{ij} = \sigma_{xx} + \sigma_{yy} + \sigma_{zz}$$

$$I_2 = \frac{1}{2}\left(\sigma_{ii}\sigma_{jj} - \sigma_{ij}\sigma_{ji}\right)$$

i.e.,

$$I_2 = \begin{vmatrix} \sigma_{xx} & \tau_{xy} \\ \tau_{yx} & \sigma_{yy} \end{vmatrix} + \begin{vmatrix} \sigma_{yy} & \tau_{yz} \\ \tau_{zy} & \sigma_{zz} \end{vmatrix} + \begin{vmatrix} \sigma_{xx} & \tau_{xz} \\ \tau_{zx} & \sigma_{zz} \end{vmatrix}$$

and

$$I_3 = \begin{vmatrix} \sigma_{xx} & \tau_{xy} & \tau_{xz} \\ \tau_{yx} & \sigma_{yy} & \tau_{yz} \\ \tau_{zx} & \tau_{zy} & \sigma_{zz} \end{vmatrix}$$

With I_1, I_2 and I_3 are known as the stress invariants.

The above cubic equation will provide the eigenvalues σ_1, σ_2 and σ_3 and these are denoted the principal stresses, where $\sigma_1 > \sigma_2 > \sigma_3$. Associated with the principal stresses there are principal planes and these are defined by the eigenvectors (l, m, n). Accordingly, for each eigenvalue, $\sigma = \sigma_1$ (say), the corresponding eigenvectors are determined from Equations (3.9) and (3.10).

3.2 Transformation of Stress State

If the complete stress field as referenced to one set of coordinate axes is known, it can be determined with respect to any other set of co-ordinate axes by means of appropriate transformation equations. For example, let us assume that it is desired to know the stress state σ_{ij} referenced to a new rotated set of axes x' y' z'. Now, consider that the x'-axis, as shown in Figure 3.3 has the respective direction cosines l_{11}, l_{12} and l_{13} between it and the xyz-axes. Similarly, the direction cosines of the angles made by y' with xyz are l_{21}, l_{22} and l_{23} and those made by z' are l_{31}, l_{32} and l_{33}. In this case, the transformation equation for a three-dimensional state of stress can be performed with the aid of the expression:

$$\left[\sigma_{i'j'}\right] = \left[l_{ij}\right]\left[\sigma_{ij}\right]\left[l_{ij}\right]^{T} \tag{3.13}$$

or written in the long form as:

$$\begin{bmatrix} \sigma_{x'x'} & \tau_{x'y'} & \tau_{x'z'} \\ \tau_{y'x'} & \sigma_{y'y'} & \tau_{y'z'} \\ \tau_{z'x'} & \tau_{z'y'} & \sigma_{z'z'} \end{bmatrix} = \begin{bmatrix} l_{11} & l_{12} & l_{13} \\ l_{21} & l_{22} & l_{23} \\ l_{31} & l_{32} & l_{33} \end{bmatrix} \begin{bmatrix} \sigma_{xx} & \tau_{xy} & \tau_{xz} \\ \tau_{yx} & \sigma_{yy} & \tau_{yz} \\ \tau_{zx} & \tau_{zy} & \sigma_{zz} \end{bmatrix} \begin{bmatrix} l_{11} & l_{21} & l_{31} \\ l_{12} & l_{22} & l_{32} \\ l_{13} & l_{23} & l_{33} \end{bmatrix} \tag{3.14}$$

Figure 3.3 Three-dimensional transformation of stress axes

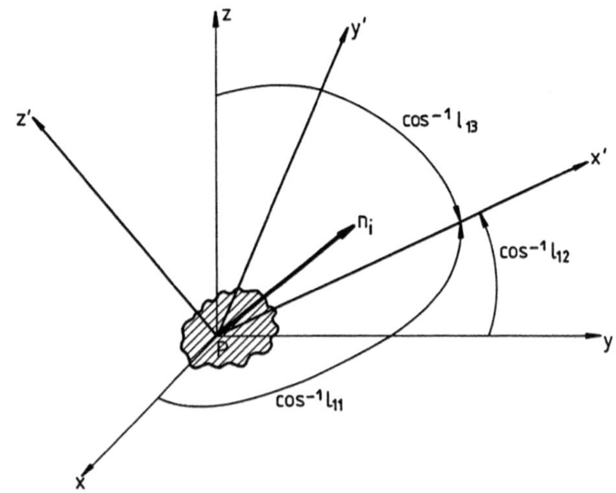

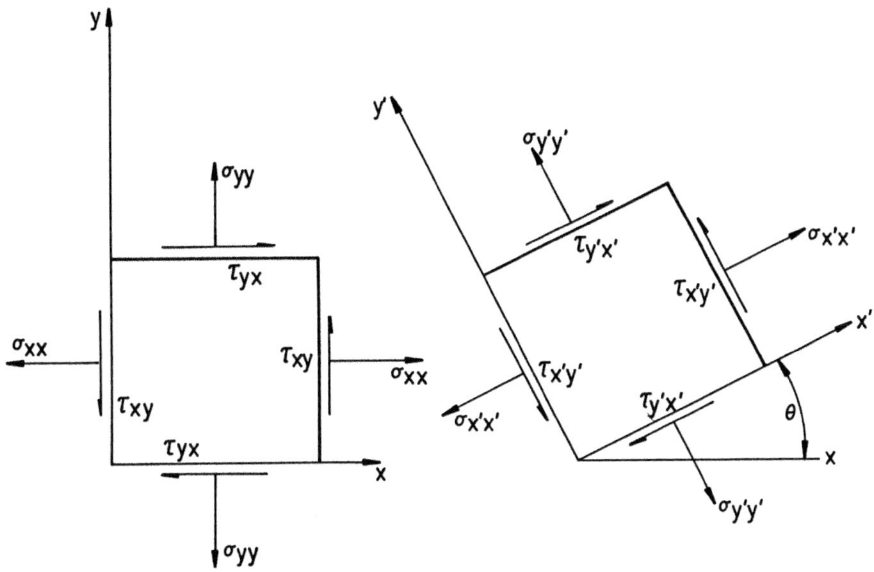

Figure 3.4 Two-dimensional transformation of rotated stress axes

For the two-dimensional state of stress shown in Figure 3.4, the above formulations reduce to:

$$\begin{bmatrix} \sigma_{x'x'} & \tau_{x'y'} \\ \tau_{y'x'} & \sigma_{y'y'} \end{bmatrix} = [\cos\theta \sin\theta - \sin\theta \cos\theta] \begin{bmatrix} \sigma_{xx} & \tau_{xy} \\ \tau_{yx} & \sigma_{yy} \end{bmatrix} [\cos\theta - \sin\theta \sin\theta \cos\theta] \tag{3.15}$$

or in their most familiar form as:

$$\begin{aligned} \sigma_{x'x'} &= \frac{\sigma_{xx} + \sigma_{yy}}{2} + \left(\frac{\sigma_{xx} - \sigma_{yy}}{2}\right) \cos 2\theta + \tau_{xy} \sin 2\theta \\ \sigma_{y'y'} &= \frac{\sigma_{xx} + \sigma_{yy}}{2} - \left(\frac{\sigma_{xx} - \sigma_{yy}}{2}\right) \cos 2\theta + \tau_{xy} \sin 2\theta \\ \tau_{x'y'} &= \tau_{xy} \cos 2\theta - \left(\frac{\sigma_{xx} - \sigma_{yy}}{2}\right) \sin 2\theta \end{aligned} \tag{3.16}$$

Alternatively, these expressions may be written as:

$$\left.\begin{aligned} \sigma_{x'x'} + \sigma_{y'y'} &= \sigma_{xx} + \sigma_{yy} \\ \sigma_{y'y'} - \sigma_{x'x'} + 2i\tau_{x'y'} &= \left(\sigma_{yy} - \sigma_{xx} + 2i\tau_{xy}\right) e^{2i\theta} \end{aligned}\right\} \tag{3.17}$$

where

$$e^{2i\theta} = \cos 2\theta + i\sin 2\theta \tag{3.18}$$

which is a suitable form for complex stress function analysis, to be discussed later.

Now, by allowing θ to vary, it is possible to identify two orthogonal planes upon which the shear stresses vanish. Again, these planes are known as the principal planes and the corresponding stresses are known as the principal stresses, and may be obtained using the expression,

$$\begin{matrix}\sigma_1 \\ \sigma_2\end{matrix} = \left(\frac{\sigma_{xx} + \sigma_{yy}}{2}\right) \pm \sqrt{\left(\frac{\sigma_{xx} - \sigma_{yy}}{2}\right)^2 + \tau_{xy}{}^2} \tag{3.19}$$

With the maximum shear stress being,

$$\tau_{max} = \sqrt{\left(\frac{\sigma_{xx} - \sigma_{yy}}{2}\right)^2 + \tau_{xy}{}^2} = \frac{\sigma_1 - \sigma_2}{2} \tag{3.20}$$

3.3 Equation of Equilibrium

In the above section, we considered stress fields that were uniform throughout the body. Such stress states cannot be found in structural analysis. The stress state usually varies from point to point in most engineering applications. This variation is depicted in Figure 3.5 in terms of the Cartesian co-ordinate system.

Since the body as a whole must be in equilibrium, the element under consideration must also be in equilibrium under the action of the stresses. By combining and factoring out the common terms, the equilibrium along the x, y and z directions reduce to,

$$\left.\begin{aligned} \frac{\partial \sigma_{xx}}{\partial x} + \frac{\partial \tau_{xy}}{\partial y} + \frac{\partial \tau_{xz}}{\partial z} + X = 0 \\ \frac{\partial \sigma_{yy}}{\partial y} + \frac{\partial \tau_{yx}}{\partial x} + \frac{\partial \tau_{yz}}{\partial z} + Y = 0 \\ \frac{\partial \sigma_{zz}}{\partial z} + \frac{\partial \tau_{zx}}{\partial x} + \frac{\partial \tau_{zy}}{\partial y} + Z = 0 \end{aligned}\right\} \tag{3.21}$$

or simply,

$$\partial \sigma_{ij}/\partial x_j + X_i = 0 \tag{3.22}$$

where X_i are the body forces per unit volume, and act along x, y and z directions.

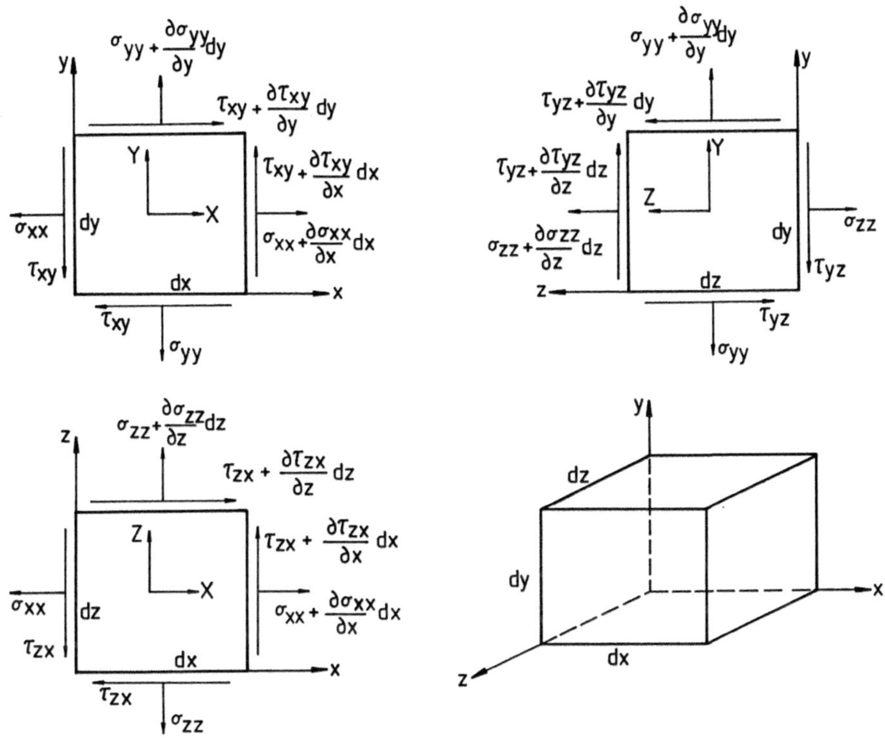

Figure 3.5 Equilibrium of forces in Cartesian frame

For the two-dimensional state of stress with no body forces, the above equations reduce to,

$$\left.\begin{aligned}\frac{\partial\sigma_{xx}}{\partial x}+\frac{\partial\tau_{xy}}{\partial y}&=0\\ \frac{\partial\sigma_{yy}}{\partial y}+\frac{\partial\tau_{yx}}{\partial x}&=0\end{aligned}\right\}\qquad(3.23)$$

For a number of engineering problems, it is more convenient to set up the equilibrium equations in terms of cylindrical polar coordinates. With the stress pattern depicted in Figure 3.6, the equilibrium equations are given by:

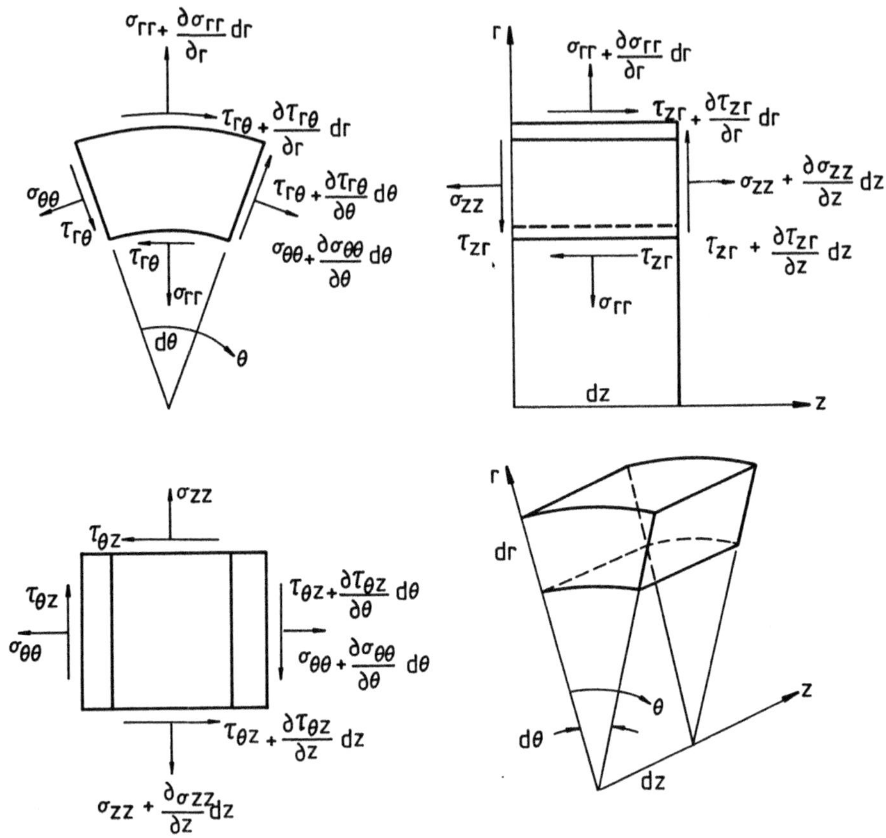

Figure 3.6 Force equilibrium in cylindrical polar frame

$$
\left.
\begin{aligned}
&\frac{\partial \sigma_{rr}}{\partial r} + \frac{\partial \tau_{r\theta}}{r\partial \theta} + \frac{\partial \tau_{rz}}{\partial z} + \frac{\sigma_{rr} - \sigma_{\theta\theta}}{r} + R = 0 \\
&\frac{\partial \sigma_{\theta\theta}}{r\partial \theta} + \frac{\partial \tau_{\theta r}}{\partial r} + \frac{\partial \tau_{\theta z}}{\partial z} + \frac{2\tau_{r\theta}}{r} + \theta = 0 \\
&\frac{\partial \sigma_{zz}}{\partial z} + \frac{\partial \tau_{z\theta}}{r\partial \theta} + \frac{\partial \tau_{zr}}{\partial r} + \frac{\tau_{zr}}{r} + Z = 0
\end{aligned}
\right\} \tag{3.24}
$$

Similarly, for two-dimensional axisymmetric bodies with no body forces, the equilibrium equations reduce to:

$$
\frac{\partial \sigma_{rr}}{\partial r} + \frac{\sigma_{rr} - \sigma_{\theta\theta}}{r} = 0 \tag{3.25}
$$

The other equations are equivalently satisfied.

3.4 Definition of Displacement and Strain

Strain is a geometric quantity which depends on the relative movement of two or three points in a body, and therefore it is governed by deformation. Since the rigid-body displacements do not produce strains, they will be neglected. In the preceding sections, two types of stress were discussed: normal stresses σ_{ij} and shear stress τ_{ij}. This same classification will be used for strains. In the small deformation theory, a normal strain ε is defined as the change in length of a line segment between two points divided by the original length of that line segment. A shearing strain γ is defined as the angular change between two line segments, which were originally perpendicular.

The relationship between strains and (deformation) displacements can be determined by considering the deformation of a two-dimensional element with dimensions d*x* and d*y* and unit thickness. This deformation is illustrated in Figure 3.7 in which a general point P is displaced a distance *u* in the *x*-direction and *v* in the *y*-direction. It is the differences in linear and angular displacements which produce normal and shear strains.

Examination of Figure 3.7 reveals that the two normal or longitudinal strains in the x and y directions are,

$$\varepsilon_{xx} = \frac{\partial u}{\partial x} \text{ and } \varepsilon_{yy} = \frac{\partial v}{\partial y}$$

A positive sign applies to elongation and a negative sign applies to contraction. Now, if we consider the change experienced by the right-angle P_1PP_2 of Figure 3.7, we obtain,

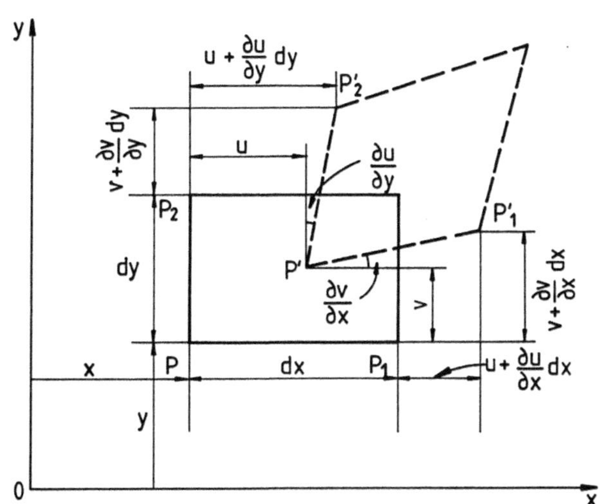

Figure 3.7 Two-dimensional representation of deformation

$$\gamma_{xy} = \frac{\partial u}{\partial y} + \frac{\partial v}{\partial x}$$

The shear strain is positive when the right-angle between two positive axes decreases; otherwise the shear strain is negative.

In the case of a three-dimensional element, a rectangular 3-D prism with sides dx, dy and dz will lead to the following normal and shear strains:

$$\left.\begin{aligned} &\varepsilon_{xx} = \frac{\partial u}{\partial x}, \varepsilon_{yy} = \frac{\partial v}{\partial y}, \varepsilon_{zz} = \frac{\partial w}{\partial z} \\ &\gamma_{xy} = \frac{\partial u}{\partial y} + \frac{\partial v}{\partial x}, \gamma_{yz} = \frac{\partial v}{\partial z} + \frac{\partial w}{\partial y}, \gamma_{zx} = \frac{\partial w}{\partial x} + \frac{\partial u}{\partial z} \end{aligned}\right\} \tag{3.26}$$

In the above formulations,

$$\gamma_{xy} = \gamma_{yx}, \gamma_{yz} = \gamma_{zy} \text{ and } \gamma_{zx} = \gamma_{xz} \tag{3.27}$$

Expressions (3.26) are known as the strain-displacement relationships. They are also referred to as the kinematic relationships. Using tensor notations, the strain displacement relations can be rewritten as,

$$\varepsilon_{ij} = \frac{1}{2}\left(\frac{\partial u_i}{\partial x_j} + \frac{\partial u_j}{\partial x_i}\right) \qquad (i, j = x, y, z) \tag{3.28}$$

The factor ½ facilitates the representation of the strain equations in indicial notation. It is apparent from Equation (3.28) that,

$$\varepsilon_{xy} = \frac{1}{2}\gamma_{xy}, \varepsilon_{yz} = \frac{1}{2}\gamma_{yz} \text{ and } \varepsilon_{zx} = \frac{1}{2}\gamma_{zx} \tag{3.29}$$

Just as the state of stress at a point is described by a tensor, the strain is also defined by a tensor; as follows,

$$\varepsilon_{ij} = \begin{bmatrix} \varepsilon_{xx} & \frac{1}{2}\gamma_{xy} & \frac{1}{2}\gamma_{xz} \\ \frac{1}{2}\gamma_{yx} & \varepsilon_{yy} & \frac{1}{2}\gamma_{yz} \\ \frac{1}{2}\gamma_{zx} & \frac{1}{2}\gamma_{zy} & \varepsilon_{zz} \end{bmatrix} \tag{3.30}$$

It is worth observing that the Cartesian coordinate system used in connection with the static equilibrium equations pertains to the deformed state, and the coordinate set was thus established in a deformed body. In this instance, the *xyz* axes are known as an Eulerian co-ordinate system. In discussing kinematics of deformation, the *xyz* set was established in the undeformed body. In this case, the *xyz* coordinates are referred to as a Lagrangian coordinate system. Though these systems are clearly not the same, the assumption of small deformation permits us to assume that the *xyz* coordinates in the undeformed body are applicable to equations of stress and strain alike. The choice of a Lagrangian system should not lead to errors of consequence, provided they are not used in cases involving large deformation. Under such circumstances, the above approximation is not valid and modified formulations must be adopted. In terms of cylindrical polar coordinates, the three-dimensional strain and displacement relations take the form:

$$\left.\begin{aligned}\varepsilon_{rr} &= \frac{\partial u}{\partial r}, \varepsilon_{\theta\theta} = \frac{1}{r}\frac{\partial v}{\partial \theta} + \frac{u}{r}, \varepsilon_{zz} = \frac{\partial w}{\partial z} \\ \gamma_{r\theta} &= \frac{\partial v}{\partial r} + \frac{1}{r}\frac{\partial u}{\partial \theta} - \frac{v}{r}, \gamma_{\theta z} = \frac{1}{r}\frac{\partial w}{\partial \theta} + \frac{\partial v}{\partial z}, \gamma_{zr} = \frac{\partial u}{\partial z} + \frac{\partial w}{\partial r}\end{aligned}\right\} \tag{3.31}$$

3.5 Compatibility of Strain

Expressions (3.26) and (3.31) of the strain displacement relationships indicate that if the displacements u, v and w are specified as continuous functions of x, y and z or r, θ and z, then one can compute the strain components uniquely. Conditions of compatibility have both mathematical and physical significance. From the mathematical viewpoint, they establish that the displacement field, as expressed by u, v and w is single-valued and continuous. Physically, it means that the body must be connected together with no gaps or overlaps.

Examination of the kinematic relations (3.26) indicates that there exist six components of strain associated with only three displacement components. It is obvious that several problems are encountered, if we consider the inverse problem; i.e. to calculate the displacements u, v and w from the strain components ε_{ij}. In the first place, it is apparent that the solution will not be unique, since strains represent pure deformations whereas displacements include rigid body displacements. This can be overcome, if we are able to specify the rigid body displacement and rotation at some point in the body. However, a more difficult problem is encountered in calculating the displacement from the strains, since Equation (3.26) or (3.31) are six equations for the three unknown displacements u, v and w. It is evident, therefore, that some restrictions must be placed on the strains in order that Equations (3.26) or (3.31) have a solution.

In two-dimensional problems, the sum of the second derivative of ε_{xx} with respect to y and ε_{yy} with respect to x is equal to the second derivative of γ_{xy} with respect to x and y, i.e.,

$$\frac{\partial^2 \varepsilon_{xx}}{\partial y^2} + \frac{\partial^2 \varepsilon_{yy}}{\partial x^2} = \frac{\partial^2 \gamma_{xy}}{\partial x \partial y} \tag{3.32}$$

This is the condition of compatibility or continuity of the two-dimensional problem, expressed in terms of strain. The three-dimensional equations of compatibility, derived in a similar manner, are:

$$\begin{aligned}
&\frac{\partial^2 \varepsilon_{xx}}{\partial y^2} + \frac{\partial^2 \varepsilon_{yy}}{\partial x^2} = \frac{\partial^2 \gamma_{xy}}{\partial x \partial y}, 2\frac{\partial^2 \varepsilon_{xx}}{\partial y \partial z} = \frac{\partial}{\partial x}\left(-\frac{\partial \gamma_{yz}}{\partial x} + \frac{\partial \gamma_{xz}}{\partial y} + \frac{\partial \gamma_{xy}}{\partial z}\right)\\
&\frac{\partial^2 \varepsilon_{yy}}{\partial z^2} + \frac{\partial^2 \varepsilon_{zz}}{\partial y^2} = \frac{\partial^2 \gamma_{yz}}{\partial y \partial z}, 2\frac{\partial^2 \varepsilon_{yy}}{\partial z \partial x} = \frac{\partial}{\partial x}\left(\frac{\partial \gamma_{yz}}{\partial x} - \frac{\partial \gamma_{xz}}{\partial y} + \frac{\partial \gamma_{xy}}{\partial z}\right)\\
&\frac{\partial^2 \varepsilon_{zz}}{\partial x^2} + \frac{\partial^2 \varepsilon_{xx}}{\partial z^2} = \frac{\partial^2 \gamma_{xz}}{\partial z \partial x}, 2\frac{\partial^2 \varepsilon_{zz}}{\partial x \partial y} = \frac{\partial}{\partial z}\left(\frac{\partial \gamma_{yz}}{\partial x} + \frac{\partial \gamma_{xz}}{\partial y} - \frac{\partial \gamma_{xy}}{\partial z}\right)
\end{aligned} \tag{3.33}$$

To gain further insight into the meaning of the above equations, let us assume a body is divided into infinitesimal cubes. We now assume that these cubes can be separated from each other and let each of them be subjected to some arbitrary deformation (strain). It is obvious that if we now try to place all the cubes back together we will, in general, no longer be able to do so and we will be violating the condition of continuity. Due to this arbitrary deformation, the cube boundaries will take a very general incompatible shape. This is because some of the cube boundaries will separate and others will overlap. This shows that there must be some relationships between the strains at different points in a body for that body to remain continuous after straining.

3.6 Strains and Strain Invariants

In the study of the stress state at a point, we have found that there exist at least three mutually orthogonal planes which have no shear stress acting on them. Either from the constitutive formulation of the behavior or the strain as a geometrical quantity, one can deduce that there exist at least three mutually perpendicular planes with no shear stress acting. By this we mean planes whose normal will not change orientation, when the body is strained. Thus, vector ***n***, which was originally normal to such a plane, will either shorten or lengthen, but will not change direction. Such planes are known as the principal planes, the associated normal ***n*** are known as the principal directions and the corresponding strains are termed the principal strains [3.1, 3.2].

From similarity between the laws of stress and strain transformation, it can be concluded that the following characteristic equation applies:

$$\left[\varepsilon_{ij} - \varepsilon\delta_{ij}\right]\left[l_j\right] = 0 \tag{3.34}$$

with

$$\left[l_j\right]\left[l_j\right]^T = 1$$

or alternatively,

$$\left.\begin{aligned} &\begin{bmatrix} \varepsilon_{xx} - \varepsilon & \frac{1}{2}\gamma_{xy} & \frac{1}{2}\gamma_{xz} \\ \frac{1}{2}\gamma_{yx} & \varepsilon_{yy} - \varepsilon & \frac{1}{2}\gamma_{yz} \\ \frac{1}{2}\gamma_{zx} & \frac{1}{2}\gamma_{zy} & \varepsilon_{zz} - \varepsilon \end{bmatrix} \begin{bmatrix} l \\ m \\ n \end{bmatrix} = 0 \\ &[l \quad m \quad n] \cdot \begin{bmatrix} l \\ m \\ n \end{bmatrix} = 1 \end{aligned}\right\} \tag{3.35}$$

Again, the first expression (3.35) provides a system of homogeneous equations in the variables, (l, m, n), and can admit a non-trivial solution if, and only if, the determinant,

$$\left|\varepsilon_{ij} - \varepsilon\delta_{ij}\right| = 0$$

Giving

$$\varepsilon^3 - I_1'\varepsilon^2 + I_2'\varepsilon - I_3' = 0$$

The invariants I_1', I_2', and I_3' appearing in the cubic equation are given by,

$$I_1' = \varepsilon_{xx} + \varepsilon_{yy} + \varepsilon_{zz}$$

$$I_2 = \left(\varepsilon_{xx}\varepsilon_{yy} + \varepsilon_{yy}\varepsilon_{zz} + \varepsilon_{zz}\varepsilon_{xx}\right) - \frac{1}{4}\left(\gamma_{xy}^2 + \gamma_{yz}^2 + \gamma_{zx}^2\right)$$

$$I_3' = \begin{vmatrix} \varepsilon_{xx} & \frac{1}{2}\gamma_{xy} & \frac{1}{2}\gamma_{xz} \\ \frac{1}{2}\gamma_{yx} & \varepsilon_{yy} & \frac{1}{2}\gamma_{yz} \\ \frac{1}{2}\gamma_{zx} & \frac{1}{2}\gamma_{zy} & \varepsilon_{zz} \end{vmatrix}$$

3.7 Elastic Stress-Strain Relationships

Unless otherwise stated, we shall assume that the material of the body is linearly elastic, isotropic and homogeneous, so that its elastic properties are completely identified by the mutually independent constants E and ν, denoting Young's modulus and Poisson's ratio, respectively. In this case, the normal strains are related to the stresses via Hooke's law; viz.,

$$\left.\begin{aligned} \varepsilon_{xx} &= \frac{1}{E}\left[\sigma_{xx} - \nu\left(\sigma_{yy} + \sigma_{zz}\right)\right] \\ \varepsilon_{yy} &= \frac{1}{E}\left[\sigma_{yy} - \nu\left(\sigma_{xx} + \sigma_{zz}\right)\right] \\ \varepsilon_{zz} &= \frac{1}{E}\left[\sigma_{zz} - \nu\left(\sigma_{xx} + \sigma_{yy}\right)\right] \end{aligned}\right\} \tag{3.36}$$

Unlike the normal strains, the shear strains are mutually independent, and related to their respective stresses through the relations,

$$\left.\begin{aligned} \gamma_{xy} &= \tau_{xy}/G \\ \gamma_{yz} &= \tau_{yz}/G \\ \gamma_{zx} &= \tau_{zx}/G \end{aligned}\right\} \tag{3.37}$$

where G is the modulus of rigidity of the component and is related to E by,

$$G = E/2(1+\nu)$$

It is convenient to write the above six strain components in a matrix form as,

$$\begin{Bmatrix} \varepsilon_{xx} \\ \varepsilon_{yy} \\ \varepsilon_{zz} \\ \gamma_{xy} \\ \gamma_{yz} \\ \gamma_{zx} \end{Bmatrix} = \frac{1}{E}\begin{bmatrix} 1 & -\nu & -\nu & 0 & 0 & 0 \\ -\nu & 1 & -\nu & 0 & 0 & 0 \\ -\nu & -\nu & 1 & 0 & 0 & 0 \\ 0 & 0 & 0 & 2(1+\nu) & 0 & 0 \\ 0 & 0 & 0 & 0 & 2(1+\nu) & 0 \\ 0 & 0 & 0 & 0 & 0 & 2(1+\nu) \end{bmatrix}\begin{Bmatrix} \sigma_{xx} \\ \sigma_{yy} \\ \sigma_{zz} \\ \sigma_{xy} \\ \sigma_{yz} \\ \sigma_{zx} \end{Bmatrix} \tag{3.38}$$

By inversion, we can express the above equations in the alternative form,

$$\left\{\begin{array}{c}\sigma_{xx}\\ \sigma_{yy}\\ \sigma_{zz}\\ \sigma_{xy}\\ \sigma_{yz}\\ \sigma_{zx}\end{array}\right\}=\begin{bmatrix}\lambda+2G & \lambda & \lambda & 0 & 0 & 0\\ \lambda & \lambda+2G & \lambda & 0 & 0 & 0\\ \lambda & \lambda & \lambda+2G & 0 & 0 & 0\\ 0 & 0 & 0 & G & 0 & 0\\ 0 & 0 & 0 & 0 & G & 0\\ 0 & 0 & 0 & 0 & 0 & G\end{bmatrix}\left\{\begin{array}{c}\varepsilon_{xx}\\ \varepsilon_{yy}\\ \varepsilon_{zz}\\ \gamma_{xy}\\ \gamma_{yz}\\ \gamma_{zx}\end{array}\right\} \tag{3.39}$$

in which λ is well known as the Lamé coefficient, given by,

$$\lambda=\frac{\nu E}{(1+\nu)(1-2\nu)} \tag{3.40}$$

Equation (3.39) state Hooke's law in three dimensions which can be written in the compact form,

$$\{\sigma\}=[D]\{\varepsilon\} \tag{3.41}$$

in which [D] is the elasticity matrix represented by the three-dimensional array of Equation (3.39):

$$\left.\begin{array}{l}\{\varepsilon\}=\{\varepsilon_{xx}\quad \varepsilon_{yy}\quad \varepsilon_{zz}\quad \gamma_{xy}\quad \gamma_{yz}\quad \gamma_{zx}\}^{T}\\ \{\sigma\}=\{\sigma_{xx}\quad \sigma_{yy}\quad \sigma_{zz}\quad \tau_{xy}\quad \tau_{yz}\quad \tau_{zx}\}^{T}\end{array}\right\} \tag{3.42}$$

3.8 Plane Stress and Plane Strain Problems

If a body consists of two parallel planes a constant thickness apart and bounded by a closed surface, it is said to be a plane body. Associated with this type of body there is a particular class of problems within the general theory of elasticity which are termed plane elastic problems, and these allow a number of simplifying assumptions in their treatment.

3.8.1 Plane Stress Problems

A plane stress condition can be used to analyze elasticity problems in which the body is subjected to in-plane forces and one of its dimensions is very small compared to the rest (typically, the thickness). A typical problem is shown in Figure 3.8 in which,

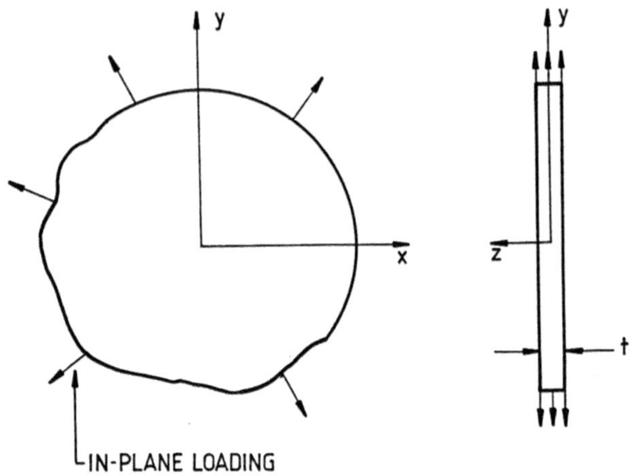

Figure 3.8 Plane stress configuration

$$\sigma_{zz} = \tau_{xz} = \tau_{yz} = 0$$

Furthermore, since t is sufficiently small compared with the other dimensions of the body, and provided that the forces applied at the edges are uniform across t, then we may assume that the above condition is also valid within the cross-section of the body. In this case, the remaining stresses σ_{xx}, σ_{yy} and τ_{xy} are assumed not to vary across the thickness t. The problem thus becomes a two-dimensional one in which the stresses along z-axis can be ignored, and the stress-strain relations reduce to:

$$\left.\begin{aligned} \varepsilon_{xx} &= \frac{1}{E}\left(\sigma_{xx} - \nu\sigma_{yy}\right) \\ \varepsilon_{yy} &= \frac{1}{E}\left(\sigma_{yy} - \nu\sigma_{xx}\right) \\ \gamma_{xy} &= \tau_{xy}/G \end{aligned}\right\} \tag{3.43}$$

and a state of plane stress is said to exist. In this case, the elasticity matrix [D] is reduced to,

$$[D] = \frac{E}{1-\nu^2}\begin{bmatrix} 1 & \nu & 0 \\ \nu & 1 & 0 \\ 0 & 0 & (1-\nu)/2 \end{bmatrix} \tag{3.44}$$

3.8.2 Plane Strain Problems

Figure 3.9 shows a solid body whose cross-section is uniform in the z-direction. Provided its length in this direction is large compared to the other dimensions, the section OACB can be regarded as being remote from the ends.

Assuming the applied loads are in the xy-plane, the resulting state of strain is reduced to two-dimensional with

$$w = 0 \text{ and } \frac{\partial}{\partial z} = 0$$

Figure 3.9 Plane strain configuration

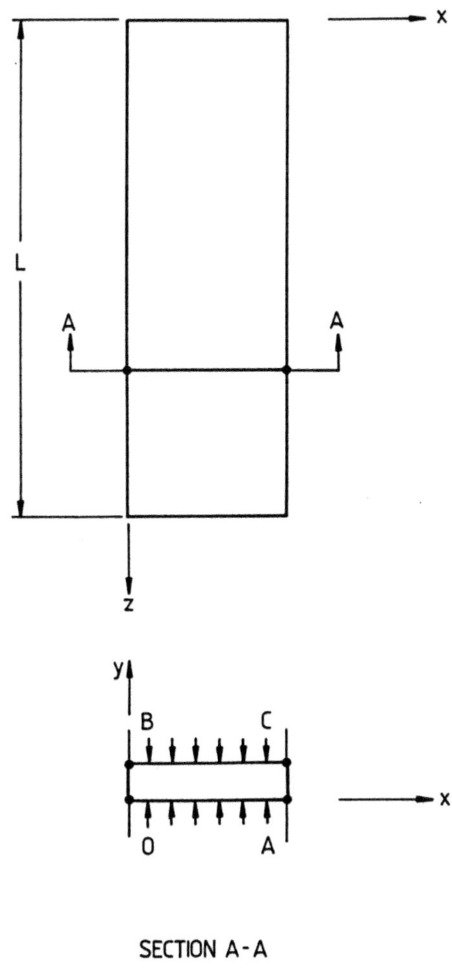

which results in,

$$\varepsilon_{zz} = \gamma_{xz} = \gamma_{yz} = 0$$

and a state of plane strain is said to exist. In this case, the elasticity matrix [D] is reduced to:

$$\begin{bmatrix} \lambda + 2G & \lambda & 0 \\ \lambda & \lambda + 2G & 0 \\ 0 & 0 & G \end{bmatrix} \tag{3.45}$$

Alternatively, the strains can be related to the stresses by the following expressions,

$$\left.\begin{aligned} \varepsilon_{xx} &= \frac{1-\nu^2}{E}\left(\sigma_{xx} - \frac{\nu}{1-\nu}\sigma_{yy}\right) \\ \varepsilon_{yy} &= \frac{1-\nu^2}{E}\left(\sigma_{yy} - \frac{\nu}{1-\nu}\sigma_{xx}\right) \\ \gamma_{xy} &= \frac{2(1+\nu)}{E}\tau_{xy} \end{aligned}\right\} \tag{3.46}$$

Note that when $\nu/(1-\nu)$ is replaced by ν and $E/(1-\nu^2)$ is replaced by E in the above expressions, plane stress equations are recovered. Conversely, if ν is replaced by $\nu/(1+\nu)$ and E is replaced by $E(1+2\nu)/(1+\nu)^2$ in the plane stress expressions (3.43), plane strain conditions are recovered. This approach will be adopted quite readily in fracture mechanics.

3.9 Strain Energy: Elastic Energy Density

Whenever an elastic body is subjected to either external or body forces, it undergoes deformations that are linearly related to the applied forces. The work done by these forces is stored within the body in the form of strain energy. In an ideal elastic body, no dissipation of energy takes place, and all of the stored energy is recoverable upon unloading. Accordingly, for a uniaxial state of stress in a linearly elastic body, the total work done W by the externally applied force F_x in producing a final relative displacement u can be expressed as,

$$W = \int_u F_x du \tag{3.47}$$

which is equal to the corresponding increase in the strain energy of the body U_l; as follows:

$$W = U_l = \frac{1}{2}\int_V \sigma_{xx}\varepsilon_{xx}dxdydz \tag{3.48}$$

Designating the strain energy per unit volume (strain energy density) as U then, for a linearly elastic body under uniform uniaxial loading condition, U is given by,

$$U = \int_{\varepsilon_{xx}} \sigma_{xx}d\varepsilon_{xx} = \frac{1}{2}E\varepsilon_{xx}^2 = \frac{1}{2}\sigma_{xx}\varepsilon_{xx} = \frac{\sigma_{xx}^2}{2E} \tag{3.49}$$

In the case of an externally applied shear force, the associated elastic strain energy density U is given by,

$$U = \frac{1}{2}G\gamma_{xy}^2 = \frac{1}{2}\tau_{xy}\gamma_{xy} = \frac{\tau_{xy}^2}{2G} \tag{3.50}$$

We can now generalize the above expressions to account for general loading conditions. In this case, the strain energy density is given by,

$$\left.\begin{aligned} U &= \int_{\varepsilon_{ij}} \sigma_{ij}d\varepsilon_{ij} \\ U &= \frac{1}{2}\begin{bmatrix}\sigma_{xx} & \sigma_{yy} & \sigma_{zz} & \tau_{xy} & \tau_{yz} & \tau_{zx}\end{bmatrix}\begin{bmatrix}\varepsilon_{xx}\\ \varepsilon_{yy}\\ \varepsilon_{zz}\\ \gamma_{xy}\\ \gamma_{yz}\\ \gamma_{zx}\end{bmatrix}\end{aligned}\right\} \tag{3.51}$$

i.e.,

$$U = \frac{1}{2}\sigma^T \cdot \varepsilon$$

Clearly, stress or strain can be eliminated from the above expressions using the stress-strain relations. For example, in terms of the stresses only, the above expression becomes,

$$U = \frac{1}{2E}\left[\begin{array}{l} \sigma_{xx}^2 + \sigma_{yy}^2 + \sigma_{zz}^2 \\ -2\nu\left(\sigma_{xx}\sigma_{yy} + \sigma_{yy}\sigma_{zz} + \sigma_{zz}\sigma_{xx}\right) \\ +2(1+\nu)\left(\tau_{xy}^2 + \tau_{yz}^2 + \tau_{zx}^2\right) \end{array}\right] \tag{3.52}$$

while in terms of the strains, it becomes,

$$U = \frac{E}{2}\left[\begin{array}{l} \frac{\nu}{(1+\nu)(1-2\nu)}\left(\varepsilon_{xx} + \varepsilon_{yy} + \varepsilon_{zz}\right)^2 \\ +\frac{1}{1+\nu}\left(\varepsilon_{xx}^2 + \varepsilon_{yy}^2 + \varepsilon_{zz}^2\right) \\ +\frac{1}{2(1+\nu)}\left(\gamma_{xy}^2 + \gamma_{yz}^2 + \gamma_{zx}^2\right) \end{array}\right] \tag{3.53}$$

An interesting relationship is deduced, if we differentiate the above energy expression (3.53) with respect to the strain components. Considering first the differential of U with respect to ε_{xx}, we get:

$$\frac{\partial U}{\partial \varepsilon_{xx}} = \frac{E}{2}\left[\frac{\nu}{(1+\nu)(1-2\nu)}2\left(\varepsilon_{xx} + \varepsilon_{yy} + \varepsilon_{zz}\right) + \frac{2}{1+\nu}\varepsilon_{xx}\right]$$

i.e.,

$$\frac{\partial U}{\partial \varepsilon_{xx}} = \lambda\varepsilon_V + 2G\varepsilon_{xx} = \sigma_{xx} \tag{3.54}$$

In a similar manner, we are able to show that:

$$\left.\begin{array}{c} \frac{\partial U}{\partial \varepsilon_{yy}} = \sigma_{yy}, \ \frac{\partial U}{\partial \varepsilon_{zz}} = \sigma_{zz} \\ \frac{\partial U}{\partial \gamma_{xy}} = \tau_{xy}, \ \frac{\partial U}{\partial \gamma_{yz}} = \tau_{yz}, \ \frac{\partial U}{\partial \gamma_{zx}} = \tau_{zx} \end{array}\right\} \tag{3.55}$$

Clearly, in order to ascertain the energy stored within an entire body, the elastic energy density is integrated over the original or undeformed volume V, viz.,

$$U_t = \int_V U dV = \int_x \int_y \int_z U dx dy dz \tag{3.56}$$

3.10 Components of Strain Energy

It is interesting to note that the elastic strain energy density is divided into two components: (i) volumetric energy and (ii) distortion energy. The first is responsible for change in volume as described by the first invariants of the stress and strain and the second is responsible for the change of shape as described by the deviation stresses. Accordingly,

$$U = U_V + U_d \tag{3.57}$$

where it can be shown that,

$$U_V = \frac{1}{2}\sigma_m \varepsilon_V$$

or equivalently,

$$U_V = \frac{1-2\nu}{6E}(\sigma_1 + \sigma_2 + \sigma_3)^2 \tag{3.58}$$

and

$$U_d = \frac{1+\nu}{3E}\left[\sigma_1^2 + \sigma_2^2 + \sigma_3^2 - (\sigma_1\sigma_2 + \sigma_2\sigma_3 + \sigma_3\sigma_1)\right] \tag{3.59}$$

The significance of the above expressions will be dealt with later in this chapter under yield criteria.

3.11 Plane Elasticity: Airy's Stress Function

In general elasticity problems, it is possible to find a stress function ϕ, which satisfies the equilibrium equation, the compatibility condition and the boundary conditions. The resulting stress field is then representative of the state of stress. This approach has been used more effectively in two-dimensional problems.

Consider the following equations of equilibrium in two dimensions with no body forces; namely,

$$\left.\begin{aligned} \frac{\partial \sigma_{xx}}{\partial x} + \frac{\partial \tau_{yx}}{\partial y} = 0 \\ \frac{\partial \sigma_{yy}}{\partial y} + \frac{\partial \tau_{xy}}{\partial x} = 0 \end{aligned}\right\} \tag{3.60}$$

Next, consider the compatibility condition provided earlier by Equation (3.32) i.e.,

$$\frac{\partial^2 \varepsilon_{xx}}{\partial y^2} + \frac{\partial^2 \varepsilon_{yy}}{\partial x^2} = \frac{\partial^2 \gamma_{xy}}{\partial x \partial y} \tag{3.32}$$

Let us now choose a function $\phi(x, y)$ such that,

$$\sigma_{xx} = \frac{\partial^2 \phi}{\partial y^2}, \sigma_{yy} = \frac{\partial^2 \phi}{\partial x^2} \text{ and } \tau_{xy} = -\frac{\partial^2 \phi}{\partial x \partial y} \tag{3.61}$$

If one substitutes the above expressions for the stresses in terms of the stress function ϕ into the equilibrium equations, one realizes that the equilibrium equations are identically satisfied for all values of ϕ. Substituting the above expressions for the stress into the compatibility condition gives:

$$\left.\begin{array}{c} \dfrac{\partial^4 \phi}{\partial x^4} + 2\dfrac{\partial^4 \phi}{\partial x^2 \partial y^2} + \dfrac{\partial^4 \phi}{\partial y^4} = 0 \\ \text{or} \\ \nabla^2\left(\nabla^2 \phi\right) = \nabla^2\left(\sigma_{xx} + \sigma_{yy}\right) = 0 \end{array}\right\} \tag{3.62}$$

where

$$\nabla^2 = \frac{\partial^2}{\partial x^2} + \frac{\partial^2}{\partial y^2} \tag{3.63}$$

and is known as the Laplacian. In the above expressions, ϕ is a bi-harmonic function known as Airy's stress function and is used in solving two-dimensional boundary-value problems. Since solutions of Equation (3.62) satisfy both compatibility and equilibrium, any solution of the equation gives a possible stress distribution in an elastic body. In order that the stress distribution corresponds to a specific problem, it is further required that the boundary condition be fulfilled. Clearly, the above expressions apply for plane stress and plane strain conditions.

3.11.1 Stress Functions with Body Forces

In this case, let us assume that the body forces can be expressed in terms of a potential function $V(x, y)$, such that:

$$X = -\frac{\partial V}{\partial x} \text{ and } Y = -\frac{\partial V}{\partial y} \tag{3.64}$$

The equilibrium equations then become,

$$\frac{\partial}{\partial x}(\sigma_{xx} - V) + \frac{\partial \tau_{yx}}{\partial y} = 0$$

$$\frac{\partial}{\partial y}(\sigma_{yy} - V) + \frac{\partial \tau_{xy}}{\partial x} = 0$$

If we now select a stress function ϕ, such that:

$$\left.\begin{aligned} \sigma_{xx} &= \frac{\partial^2 \phi}{\partial y^2} + V \\ \sigma_{yy} &= \frac{\partial^2 \phi}{\partial x^2} + V \\ \tau_{xy} &= -\frac{\partial^2 \phi}{\partial x \partial y} \end{aligned}\right\} \tag{3.65}$$

then the equilibrium equations are satisfied. For the complete solution, we must consider the compatibility condition (3.32).

If we now make use of the stress-strain relations, we arrive at the following differential equation:

$$\left.\begin{gathered} \frac{\partial^4 \phi}{\partial x^4} + 2\frac{\partial^4 \phi}{\partial x^2 \partial y^2} + \frac{\partial^4 \phi}{\partial y^4} + \frac{1-2\nu}{1-\nu}\left(\frac{\partial^2 V}{\partial x^2} + \frac{\partial^2 V}{\partial y^2}\right) = 0 \\ \text{or simply} \\ \nabla^2(\nabla^2 \phi) = -\frac{1-2\nu}{1-\nu}(\nabla^2 V) \end{gathered}\right\} \tag{3.66}$$

for plane strain conditions, and

$$\nabla^2(\nabla^2 \phi) = -(1-\nu)(\nabla^2 V) \tag{3.67}$$

for plane stress conditions [3.2].

3.11.2 Stress Function in Cylindrical Coordinates

Using the appropriate transformation, the following expressions for cylindrical polar stresses are obtained in terms of the stress function $\phi(r, \theta)$:

$$\left.\begin{aligned} \sigma_{rr} &= \frac{1}{r}\frac{\partial \phi}{\partial r} + \frac{1}{r^2}\frac{\partial^2 \phi}{\partial \theta^2} \\ \sigma_{\theta\theta} &= \frac{\partial^2 \phi}{\partial r^2} \\ \tau_{r\theta} &= -\frac{1}{r}\frac{\partial \phi}{\partial \theta} - \frac{1}{r}\frac{\partial^2 \phi}{\partial r \partial \theta} = -\frac{\partial}{\partial r}\left(\frac{1}{r}\frac{\partial \phi}{\partial \theta}\right) \end{aligned}\right\} \tag{3.68}$$

These expressions satisfy the following equilibrium equations with no body forces:

$$\left.\begin{aligned} \frac{\partial \sigma_{rr}}{\partial r} + \frac{\partial \tau_{r\theta}}{r\partial \theta} + \frac{\sigma_{rr} - \sigma_{\theta\theta}}{r} = 0 \\ \frac{\partial \sigma_{\theta\theta}}{r\partial \theta} + \frac{\partial \tau_{\theta r}}{\partial r} + \frac{2\tau_{r\theta}}{r} = 0 \end{aligned}\right\} \tag{3.69}$$

and will identically satisfy the compatibility condition, thus giving,

$$\left(\frac{\partial^2}{\partial r^2} + \frac{1}{r}\frac{\partial}{\partial r} + \frac{1}{r^2}\frac{\partial^2}{\partial \theta^2}\right)\left(\frac{\partial^2 \phi}{\partial r^2} + \frac{1}{r}\frac{\partial \phi}{\partial r} + \frac{1}{r^2}\frac{\partial^2 \phi}{\partial \theta^2}\right) = 0 \tag{3.70}$$

i.e.,

$$\left.\begin{aligned} \nabla^2(\nabla^2 \phi) = 0 \\ \text{or} \\ \nabla^2(\sigma_{rr} + \sigma_{\theta\theta}) = 0 \end{aligned}\right\} \tag{3.71}$$

The above expressions can be modified to account for body forces. For plane strain conditions, the above expressions reduce to:

$$\nabla^2(\nabla^2 \phi) = -\frac{1-2\nu}{1-\nu}\left(\frac{\partial^2 V}{\partial r^2} + \frac{1}{r}\frac{\partial V}{\partial r}\right) \tag{3.72}$$

and for plane stress conditions they reduce to:

$$\nabla^2(\nabla^2 \phi) = -(1-\nu)\left(\frac{\partial^2 V}{\partial r^2} + \frac{1}{r}\frac{\partial V}{\partial r}\right) \tag{3.73}$$

3.11.3 Stress Concentration Around a Hole

Even a superficial look at the history of engineering failures indicates that fatigue failures inevitably start at a geometric discontinuity or from stress concentration features. The stress concentration features include material inhomogeneities, sharp notches, holes, machining marks, corrosion pits or defective welds. This, together with the development of new materials working in aggressive environment and high temperatures, has resulted in numerous catastrophic failures such as the Comet catastrophic failure, which is covered in a later chapter. For the moment, let us content ourselves with the study of stress concentration due to the presence of a circular hole in a very large thin plate (plane stress) subjected to a remote uniform uniaxial boundary loading.

Let us now use Airy's stress function to develop the stress field in the vicinity of a circular hole. We commence by initially considering that the plate was subjected to a uniform tensile stress σ_0, as shown in Figure 3.10a. In terms of Cartesian coordinates, the uniform stress field can be expressed by the stress function,

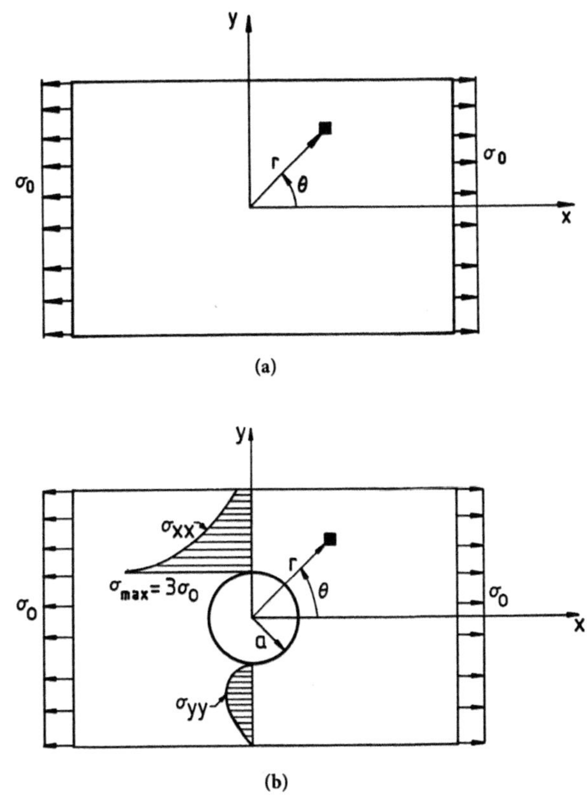

Figure 3.10 (**a**) Large thin plate subjected to a uniform tensile stress, σ_0. (**b**) Circular hole in the loaded plate

$$\phi_1 = \frac{\sigma_0}{2} y^2 \tag{a}$$

from which

$$\sigma_{xx} = \sigma_0, \sigma_{yy} = \tau_{xy} = 0 \tag{b}$$

In terms of cylindrical polar coordinates, the above expressions take the form,

$$\left.\begin{aligned}
\phi_1 &= \frac{\sigma_0}{4} r^2 (1 - \cos 2\theta) \\
\sigma_{rr}^{(1)} &= \frac{\sigma_0}{2} (1 + \cos 2\theta) \\
\sigma_{\theta\theta}^{(1)} &= \frac{\sigma_0}{2} (1 - \cos 2\theta) \\
\tau_{r\theta}^{(1)} &= -\frac{\sigma_0}{2} \sin 2\theta
\end{aligned}\right\} \tag{c}$$

If a hole of radius a is now introduced into the plate (Figure 3.10b), the inner boundary must satisfy the following boundary conditions for all values of θ:

$$\sigma_{rr} = \tau_{r\theta} = 0 \text{ at } r = a \tag{d}$$

In other words, we must introduce another stress function, call it ϕ_2, to provide the additional terms to give values of σ_{rr} and $\tau_{r\theta}$ that are equal and opposite to those of the expressions given in (3.73c). This can be satisfied by a stress function of the form:

$$\phi_2 = \frac{\sigma_0 a^2}{2} \left[-\log r + \left(1 - \frac{1}{2} \left(\frac{a}{r} \right)^2 \right) \cos 2\theta \right] \tag{e}$$

which, upon the appropriate differentiation, yields the following stress expressions,

$$\left.\begin{aligned}
\sigma_{rr}^{(2)} &= \frac{\sigma_0 a^2}{2r^2} \left[-1 + \left(-4 + 3\left(\frac{a}{r} \right)^2 \right) \cos 2\theta \right] \\
\sigma_{\theta\theta}^{(2)} &= \frac{\sigma_0 a^2}{2r^2} \left[1 - 3\left(\frac{a}{r} \right)^2 \cos 2\theta \right] \\
\tau_{r\theta}^{(2)} &= \frac{\sigma_0 a^2}{2r^2} \left[-2 + 3\left(\frac{a}{r} \right)^2 \right] \sin 2\theta
\end{aligned}\right\} \tag{f}$$

which at $r = a$ become,

$$\left.\begin{aligned} \sigma_{rr}^{(2)} &= -\frac{\sigma_0}{2}(1+\cos 2\theta) = -\sigma_{rr}^{(1)} \\ \tau_{r\theta}^{(2)} &= +\frac{\sigma_0}{2}\sin 2\theta = -\tau_{r\theta}^{(1)} \end{aligned}\right\} \quad \text{(g)}$$

The other condition to be satisfied is that ϕ_2 must have a negligible effect upon the stress pattern at large values of *r*. Inserting a large value of *r* in Equation (3.80) yields,

$$\left.\begin{aligned} \sigma_{rr}^{(2)} &= +\frac{\sigma_0}{2}\left[-\frac{a^2}{r^2}+\left(3\frac{a^4}{r^4}-4\frac{a^2}{r^2}\right)\cos 2\theta\right] \simeq 0 \\ \sigma_{\theta\theta}^{(2)} &= +\frac{\sigma_0}{2}\left[\frac{a^2}{r^2}-3\frac{a^4}{r^4}\cos 2\theta\right] \simeq 0 \\ \tau_{r\theta}^{(2)} &= +\frac{\sigma_0}{2}\left[\left(3\frac{a^4}{r^4}-2\frac{a^2}{r^2}\right)\sin 2\theta\right] \simeq 0 \end{aligned}\right\} \quad \text{(h)}$$

since $a \ll r$; accordingly $(a/r)^2$ and $(a/r)^4$ are very small values.

The resulting stress system is therefore given by the superposition of the stress systems provided by ϕ_1 and ϕ_2, i.e.

$$\left.\begin{aligned} \sigma_{rr} &= \sigma_{rr}^{(1)}+\sigma_{rr}^{(2)} = \frac{\sigma_0}{2}\left\{\left(1-\left(\frac{a}{r}\right)^2\right)\left[1+\left(1-3\left(\frac{a}{r}\right)^2\right)\cos 2\theta\right]\right\} \\ \sigma_{\theta\theta} &= \sigma_{\theta\theta}^{(1)}+\sigma_{\theta\theta}^{(2)} = \frac{\sigma_0}{2}\left[\left(1+\left(\frac{a}{r}\right)^2\right)-\left(1+3\left(\frac{a}{r}\right)^4\right)\cos 2\theta\right] \\ \tau_{r\theta} &= \tau_{r\theta}^{(1)}+\tau_{r\theta}^{(2)} = -\frac{\sigma_0}{2}\left[\left(1+3\left(\frac{a}{r}\right)^2\right)\left(1-\left(\frac{a}{r}\right)^2\right)\sin 2\theta\right] \end{aligned}\right\} \quad \text{(i)}$$

Equation (3.73i) give the stress components at any point in the plate. For example, the stresses along the *y*-axis can be obtained by setting $\theta = \pm\pi/2$ and $r = y$ in the above equations. This yields:

$$\left.\begin{aligned} \sigma_{rr} = \sigma_{yy} &= \frac{\sigma_0}{2}\left[1-\left(\frac{a}{r}\right)^2\right]3\left(\frac{a}{y}\right)^2 \\ \sigma_{\theta\theta} = \sigma_{xx} &= \frac{\sigma_0}{2}\left[2+\left(\frac{a}{y}\right)^2+3\left(\frac{a}{y}\right)^4\right] \\ \tau_{r\theta} = \tau_{xy} &= 0 \end{aligned}\right\} \quad \text{(j)}$$

The above normal stresses equations are normalized (σ_{xx}/σ_0 and σ_{yy}/σ_0) and plotted as a function of the normalized position along the *y*-axis in Figure 3.11a. An examination of the figure indicates that the presence of a hole in an infinite plate

Figure 3.11 Distribution of σ_{xx}/σ_0 and σ_{yy}/σ_0 (**a**) along the horizontal y-axis and (**b**) along the vertical x-axis

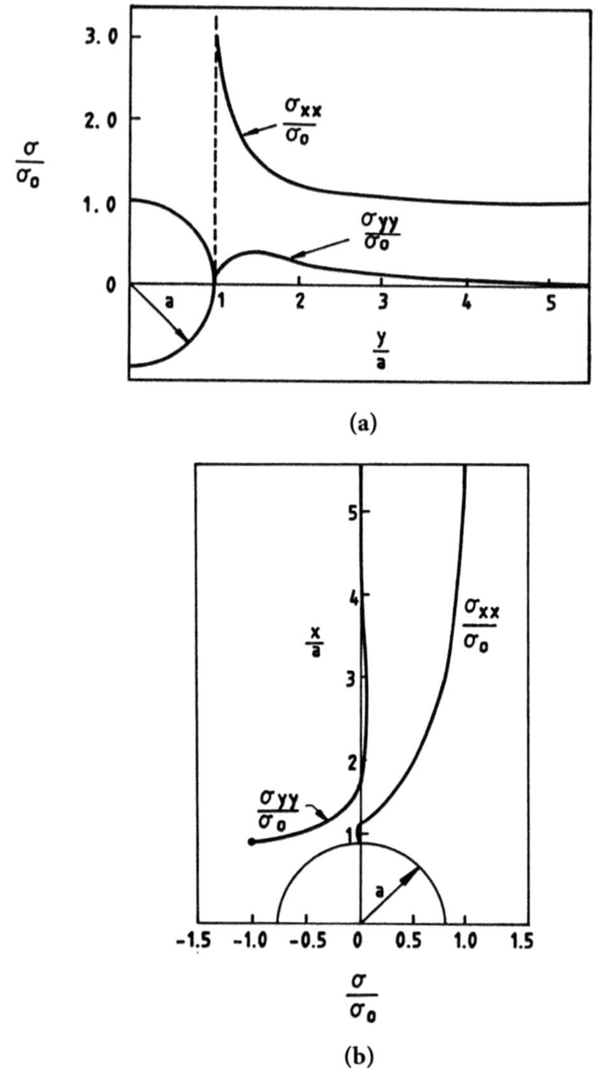

under uniaxial tension increases the σ_{xx} stress by a factor of 3. This factor is known as the *stress concentration factor*. It is also clear from the figure that, in spite of the fact that the applied stress is uniaxial tension, the resulting state of stress along the x-axis is two-dimensional. This again has been created by the presence of the hole.

In a similar manner, the stresses along the x-axis can be obtained by setting $\theta = 0$ and $r = x$ in Equation (3.73i), thus giving:

$$\left.\begin{aligned} \sigma_{rr} = \sigma_{xx} &= \frac{\sigma_0}{2}\left[2 - 5\left(\frac{a}{x}\right)^2 + 3\left(\frac{a}{x}\right)^4\right] \\ \sigma_{\theta\theta} = \sigma_{yy} &= \frac{\sigma_0}{2}\left[\left(\frac{a}{x}\right)^2 - 3\left(\frac{a}{x}\right)^4\right] \\ \tau_{r\theta} &= \tau_{xy} = 0 \end{aligned}\right\} \tag{k}$$

Again, the normalized distributions of the stresses against position are given in Figure 3.11b. It can be seen that $\sigma_{yy}/\sigma_0 = -1$ at the boundary of the hole, thus indicating that the presence of the hole produces a change in the sign of the stress in the body.

The above analysis clearly indicates that the presence of a hole in a plate has a number of important features:

(i) it produces a concentration of stress,
(ii) it induces stress biaxiality,
(iii) it produces a change in signs of stress, and
(iv) it creates non-uniform stress distribution in the body with the consequence that certain regions of the body are under-stresses and others are over-stressed.

As a matter of fact, since fatigue failure involves both the initiation and propagation of cracks from a stress concentration feature, the presence of a hole in a plate will promote initiation and enhance the propagation of these cracks.

3.12 Complex Stress Functions Approach

In the previous section, we adopted Airy's stress function for the determination of the stress field in a plate containing a hole. Unfortunately, this approach is limited to 'smooth' boundaries and represents a special case of the more general case of the complex stress functions due to Muskhelishvili [3.3–3.5].

Muskhelishvili's complex stress function approach enables the Airy stress function ϕ to be written in terms of two complex potentials $\psi(z)$ and $\chi(z)$. Indeed, considerable flexibility can be gained by adopting the complex stress function approach, since they are generally more concise and permit the solution of problems containing inhomogeneities, sharp corners, notches and cracks. This may be partly due to the fact that these functions allow the use of conformal mapping, and partly due to the fact that they permit the use of sophisticated analytical and numerical techniques, discussed later in this text.

Before we begin with the concept and the development of complex stress functions, let us examine some of the pertinent aspects of analytic and harmonic functions.

An analytical function $f(z)$ can be regarded as a function of x and y having single-valued continuous partial derivatives in a given region R. Thus,

$$\left.\begin{aligned} \frac{\partial}{\partial x} f(z) &= f'(z) \\ \frac{\partial}{\partial y} f(z) &= i f'(z) \end{aligned}\right\} \tag{3.74}$$

where

$$z = x + iy$$

Now, let us assume the function $f(z)$ takes the form,

$$\left.\begin{array}{c} f(z) = P(x, y) + iQ(x, y) \\ \text{or simply} \\ f(z) = P + iQ \end{array}\right\} \tag{3.75}$$

Accordingly,

$$\left.\begin{array}{c} \dfrac{\partial f(z)}{\partial x} = \dfrac{\partial P}{\partial x} + i\dfrac{\partial Q}{\partial x} \\ \text{and} \\ \dfrac{\partial f(z)}{\partial y} = \dfrac{\partial P}{\partial y} + i\dfrac{\partial Q}{\partial y} \end{array}\right\} \tag{3.76}$$

Since

$$i\frac{\partial f(z)}{\partial x} = \frac{\partial f(z)}{\partial y}$$

then

$$i\left(\frac{\partial P}{\partial x} + i\frac{\partial Q}{\partial x}\right) = \left(\frac{\partial P}{\partial y} + i\frac{\partial Q}{\partial y}\right)$$

Equating the real and imaginary parts of the above expression yields,

$$\frac{\partial P}{\partial x} = \frac{\partial Q}{\partial y} \text{ and } \frac{\partial P}{\partial y} = -\frac{\partial Q}{\partial x} \tag{3.77}$$

Expressions (3.77) are known as the Cauchy-Riemann equations.

Differentiating the first of Equation (3.77) with respect to x and the second with respect to y, and adding the terms, yields,

$$\left.\begin{array}{c}\left(\frac{\partial^2}{\partial x^2}+\frac{\partial^2}{\partial y^2}\right)P=0\\ \text{or}\\ \nabla^2 P=0\end{array}\right\}\tag{3.78}$$

As explained earlier, an equation of this form is known as Laplace's equation and any solution is called a *harmonic function*. In the same way, we can eliminate *P* from expression (3.77), thus giving,

$$\left.\begin{array}{c}\left(\frac{\partial^2}{\partial x^2}+\frac{\partial^2}{\partial y^2}\right)Q=0\\ \text{or}\\ \nabla^2 Q=0\end{array}\right\}\tag{3.79}$$

Thus, if two functions *P* and *Q* of *x* and *y* are derived as the real and imaginary parts of an analytic function *f*(*z*), then each could be a valid solution to Laplace's equation. In this case, the functions *P* and *Q* are termed *conjugate harmonic functions*. It is evident that both functions are real and that the knowledge of one will enable the determination of the conjugate via the use of the Cauchy-Riemann equations.

3.12.1 Development of Complex Stress Function

Airy's stress function Equation (3.62) indicates that,

$$\left(\frac{\partial^2}{\partial x^2}+\frac{\partial^2}{\partial y^2}\right)\left(\frac{\partial^2\phi}{\partial x^2}+\frac{\partial^2\phi}{\partial y^2}\right)=\left(\frac{\partial^2}{\partial x^2}+\frac{\partial^2}{\partial y^2}\right)(\sigma_{xx}+\sigma_{yy})=0$$

If we use the notations adopted in Ref. [3.1], and write,

$$P=\sigma_{xx}+\sigma_{yy}\tag{a}$$

we observe that $f(z) = P + iQ$ is an analytic function. The integral of the above function with respect to *z* is also another analytic function. Therefore, according to Muskhelishvili.

$$\int f(z)dz=\int(P+iQ)dz=4\psi(z)(\text{say})\tag{b}$$

or alternatively,

$$\psi(z) = p + iq = \frac{1}{4}\int f(z)dz \tag{c}$$

If we now differentiate expression (c) with respect to x, we get:

$$\psi'(z) = \frac{\partial p}{\partial x} + i\frac{\partial q}{\partial x} = \frac{1}{4}f(z) \tag{d}$$

i.e.,

$$\frac{\partial p}{\partial x} + i\frac{\partial q}{\partial x} = \frac{1}{4}(P + iQ) \tag{e}$$

Accordingly,

$$\frac{\partial p}{\partial x} = \frac{P}{4} \text{ and } \frac{\partial q}{\partial x} = \frac{Q}{4} \tag{f}$$

Now, since p and q are conjugate analytic functions, they satisfy the Cauchy-Riemann equation, i.e.

$$\frac{\partial p}{\partial x} = \frac{\partial q}{\partial y} = \frac{P}{4} \text{ and } \frac{\partial p}{\partial y} = -\frac{\partial q}{\partial x} = -\frac{Q}{4} \tag{g}$$

Then, it is possible to express Equation (a) as,

$$\nabla^2\phi = 4\frac{\partial p}{\partial x} = 2\frac{\partial p}{\partial x} + 2\frac{\partial q}{\partial y} \tag{h}$$

i.e.,

$$\nabla^2\phi - 2\frac{\partial p}{\partial x} - 2\frac{\partial q}{\partial y} = 0 \tag{i}$$

Equivalently, the above expression can be written as

$$\nabla^2(\phi - xp - yq) = 0 \tag{j}$$

i.e.,

$$\phi - xp - yq = p_1$$

where p_1 is some harmonic function. Consequently, one can obtain an expression for the stress function ϕ as being,

$$\phi = xp + yq + p_1 \quad \text{(k)}$$

Clearly, p_1 has its conjugate q_1, such that,

$$\chi(z) = p_1 + iq_1 \quad \text{(l)}$$

It is then easily verified that the real part of,

$$(x - iy)(p + iq) + p_1 + iq_1$$

is identical to the right-hand side of Equation (k). Accordingly, any stress function can be written as,

$$\phi = \text{Re}\ [\bar{z}\psi(z) + \chi(z)] \quad (3.80)$$

where $\psi(z)$ and $\chi(z)$ are suitably chosen analytic functions.

It is worth noting that the general expression for the stress function $F(z)$ is:

$$F(z) = \phi + i\Omega = \bar{z}\psi(z) + \chi(z) \quad (3.81)$$

and it is always possible to obtain the equivalent Airy's stress function ϕ by this method.

3.12.2 Determination of Displacement Field

Let us consider the case for plane stress conditions, where $\sigma_{zz} = \tau_{xz} = \tau_{yz} = 0$, and according to Equations (3.43) and (3.26), we have,

$$\left.\begin{aligned} E\frac{\partial u}{\partial x} &= \sigma_{xx} - \nu\sigma_{yy} \\ E\frac{\partial v}{\partial y} &= \sigma_{yy} - \nu\sigma_{xx} \\ G\left(\frac{\partial u}{\partial y} + \frac{\partial v}{\partial x}\right) &= \tau_{xy} \end{aligned}\right\} \quad (3.82)$$

The first of Equation (3.82) can be rewritten in terms of the stress function ϕ and $P = \sigma_{xx} + \sigma_{yy}$, such that,

$$E\frac{\partial u}{\partial x} = -(1 + \nu)\frac{\partial^2 \phi}{\partial x^2} + P \quad \text{(a)}$$

and similarly,

$$E\frac{\partial v}{\partial x} = -(1+\nu)\frac{\partial^2 \phi}{\partial y^2} + P \tag{b}$$

Equations (a) and (b) can be rewritten as,

$$2G\frac{\partial u}{\partial x} = -\frac{\partial^2 \phi}{\partial x^2} + \left(\frac{4}{1+\nu}\right)\frac{\partial p}{\partial x} \tag{c}$$

$$2G\frac{\partial u}{\partial x} = -\frac{\partial^2 \phi}{\partial x^2} + \left(\frac{4}{1+\nu}\right)\frac{\partial q}{\partial y} \tag{d}$$

which upon integration yield the following expressions for u and v.

$$2Gu = -\frac{\partial \phi}{\partial x} + \left(\frac{4}{1+\nu}\right)p + f_1(y) \tag{e}$$

$$2Gv = -\frac{\partial \phi}{\partial y} + \left(\frac{4}{1+\nu}\right)q + f_2(x) \tag{f}$$

where $f_1(y)$ and $f_2(x)$ are arbitrary functions. However, the displacement equations given by expression (e) and (f) must satisfy the third equation of (3.82), i.e.

$$-\frac{\partial^2 \phi}{\partial x \partial y} + \frac{2}{1+\nu}\left(\frac{\partial p}{\partial y} + \frac{\partial q}{\partial x}\right) + \frac{1}{2}\frac{\partial f_1}{\partial y} + \frac{1}{2}\frac{\partial f_2}{\partial x} = \tau_{xy} \tag{g}$$

Clearly, the first term on the left-hand side, $-\partial^2 \phi/\partial x\partial y$, is the shear stress τ_{xy}. Now, since p and q are conjugate harmonic functions, they satisfy the Cauchy-Riemann equations and as a result $(\partial p/\partial y + \partial q/\partial x) = 0$. Accordingly, Equations (e) and (f) represent expressions for the displacement, with the condition that

$$\frac{\partial f_1}{\partial y} + \frac{\partial f_2}{\partial x} = 0$$

which implies that

$$\frac{\partial f_1}{\partial y} = -\frac{\partial f_2}{\partial x} = \text{constant} \tag{h}$$

representing rigid body displacement.

Accordingly, the displacement expressions for plane stress conditions now reduce to,

$$\left.\begin{aligned} 2Gu &= -\frac{\partial \phi}{\partial x} + \left(\frac{4}{1+\nu}\right)p \\ 2Gv &= -\frac{\partial \phi}{\partial y} + \left(\frac{4}{1+\nu}\right)q \end{aligned}\right\} \tag{3.83}$$

These expressions were first stated by Love [3.2]. In the case of plane strain, the above expressions can be modified according to the approach described earlier.

3.12.3 Stresses and Displacements Using Complex Potentials

It is now our intention to generalize the above expressions for the stresses and displacements in terms of the complex $\psi(z)$ and $\chi(z)$. To this end, let us assume that $f(z) = \alpha + i\beta$ are real functions. Let us now define $f(\overline{z})$ as being the conjugate of $f(z)$, in which i is replaced by $-i$. The long term conjugate here is quite different from that used earlier in relation to conjugate harmonic functions. Accordingly,

$$\overline{f}(\overline{z}) = \alpha - i\beta \tag{a}$$

Evidently, the use of $\overline{f}(x)$ and $f(\overline{z})$ enables us to write,

$$f(z) + \overline{f}(\overline{z}) = 2\alpha = 2\ \mathrm{Re}(\ f(z)) \tag{b}$$

If we adopt the same approach with the bracketed complex stress function described in Equation (3.80), we get,

$$2\phi = \overline{z}\psi(z) + \chi(z) + z\overline{\psi}(\overline{z}) + \overline{\chi}(\overline{z}) \tag{3.84}$$

which upon differentiation yields

$$2\frac{\partial \phi}{\partial x} = \overline{z}\psi'(z) + \psi(z) + \chi'(z) + z\overline{\psi}'(\overline{z}) + \overline{\psi}(\overline{z}) + \overline{\chi}'(\overline{z}) \tag{c}$$

$$2\frac{\partial \phi}{\partial y} = i[\overline{z}\psi'(z) - \psi(z) + \chi'(z) - z\overline{\psi}'(\overline{z}) + \overline{\psi}(\overline{z}) - \overline{\chi}'(\overline{z})] \tag{d}$$

The above expressions are combined to give,

$$\frac{\partial \phi}{\partial x} + i\frac{\partial \phi}{\partial y} = \psi(z) + z\overline{\psi}'(\overline{z}) + \overline{\chi}'(\overline{z}) \tag{e}$$

Now, combining Equation (3.83), we find,

$$2G(u + iv) = -\left(\frac{\partial \phi}{\partial x} + i\frac{\partial \phi}{\partial y}\right) + \frac{4}{1+\nu}(p + iq) \tag{f}$$

Replacing $-\left(\left(\frac{\partial \phi}{\partial x}\right) + i\left(\frac{\partial \phi}{\partial y}\right)\right)$ by its equivalent from expression (e) and noting that $p + iq = \psi(z)$, we get the following expression for plane stress displacements u and v in terms of complex potentials $\psi(z)$ and $\chi(z)$:

$$2G(u + iv) = \frac{3-\nu}{1+\nu}\psi(z) - z\overline{\psi}'(\overline{z}) - \overline{\chi}'(\overline{z}) \tag{3.85}$$

Again, for plane strain conditions, replace ν by $\nu/(1 - \nu)$ to obtain the appropriate expressions.

Now, let us turn our attention to the derivation of the stress components σ_{xx}, σ_{yy} and τ_{xy}. Clearly, these can be obtained directly from the second derivatives of the stress function expression given by Equation (3.61). However, in view of the latter applications, we intend to use the concept of the complex potentials. In this approach, it is suggested to differentiate expression (e) once with respect to x to yield:

$$\frac{\partial^2 \phi}{\partial x^2} + i\frac{\partial^2 \phi}{\partial x \partial y} = \psi'(z) + z\overline{\psi}''(\overline{z}) + \overline{\psi}'(\overline{z}) + \overline{\chi}''(\overline{z}) \tag{g}$$

and another with respect to y and multiplying by i, will yield,

$$i\frac{\partial^2 \phi}{\partial x \partial y} - \frac{\partial^2 \phi}{\partial y^2} = -\psi'(z) + z\overline{\psi}''(\overline{z}) - \overline{\psi}'(\overline{z}) + \overline{\chi}''(\overline{z}) \tag{h}$$

Subtracting Equation (h) from (g) yields,

$$\sigma_{xx} + \sigma_{yy} = 2\psi'(z) + 2\overline{\psi}'(\overline{z}) = 4\ \mathrm{Re}\ \psi'(z) \tag{3.86}$$

and adding Equation (h) to (g) yields,

$$\sigma_{yy} - \sigma_{xx} - 2i\tau_{xy} = 2[z\overline{\psi}''(\overline{z}) + \overline{\chi}''(\overline{z})] \tag{3.87}$$

Changing i to $-i$ on both sides of Equation (3.87) gives an alternative expression, i.e.,

$$\sigma_{yy} - \sigma_{xx} + 2i\tau_{xy} = 2[\overline{z}\psi''(z) + \chi''(z)] \tag{3.88}$$

With the knowledge of the complex potentials $\psi(z)$ and $\chi(z)$, separating real and imaginary parts of Equation (3.88) yields expression for $(\sigma_{yy} - \sigma_{xx})$ and τ_{xy}. Combining these results with Equation (3.85) gives the required expressions for σ_{xx}, σ_{yy} and τ_{xy}.

3.12.4 Curvilinear Coordinates

These are often used in association with certain geometries in order to simplify the calculations of stresses and displacements. Typically, they are used in cases involving notches, circular or elliptic holes, and cracks. They allow conformal mapping and accordingly are very useful for handling complex problems [3.1, 3.5].

In general, one can define the position of a point in two-dimensional space via the use of the Cartesian co-ordinate system (*x*,*y*). Alternatively, point P may also be intrinsically defined by the intersection of two curves via the use of the curvilinear o, $\alpha = C_1$ and $\beta = C_2$, as depicted in Figure 3.12, α and *β* may intersect orthogonally, in which case the system is called orthogonal. It is worth noting that α and *β* represent some intrinsic coordinates but do not necessarily represent the length directly.

Let us now consider the rectangular coordinates (*x*,*y*) of a point expressed in terms of α and *β*, and that:

$$F_1(x, y) = \alpha, F_2(x, y) = \beta \tag{a}$$

Giving definite constant values to α and *β*, the above expressions will represent two curves which will intersect, when $F_1(x,y)$ and $F_2(x,y)$ are suitable functions. In the case of polar coordinates, the relationship is evidently clear:

$$\left.\begin{aligned} F_1(x, y) &= \sqrt{x^2 + y^2} \text{ and } \alpha = r \\ F_2(x, y) &= \arctan\left(\frac{y}{x}\right) \text{ and } \beta = \theta \end{aligned}\right\} \tag{b}$$

Suppose that expression (a) can be solved for α and *β* in terms of *x* and *y*, i.e.,

Figure 3.12 Generalized curvilinear co-ordinate system

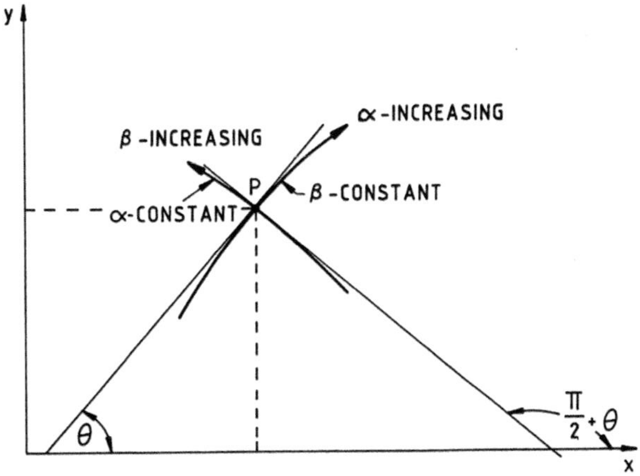

$$x = f_1(\alpha, \beta), y = f_2(\alpha, \beta) \tag{c}$$

In this case, the functions in Equations (a) and (c) are assumed to be single-valued and to have continuous derivatives so that the correspondence between (x,y) and (α,β) is unique. Effectively, given a point P in terms of the rectangular coordinates (x,y), we can associates with P a unique set of curvilinear coordinates (α,β). The set of Equations (a) and (c) define a transformation of coordinates.

Let us now consider the case corresponding to the equations,

$$x = c\cosh\alpha\cos\beta, y = c \ \sinh\alpha\sin\beta \tag{d}$$

where c is a constant. Elimination of β gives,

$$\frac{x^2}{c^2\cosh^2\alpha} + \frac{y^2}{c^2\sinh^2\alpha} = 1 \tag{e}$$

For a constant value of α (α_0, say) Equation (e) describes an ellipse with semi-axes,

$$\left.\begin{aligned} c\cosh\alpha_0 &= a \\ c\sinh\alpha_0 &= b \end{aligned}\right\} \tag{f}$$

with foci at $\pm C$. It is worth noting that, for different values of α, we obtain different ellipses with the same foci as depicted in Figure 3.13. In any of these ellipses α is constant and β varies through a range of 2π.

Figure 3.13 An elliptical co-ordinate system

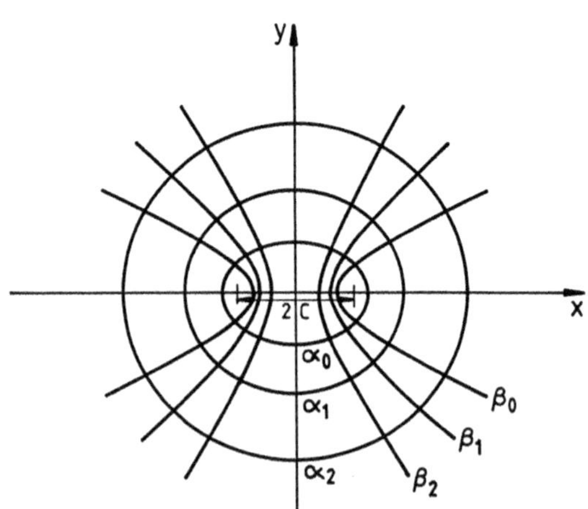

Alternatively, if we eliminate α, we get,

$$\frac{x^2}{c^2\cos^2\beta} - \frac{y^2}{c^2\sin^2\beta} = 1 \tag{g}$$

For a constant value of $\beta = \beta_0$, the above equation represents a hyperbola having the same foci as the ellipse, as shown in Figure 3.13. This Equation (g) represents a family of confocal hyperbolae.

3.12.5 *Stress and Displacement in Curvilinear Coordinates*

From the foregoing, it is now possible to express Equations (3.85), (3.86) and (3.88) in terms of the curvilinear coordinates α and β. In this case the complex potentials $\psi(z)$ and $\chi(z)$ are taken to be functions of α and β instead of x and y.

Figure 3.14 shows the notation used for the stresses. In this case, it is possible to use the transformation relations described earlier in this chapter to evaluate σ_α, σ_β and $\tau_{\alpha\beta}$. These give,

$$\left.\begin{aligned} \sigma_\alpha + \sigma_\beta &= \sigma_{xx} + \sigma_{yy} \\ \sigma_\beta - \sigma_\alpha + 2i\tau_{\alpha\beta} &= \left(\sigma_{yy} - \sigma_{xx} + 2i\tau_{xy}\right)e^{2i\theta} \end{aligned}\right\} \tag{3.89}$$

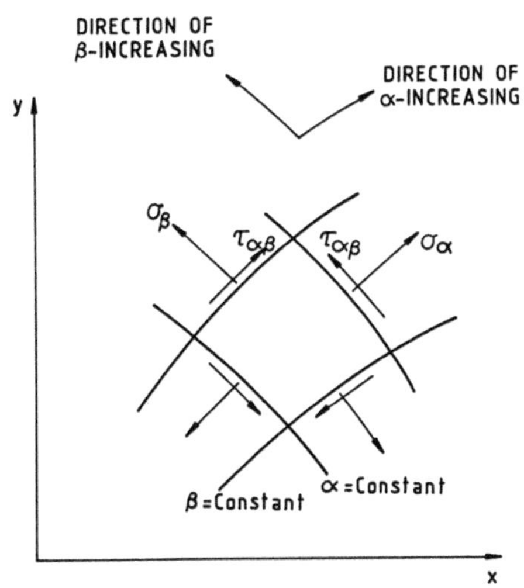

Figure 3.14 Notation used for the stresses in the generalized curvilinear coordinate system

where the factor $e^{2i\theta}$ for curvilinear coordinates can be obtained from expression [3.3],

$$e^{2i\theta} = \frac{f'(\xi)}{\overline{f}(\xi)} \tag{3.90}$$

where

$$\xi = \alpha + i\beta$$

Separating real and imaginary parts of the second of Equation (3.89), one obtains expressions for $(\sigma_\beta - \sigma_\alpha)$ and $\tau_{\alpha\beta}$ in terms of the original components σ_{xx}, σ_{yy} and τ_{xy}. Combining these results with the first of Equation (3.89) yields expressions for σ_α σ_β and $\tau_{\alpha\beta}$ which obviously agree with those using the transformation laws.

Combining Equations (3.86) and (3.88) with (3.89) yields the following expressions for the stresses in terms of the curvilinear coordinates α and β and the complex potentials $\psi(z)$ and $\chi(z)$:

$$\sigma_\alpha + \sigma_\beta = 2[\psi'(z) + \overline{\psi}'(\overline{z})] = 4 \text{ Re } \psi'(z) \tag{3.91}$$

$$\sigma_\beta - \sigma_\alpha + 2i\tau_{\alpha\beta} = 2[\overline{z}\psi''(z) + \chi''(z)]e^{2i\theta} \tag{3.92}$$

It must be remembered that in the above expressions $z = f(\xi)$.

In a somewhat similar fashion, the displacements in curvilinear coordinates u_α and u_β in the increasing directions of α and β, respectively, are specified in terms of the Cartesian components u and v. In this case, the relationships between the two sets are given as,

$$u_\alpha = u\cos\theta + v\sin\theta$$

$$u_\beta = v\cos\theta - u\sin\theta$$

and accordingly,

$$u_\alpha + iu_\beta = (u + iv)e^{-i\theta} \tag{3.93}$$

The expression for the curvilinear displacement components then becomes:

$$2G(u_\alpha + iu_\beta) = \left[\frac{3-\nu}{1+\nu}\psi(z) - z\overline{\psi}'(\overline{z}) - \overline{\chi}'(\overline{z})\right]e^{-i\theta} \tag{3.94}$$

We intend to use expressions (3.91), (3.92) and (3.94) in the solution of an elliptic hole in a plate under simple tension.

3.13 Stress Concentration around an Elliptic Hole

The detailed solution of this problem was first given by Inglis [3.4]. In this sub-section we provide a summary of the approach adopted in handling this problem. According to the notations used in Figure 3.15a, it is clear that,

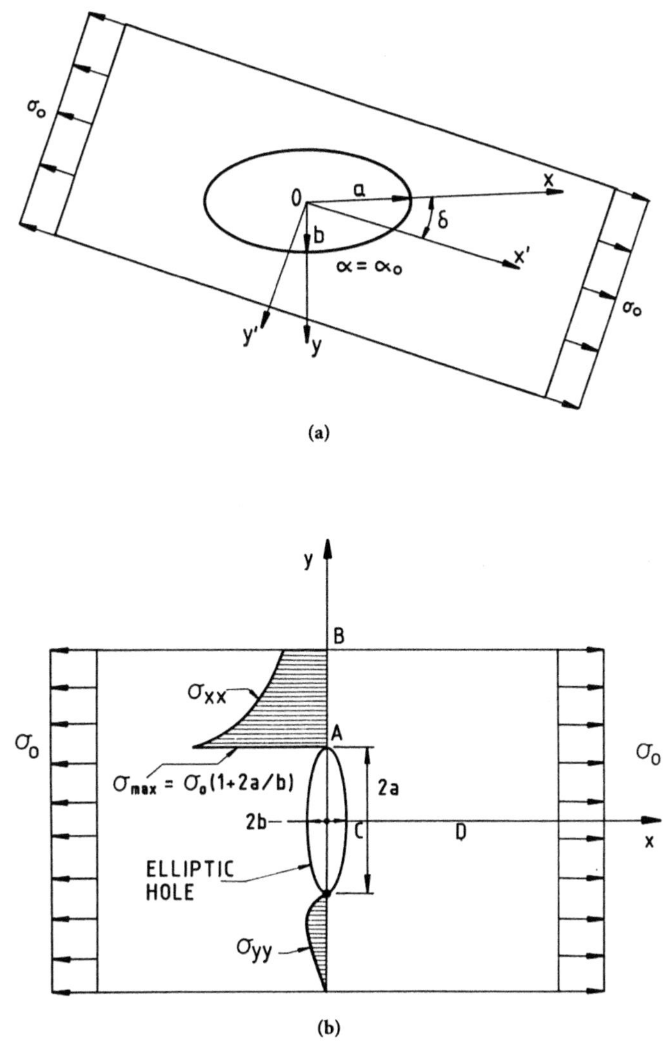

Figure 3.15 Stress analysis of an elliptical hole in a plate (**a**) Notation used, and (**b**) Resulting stress concentration and biaxiality

$$\left.\begin{array}{r}\sigma_{x'x'} + \sigma_{y'y'} = \sigma_{xx} + \sigma_{yy} \\ \sigma_{y'y'} - \sigma_{x'x'} + 2i\tau_{x'y'} = \left(\sigma_{yy} - \sigma_{xx} + 2i\tau_{xy}\right)e^{2i\theta}\end{array}\right\} \quad \text{(a)}$$

with the boundary conditions being,

$$\sigma_{x'x'} = \sigma_0 \text{ and } \sigma_{y'y'} = \tau_{x'y'} = 0 \text{ at infinity} \quad \text{(b)}$$

Accordingly,

$$\left.\begin{array}{r}\sigma_{xx} + \sigma_{yy} = \sigma_0 \\ \sigma_{yy} - \sigma_{xx} + 2i\tau_{xy} = -\sigma_0 e^{-2i\theta}\end{array}\right\} \text{at infinity} \quad \text{(c)}$$

Using expressions (3.86) and (3.88), we get,

$$\left.\begin{array}{r}4 \text{ Re } \psi'(z) = \sigma_0 \\ 2[\bar{z}\psi''(z) + \chi''(z)] = -\sigma_0 e^{-2i\theta}\end{array}\right\} \text{at infinity} \quad \text{(d)}$$

At the boundary of the elliptic hole, $\alpha = \alpha_0$, we must have $\sigma_\alpha = \tau_{\alpha\beta} = 0$.

The above boundary conditions can be satisfied by taking the complex potentials $\Psi(z)$ and $\chi(z)$ in the form:

$$\left.\begin{array}{r}\psi(z) = \frac{1}{4}[Ac\cosh\alpha + Bc\sinh\alpha] \\ \chi(z) = \frac{1}{4}\left[Cc^2\alpha + Dc^2\cosh 2\alpha + Ec^2\sinh 2\alpha\right]\end{array}\right\}$$

where A, B, C, D and E are constants, which are determined from the boundary conditions. The determination of these constants allows the evaluation of the complex potentials $\Psi(z)$ and $\chi(z)$. The stress σ_β at the hole can be obtained from Equation (3.90), since at the boundary of the ellipse $\sigma_0 = 0$, thus,

$$\left(\sigma_\beta\right)_{\alpha=\alpha_0} = \sigma_0 \frac{\sinh 2\alpha_0 + \cos 2\delta - e^{2\alpha_0}\cos 2(\delta - \beta)}{\cosh 2\alpha_0 - \cos 2\beta} \quad \text{(e)}$$

When the tensile boundary stress σ_0 is applied at right-angles to the major axis ($\delta = \pi/2$), then,

$$\left(\sigma_\beta\right)_{\alpha=\alpha_0} = \sigma_0 e^{2\alpha_0}\left[\frac{\sinh 2\alpha_0\left(1 + e^{-2\alpha_0}\right)}{\cosh 2\alpha_0 - \cos 2\beta} - 1\right] \quad (3.95)$$

and the greatest value, occurring at the ends of the major axis with $\cos 2\beta = 1$, reduces to:

$$\sigma_{\max} = \sigma_0(1 + 2a/b) \tag{3.96}$$

One can easily observe that, as the elliptic hole becomes more and more slender ($b \rightarrow 0$), the maximum stress $\sigma_{\max} \rightarrow \infty$. Also as $b \rightarrow a$, $(\sigma_\beta)_{\max} \rightarrow 3\sigma_0$, thus confirming the circular hole analysis. Figure 3.15b shows a schematic of the effect of the presence of an elliptic hole upon the resulting concentration and biaxiality of stress due to the applied tensile stress σ_0 at the boundary of the plate.

References

3.1 J. Scheel, D. Wallenta, A. Ricoeur, A critical review of the complex potentials in linear elastic fracture mechanics. J. Elast. **147**, 291–308 (2021). https://doi.org/10.1007/s10659-021-09873-1
3.2 L.W. Ratner, Part II – linear theory of infinitesimal deformations, in *Non-Linear Theory of Elasticity and Optimal Design*, (Elsevier, 2003), pp. 91–106. https://doi.org/10.1016/B978-044451427-1/50005-0
3.3 L. Anand, S. Govindjee, *Continuum Mechanics of Solids* (Oxford University Press, Oxford, 2020)
3.4 A. Ugural, S. Fenster, *Advanced Mechanics of Materials and Applied Elasticity* (Pearson, Boston, 2020)
3.5 W.S. Slaughter, *The Linearized Theory of Elasticity* (Birkhäuser, Boston, 2012)

Chapter 4
Fundamentals of the Finite Element Method

Summary Finite element method (FEM) is a very powerful numerical tool that has wide applications in a multitude of real engineering applications involving continuum modeling of coupled and uncoupled field problems. The main focus of this chapter is to provide a fundamental understanding of the principles upon which FEM is based. FEM is used extensively as a component of multiscale modeling.

4.1 Introduction

The finite element method is a powerful numerical tool used to treat problems involving coupled fields, intricate geometries, varied applied loads and complex boundary conditions. In the past, it was common practice to over-simplify such problems so that an analytical solution could be obtained. In view of the uncertainties associated with such ad hoc approximations, large safety factors were introduced in engineering design.

All finite element methods involve dividing the physical system into small sub-regions known as elements. Each element is essentially a simple unit, the behavior of which can be readily analyzed. The features of the overall system are accommodated for by using a large number of elements. Indeed, one of the attractions of the finite element method is the ease with which it can be applied to real engineering problems involving complex geometrical features and loads. The price that must be paid for flexibility and simplicity of individual elements is in the amount of numerical computations required to solve the resulting sets of simultaneous algebraic equations. For a more detailed account of the finite element method, see, e.g., Refs. [4.1, 4.2].

Unfortunately, this simple, elegant and powerful method has so far been confined mainly to postgraduate research and teaching. This, together with the purely mathematical focus, demanded by most texts available at present, is responsible for creating a curious mystique about the subject which inhibits practising engineers from utilizing what is after all a simple and a practical numerical method of analysis.

S. A. Meguid, *Atomistic and Continuum Fracture Mechanics of Solids*,
https://doi.org/10.1007/978-3-031-56085-9_4

4.2 Discretization of a Continuum

The fundamental concept of the finite element method is that any continuous field variable, such as displacement, velocity, stress, pressure or temperature, can be approximated by a discrete model composed of a set of piecewise continuous field variables defined over a finite number of subdomains, known as elements. These elements are interconnected at specified joints which are called nodes. Since the actual variation of the field variable inside the continuum is not known, some approximating functions are needed to describe its variation.

These approximating functions, which are also known as the interpolating or shape functions, are defined in terms of the discrete values of the field variable at the nodal points. When field equations, such as equilibrium or heat balance, for the whole body are written, the new unknowns will be the nodal values of that field variable. By solving the field equations, which are generally in the form of banded matrices, the nodal value of the field variable can be obtained throughout the assemblage of elements. A large and more comprehensive coverage of the method is contained in numerous textbooks. In this chapter, we provide the essential features of the finite element method rather than a complete coverage of the method.

The general solution of an engineering problem can be outlined in a step-by-step procedure. This sequence of steps describes the actual solution process, which is followed in setting up and solving structural mechanics, fluid flow or heat balance problems. In this work, we concentrate on the approach adopted in structural mechanics problems as it relates to fracture mechanics of a continuum; a summary of which is as follows:

(i) **Idealization of structure:** The geometrical features of the structure are simplified in order to accommodate sensible Discretization.
(ii) **Discretization of the structure:** In this case, the body is divided into an equivalent system of finite elements. The type, size and number of elements is dictated by the geometrical features of the component, applied loads, restraints, accuracy needed and CPU of the processor.
(iii) **Choice of interpolation or Shape function:** In the direct stiffness method, the assumed displacement function approximates the actual distribution of the displacement field within the continuum. In general, the interpolation function is taken in the form of a polynomial, and practical considerations limit the number of terms that can be retained in that polynomial.
(iv) **Derivation of element stiffness matrix:** The element stiffness matrix is composed of the coefficients of the equilibrium and compatibility equations derived from the material and the geometric properties of an element and obtained by the use of the principle of minimum potential energy (equilibrium condition) of the system. The stiffness $[\boldsymbol{K}^{(e)}]$ relates the displacements at the nodal points $\{\boldsymbol{u}^{(e)}\}$ to the applied forces at the nodal points $\{\boldsymbol{F}^{(e)}\}$, where the superscript (e) denotes the element number.

(v) **Assembly of element equations for the continuum:** This process includes the assembly of the global stiffness matrix [***K***] for the entire body from the individual element stiffness matrices [$\boldsymbol{K}^{(e)}$] and the global load vector {***F***} from the element nodal force vectors {$\boldsymbol{F}^{(e)}$}.

(vi) **Solution for the unknown nodal displacements:** the overall equilibrium equations have to be modified to account for the boundary conditions of the problem. After the incorporation of the boundary conditions, the global equilibrium equations can be expressed as

$$[\mathrm{K}]\{\mathrm{u}\} = \{\mathrm{F}\} \tag{4.1}$$

For linear elastic problems, the displacement vector can be easily obtained. But for non-linear problems, the solution is obtained in a sequence of steps, each step involving the updating of the nonlinear (geometrical, loading and coupling) stiffness matrix [***K***].

(vii) **Computation of element strains and stresses from nodal displacements:** Having determined the primary unknowns (in our case, the nodal displacements), it is often necessary to use these nodal displacements to determine the element strains and stresses.

The application of the above-mentioned seven steps of the finite element simulations is best demonstrated by the following simple example, which is concerned with the determination of the displacement, strain and stress fields in the tapered bar depicted in Figure 4.1a. In this case, we will only use a linear bar element of uniform cross section area with one degree of freedom per node; i.e. Poisson's effects are ignored. This effectively means that the stiffness matrix will have 4 components (2 × 2). In this case, we must follow the following steps:

(a) **Idealization:** The tapered bar is approximated by a limited number of one-dimensional bar elements of uniform cross section area, as shown in Figure 4.1b.

(b) **Discretization:** The bar can be approximated to an assemblage of three discretized elements, as shown in Figure 4.1c. By assuming the bar to be a one-dimensional structure, only axial displacement at any point in the element is considered. Since there are four nodes, the axial nodal displacements u_1, u_2, u_3 and u_4 will be taken as unknown.

(c) **Trial Function:** In each of this linear 2-nodes element, we assume a linear variation of axial displacement $u(x)$ such that the trial function is represented by a straight line connecting the two discrete displacements, i.e.,

$$\mathrm{u}(\mathrm{x}) = \mathrm{a} + \mathrm{bx} \tag{4.1a}$$

where a and b are coefficients to be determined from the element boundary conditions. If we consider the respective end displacements $\mathrm{u}_1^{(\mathrm{e})}$ and $\mathrm{u}_2^{(\mathrm{e})}$ $at\, x = 0$ and $x = l^{(e)}$ as unknowns, we obtain,

$$a = u_1^{(e)} \text{ and } b = \left(u_2^{(e)} - u_1^{(e)}\right)/l^{(e)} \tag{4.1b}$$

where the superscript (*e*) denotes the element number.

Thus,

$$u_{(x)}^{(e)} = u_1^{(e)} + \frac{\left(u_2^{(e)} - u_1^{(e)}\right)x}{l^{(e)}}$$

or

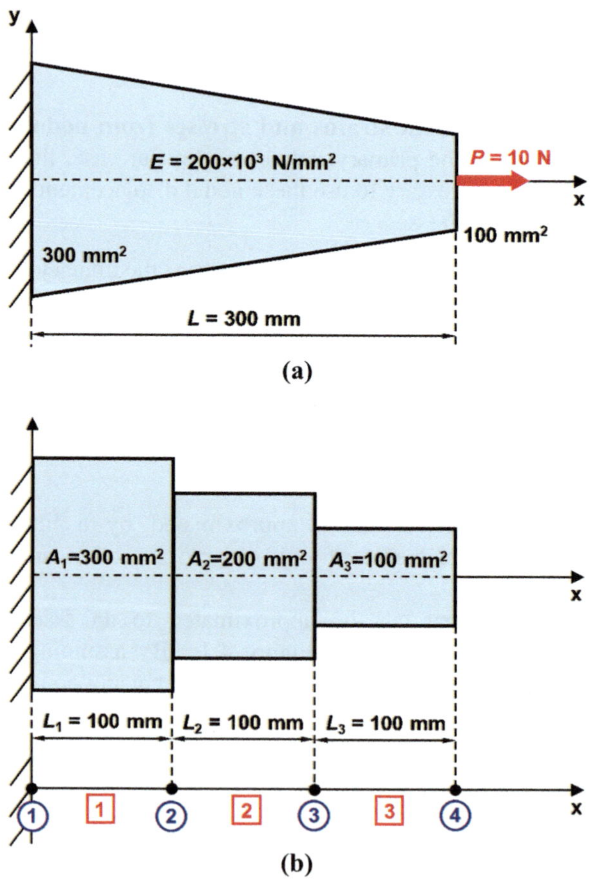

Figure 4.1 Finite element analysis of (**a**) a tapered bar, (**b**) idealization of tapered bar, (**c**) discretization of tapered bar into 3 bar elements, (**d**) description of a general one-dimensional bar element

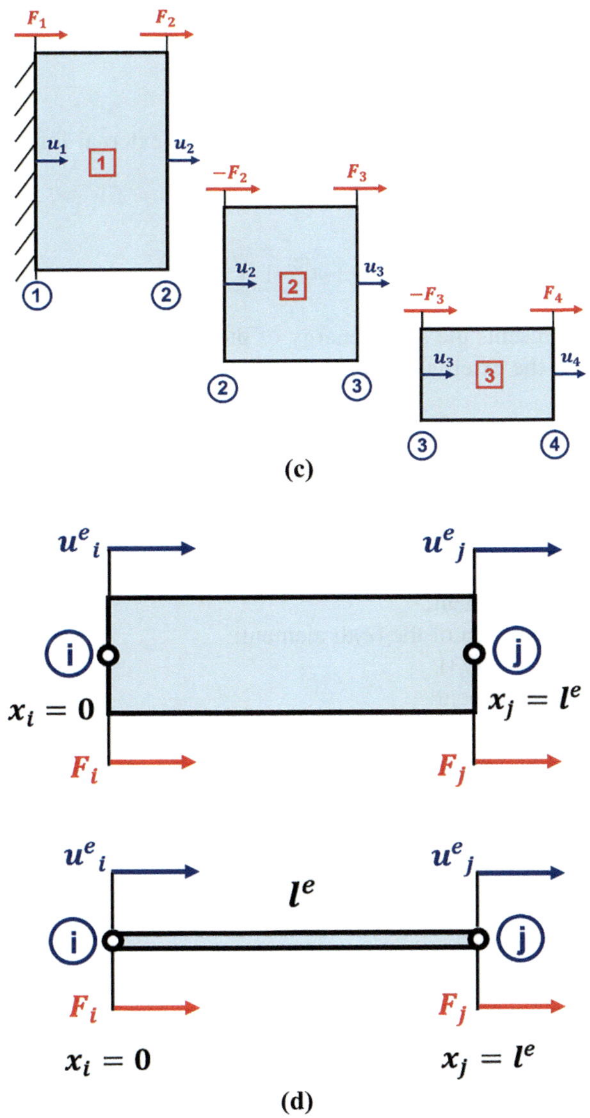

Figure 4.1 (continued)

$$u_{(x)}^{(e)} = u_1^{(e)} N_1 + u_2^{(e)} N_2 \tag{4.1c}$$

where $N_1 = 1 - \frac{x}{l^{(e)}}$ and $N_2 = \frac{x}{l^{(e)}}$ are known as the shape functions.

(d) **Element stiffness matrix:** The element stiffness matrices can be derived from the principle of minimum potential energy (MPE), principle of virtual work or weighted residual methods. For example, applying the MPE approach, will enable one to obtain the following expression for the potential energy of the bar (U) under point load,

$$U = \text{strain energy} - \text{Work done by the external forces}$$

i.e.,

$$\mathrm{U} = \pi^{(1)} + \pi^{(2)} + \pi^{(3)} - \mathrm{W_p} \tag{4.1d}$$

where $\pi^{(e)}$ represents the strain energy of the (e)th element and W_p denotes the work done by the external forces. For the element shown in Figure 4.1d:

$$\pi^{(e)} = \int_0^{l^{(e)}} (1/2)\sigma^{(e)}\varepsilon^{(e)}\mathrm{dV} = \frac{\mathrm{A}^{(e)}\mathrm{E}^{(e)}}{2}\int_0^{l^{(e)}} \left(\varepsilon^{(e)}\right)^2 \mathrm{dx} \tag{4.1e}$$

where

V = volume of (e)th element;
$A^{(e)}$ = cross-sectional area of the (e)th element;
$l^{(e)}$ = element length (=$L/3$);
$\sigma^{(e)}$ = stress in (e)th element;
$\varepsilon^{(e)}$ = strain in (e)th element;
$E^{(e)}$ = Young's Modulus of (e)th element.

Now, since $\varepsilon_{xx} = \frac{\partial u}{\partial x}$, then

$$\varepsilon^{(e)} = \frac{u_2^{(e)} - u_1^{(e)}}{l^{(e)}} \tag{4.1f}$$

therefore,

$$\pi^{(e)} = \frac{\mathrm{A}^{(e)}\mathrm{E}^{(e)}}{2l^{(e)}}\left[\left(u_1^{(e)}\right)^2 + \left(u_2^{(e)}\right)^2 - 2u_1^{(e)}u_2^{(e)}\right]$$

According to the quadratic form of matrices, this expression can be written in a matrix form as,

$$\pi^{(e)} = (1/2)\left\{u^{(e)}\right\}^{\mathrm{T}}\left[\mathrm{K}^{(e)}\right]\left\{u^{(e)}\right\} \tag{4.1g}$$

where $\{\boldsymbol{u}^{(e)}\}^{\mathrm{T}} = \{u_1{}^{(e)}\ u_2{}^{(e)}\}$ which is the transpose of the nodal displacement vector of the (e)th element; giving,

$$\left[K^{(e)}\right] = \frac{A^{(e)}E^{(e)}}{l^{(e)}}\begin{bmatrix} 1 & -1 \\ -1 & 1 \end{bmatrix} \tag{4.1h}$$

which is called the element stiffness matrix; and

$$\left\{u^{(e)}\right\} = \begin{Bmatrix} u_i^{(e)} \\ u_j^{(e)} \end{Bmatrix}$$

which is the vector of the nodal displacements:

$$= \begin{Bmatrix} u_1 \\ u_2 \end{Bmatrix} \text{for } e = 1,$$

$$= \begin{Bmatrix} u_2 \\ u_3 \end{Bmatrix} \text{for } e = 2, \text{ and}$$

$$= \begin{Bmatrix} u_3 \\ u_4 \end{Bmatrix} \text{for } e = 3.$$

The work done by the external forces can be expressed as,

$$W_p = u_1F_1 + u_2F_2 + u_3F_3 + u_4F_4 \tag{4.1i}$$

In the present example, F_1 = reaction at the fixed node, F_2 and $F_3 = 0$ according to internal force equilibrium rules and $F_4 = P = +10$ N.

If the entire bar is in equilibrium under the load vector

$$\{F\} = \begin{Bmatrix} F_1 \\ F_2 \\ F_3 \\ F_4 \end{Bmatrix} \tag{4.1j}$$

Then principle of minimum potential energy gives

$$\frac{\partial U}{\partial u_i} = 0, i = 1, 2, 3 \text{ and } 4$$

This equation can be written as

$$\frac{\partial U}{\partial u_i} = \frac{\partial}{\partial u_i}\left(\sum\nolimits_{e=1}^{3} \pi^{(e)} - W_p\right) = 0, i = 1, 2, 3 \text{ and } 4$$

i.e.,

$$\sum_{e=1}^{3}\left(\left[K^{(e)}\right]\left\{u^{(e)}\right\}-\left\{F^{(e)}\right\}\right)=0 \tag{4.1k}$$

(e) **Assemblage of element stiffness matrices:** The above equation can be written as:

$$[K^{g}]\,\{u^{g}\}=\{F^{g}\} \tag{4.1l}$$

where $[K^g]$ is the global stiffness matrix $= \sum_{e=1}^{3}\left[K^{(e)}\right]$ and $\{u^g\}$ is the vector containing the unknown global nodal displacements. For the data given, the element stiffness matrices would be:

$$\left[K^{(1)}\right]=\frac{A^{(1)}E^{(1)}}{l^{(1)}}\begin{bmatrix}1 & -1\\ -1 & 1\end{bmatrix}=10^5\begin{array}{c}\begin{array}{cc}u_1 & u_2\end{array}\\ \begin{bmatrix}6 & -6\\ -6 & 6\end{bmatrix}\end{array}\begin{array}{c}u_1\\ u_2\end{array}$$

$$\left[K^{(2)}\right]=\frac{A^{(2)}E^{(2)}}{l^{(2)}}\begin{bmatrix}1 & -1\\ -1 & 1\end{bmatrix}=10^5\begin{array}{c}\begin{array}{cc}u_2 & u_3\end{array}\\ \begin{bmatrix}4 & -4\\ -4 & 4\end{bmatrix}\end{array}\begin{array}{c}u_2\\ u_3\end{array}$$

and

$$\left[K^{(3)}\right]=\frac{A^{(3)}E^{(3)}}{l^{(3)}}\begin{bmatrix}1 & -1\\ -1 & 1\end{bmatrix}=10^5\begin{array}{c}\begin{array}{cc}u_3 & u_4\end{array}\\ \begin{bmatrix}2 & -2\\ -2 & 2\end{bmatrix}\end{array}\begin{array}{c}u_3\\ u_3\end{array}$$

Since the displacements of the left and right nodes of the first element are u_1, and u_2, the rows and columns of the stiffness matrix corresponding to these unknowns are identified as shown above. Similarly, the rows and columns of the stiffness matrices of the other elements are identified with the unknown displacements u_2, u_3 and u_4.

The overall or global stiffness of the bar can be obtained by adding the three element stiffness matrices. Since there are four nodal unknown displacements (u_1, u_2, u_3 and u_4) the global stiffness matrix **[*K*]** will be 4 × 4. The elements of the global stiffness matrix corresponding to the unknown displacements u_1, u_2, u_3 and u_4, are obtained by adding the corresponding terms as shown below:

$$[K^g] = 10^5 \begin{array}{c} \begin{matrix} u_1 & u_2 & u_3 & u_4 \end{matrix} \\ \begin{bmatrix} 6 & -6 & 0 & 0 \\ -6 & 6+4 & -4 & 0 \\ 0 & -4 & 4+2 & -2 \\ 0 & 0 & -2 & 2 \end{bmatrix} \end{array} \begin{matrix} u_1 \\ u_2 \\ u_3 \\ u_4 \end{matrix}$$

$$= 10^5 \begin{bmatrix} 6 & -6 & 0 & 0 \\ -6 & 10 & -4 & 0 \\ 0 & -4 & 6 & -2 \\ 0 & 0 & -2 & 2 \end{bmatrix}$$

The global load vector can be written as

$$\{F^g\} = \begin{Bmatrix} F_1 \\ F_2 \\ F_3 \\ F_4 \end{Bmatrix} = \begin{Bmatrix} -R \\ 0 \\ 0 \\ 10 \end{Bmatrix}$$

where R denotes the reaction at node 1. Thus the overall equilibrium equations become,

$$10^5 \begin{bmatrix} 6 & -6 & 0 & 0 \\ -6 & 10 & -4 & 0 \\ 0 & -4 & 6 & -2 \\ 0 & 0 & -2 & 2 \end{bmatrix} \begin{Bmatrix} u_1 \\ u_2 \\ u_3 \\ u_4 \end{Bmatrix} = \begin{Bmatrix} -R \\ 0 \\ 0 \\ 10 \end{Bmatrix}$$

(f) **Solving for displacements:** In order to avoid a singular global stiffness matrix, we incorporate the known boundary condition; $u_1 = 0$. The reduced equilibrium equations can be rewritten as:

$$10^5 \begin{bmatrix} 10 & -4 & 0 \\ -4 & 6 & -2 \\ 0 & -2 & 2 \end{bmatrix} \begin{Bmatrix} u_2 \\ u_3 \\ u_4 \end{Bmatrix} = \begin{Bmatrix} 0 \\ 0 \\ 10 \end{Bmatrix}$$

The solution of the above linear equations gives.

$$u_2 = 1.66 \times 10^{-5} \text{mm}$$
$$u_3 = 4.166 \times 10^{-5} \text{mm}$$
$$u_4 = 9.166 \times 10^{-5} \text{mm}$$

(g) **Element strains and stresses:** From the known displacements, the strains in the elements can be computed as follows,

$$\varepsilon_{xx}^{(1)} = \frac{\partial u}{\partial x} \text{ for element } (1) = \frac{u_2^{(1)} - u_1^{(1)}}{l^{(1)}} = 1.66 \times 10^{-7}$$

$$\varepsilon_{xx}^{(2)} = \frac{\partial u}{\partial x} \text{ for element } (2) = \frac{u_3^{(2)} - u_2^{(2)}}{l^{(2)}} = 3 \times 10^{-7}$$

$$\varepsilon_{xx}^{(3)} = \frac{\partial u}{\partial x} \text{ for element } (3) = \frac{u_4^{(3)} - u_3^{(3)}}{l^{(3)}} = 5 \times 10^{-7}$$

The stresses in the three elements are given by

$$\sigma_{xx}^{(1)} = E^{(1)} \varepsilon_{xx}^{(1)} = 332 \text{ N/m}^2$$
$$\sigma_{xx}^{(2)} = E^{(2)} \varepsilon_{xx}^{(2)} = 600 \text{ N/m}^2$$
$$\sigma_{xx}^{(3)} = E^{(3)} \varepsilon_{xx}^{(3)} = 1000 \text{ N/m}^2$$

4.3 Elastic Behavior of Solids and D-Matrix

The derivation of the stiffness matrix of bar elements discussed in the above example were deliberately restricted within the confines of elementary structural mechanics with which the beginner is familiar. In the following section, we shall present a general procedure, based upon the principle of virtual work, from which the stiffness and other characteristics of any finite element can be derived. As the understanding and application of this procedure is facilitated by some knowledge of elasticity theory, we shall discuss briefly some of the essential aspects of linear elasticity which are of interest in the general context of finite elements. For simplicity, we devote our attention to Cartesian coordinates and for the sake of convenience and ease of understanding, we will repeat some of the formulations discussed in the previous chapter.

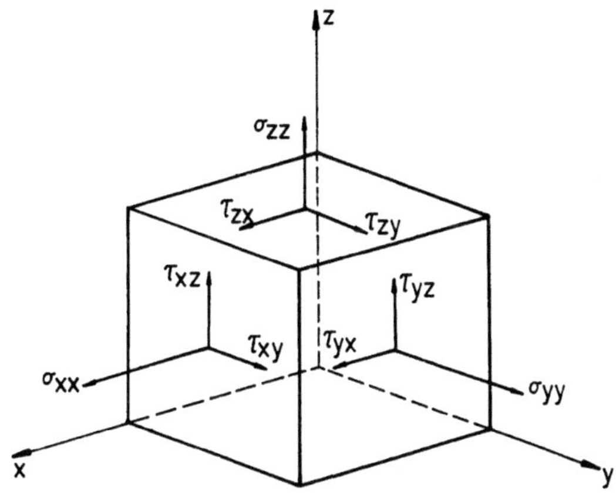

Figure 4.2 Stress components referred to Cartesian coordinates

4.3.1 Stresses and Strains

Throughout this text the symbols σ and τ are used to denote stress. Individual components of stress are indicated by double subscripts, as follows:

Direct stresses σ_{xx}, σ_{yy}, σ_{zz}
Shear stresses τ_{xy}, τ_{yz}, τ_{zx}, τ_{yx}, τ_{zy}, τ_{xz}

The first subscript defines the direction of the outward normal to the surface on which it acts and the second denotes the direction of the stress, as shown in Figure 4.2.

For rotational equilibrium to be maintained, the shear stresses must be complementary, i.e.

$$\tau_{xy} = \tau_{yx}, \tau_{yz} = \tau_{zy}, \tau_{zx} = \tau_{xz} \tag{4.2}$$

The components of the displacement for a solid are denoted by u, υ and w in the coordinates directions x, y and z, respectively. Using the same double subscript notation, the direct components of strain can be determined as

$$\varepsilon_{xx} = \frac{\partial u}{\partial x}, \varepsilon_{yy} = \frac{\partial \upsilon}{\partial y}, \varepsilon_{zz} = \frac{\partial w}{\partial z} \tag{4.3}$$

and the engineering shear strains as

$$\gamma_{xy} = \gamma_{yx} = \frac{\partial u}{\partial y} + \frac{\partial \upsilon}{\partial x} \tag{4.4}$$

$$\gamma_{yz} = \gamma_{zy} = \frac{\partial \upsilon}{\partial z} + \frac{\partial w}{\partial y} \tag{4.5}$$

$$\gamma_{zx} = \gamma_{xz} = \frac{\partial u}{\partial z} + \frac{\partial w}{\partial x} \tag{4.6}$$

4.3.2 Stress-Strain Relationships

Unless otherwise stated, we shall assume that the material of the body is linearly elastic, isotropic and homogeneous, so that its elastic properties are completely identified by the mutually independent constants E and v, denoting Young's modulus and Poisson's ratio, respectively. In this case, the normal strains are related to the stresses via Hooke's law as follows:

$$\varepsilon_{xx} = \frac{1}{E}\left[\sigma_{xx} - v\left(\sigma_{yy} + \sigma_{zz}\right)\right] \tag{4.7}$$

$$\varepsilon_{yy} = \frac{1}{E}\left[\sigma_{yy} - v(\sigma_{xx} + \sigma_{zz})\right] \tag{4.8}$$

and

$$\varepsilon_{zz} = \frac{1}{E}\left[\sigma_{zz} - v\left(\sigma_{xx} + \sigma_{yy}\right)\right] \tag{4.9}$$

Unlike the normal strains, the shear strains are mutually independent, and related to their respective stresses through the following relations:

$$\gamma_{xy} = \frac{\tau_{xy}}{G} \tag{4.10}$$

$$\gamma_{yz} = \frac{\tau_{yz}}{G} \tag{4.11}$$

$$\gamma_{zx} = \frac{\tau_{zx}}{G} \tag{4.12}$$

where G is the modulus of rigidity of the component and is given by:

$$G = \frac{E}{2(1 + v)} \tag{4.13}$$

It is convenient to write the above six strain components in a matrix form, as follows:

$$\begin{Bmatrix} \varepsilon_{xx} \\ \varepsilon_{yy} \\ \varepsilon_{zz} \\ \gamma_{xy} \\ \gamma_{yz} \\ \gamma_{zx} \end{Bmatrix} = \frac{1}{E} \begin{bmatrix} 1 & -v & -v & 0 & 0 & 0 \\ -v & 1 & -v & 0 & 0 & 0 \\ -v & -v & 1 & 0 & 0 & 0 \\ 0 & 0 & 0 & 2(1+v) & 0 & 0 \\ 0 & 0 & 0 & 0 & 2(1+v) & 0 \\ 0 & 0 & 0 & 0 & 0 & 2(1+v) \end{bmatrix} \begin{Bmatrix} \sigma_{xx} \\ \sigma_{yy} \\ \sigma_{zz} \\ \tau_{xy} \\ \tau_{yz} \\ \gamma_{zx} \end{Bmatrix} \tag{4.14}$$

By inversion, we can see the above equations in the following alternative form:

$$\begin{Bmatrix} \sigma_{xx} \\ \sigma_{yy} \\ \sigma_{zz} \\ \tau_{xy} \\ \tau_{yz} \\ \tau_{zx} \end{Bmatrix} = \begin{bmatrix} \lambda+2G & \lambda & \lambda & 0 & 0 & 0 \\ \lambda & \lambda+2G & \lambda & 0 & 0 & 0 \\ \lambda & \lambda & \lambda+2G & 0 & 0 & 0 \\ 0 & 0 & 0 & G & 0 & 0 \\ 0 & 0 & 0 & 0 & G & 0 \\ 0 & 0 & 0 & 0 & 0 & G \end{bmatrix} \begin{Bmatrix} \varepsilon_{xx} \\ \varepsilon_{yy} \\ \varepsilon_{zz} \\ \gamma_{xy} \\ \gamma_{yz} \\ \gamma_{zx} \end{Bmatrix} \tag{4.15}$$

in which λ. is the well-known Lamé coefficient, and is given by

$$\lambda = \frac{vE}{(1+v)(1-2v)} \tag{4.16}$$

Equation (4.15) state Hooke's law in three dimensions which can be written in the following compact form:

$$\{\sigma\} = [D]\{\varepsilon\} \tag{4.17}$$

in which [$\boldsymbol{D}$] is the elasticity matrix represented by the three-dimensional array of Eq. (4.15),

$$\{\varepsilon\} = \left\{\varepsilon_{xx}\varepsilon_{yy}\varepsilon_{zz}\gamma_{xy}\gamma_{yz}\gamma_{zx}\right\}^T \tag{4.18}$$

$$\text{and } \{\sigma\} = \left\{\sigma_{xx}\sigma_{yy}\sigma_{zz}\tau_{xy}\tau_{yz}\tau_{zx}\right\}^T$$

Figure 4.3 Plane stress configuration

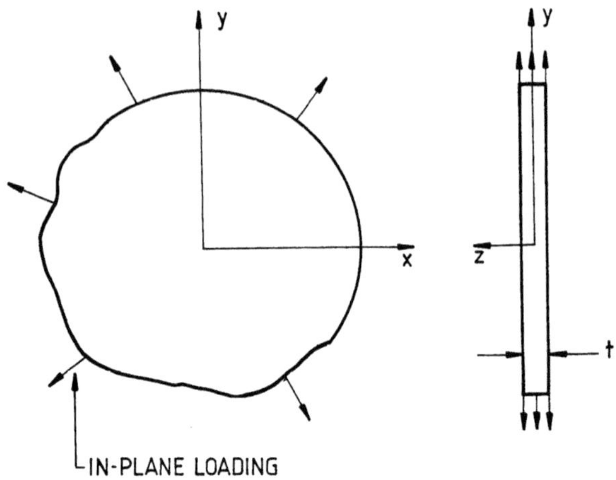

4.3.3 *Plane Stress and Plane Strain*

If a body consists of two parallel planes a constant thickness apart and bounded by any closed surface, it is said to be a plane body. Associated with this type of body there is a particular class of problems within the general theory of elasticity which are termed plane elastic problems, and these allow a number of simplifying assumptions in their treatment.

Plane Stress Problems

A plane stress condition can be used to analyze elasticity problems in which the body is subjected to in-plane forces and one of its dimensions is very small compared to the rest. A typical problem is shown in Figure 4.3, in which

$$\sigma_{zz} = \tau_{xz} = \tau_{yz} = 0 \tag{4.19}$$

Furthermore, since t is sufficiently small compared with the other dimensions of the body, and provided that the forces applied at the edges are uniform across t, then we may assume that the above condition is also valid within the cross-section of the body. In this case, the remaining stresses σ_{xx}, σ_{yy} and τ_{xy} are assumed not to vary across the thickness t. The problem thus becomes a two-dimensional one in which the z-axis can be ignored, and the stress–strain relations reduce to

$$\varepsilon_{xx} = \frac{1}{E}\left(\sigma_{xx} - v\sigma_{yy}\right) \tag{4.20}$$

$$\varepsilon_{yy} = \frac{1}{E}\left(\sigma_{yy} - v\sigma_{xx}\right) \tag{4.21}$$

and

$$\gamma_{xy} = \frac{\tau_{xy}}{G} \tag{4.22}$$

and a state of plane stress is said to exist. In this case, the elasticity matrix [***D***] is reduced to

$$[D] = \frac{E}{1 - v^2} \begin{bmatrix} 1 & v & 0 \\ v & 1 & 0 \\ 0 & 0 & \frac{1-v}{2} \end{bmatrix} \tag{4.23}$$

Plane Strain Problems

Figure 4.4 shows a solid body whose cross-section is uniform in the *z*-direction. Provided its length in this direction is large, the section OACB can be regarded as being remote from the ends.

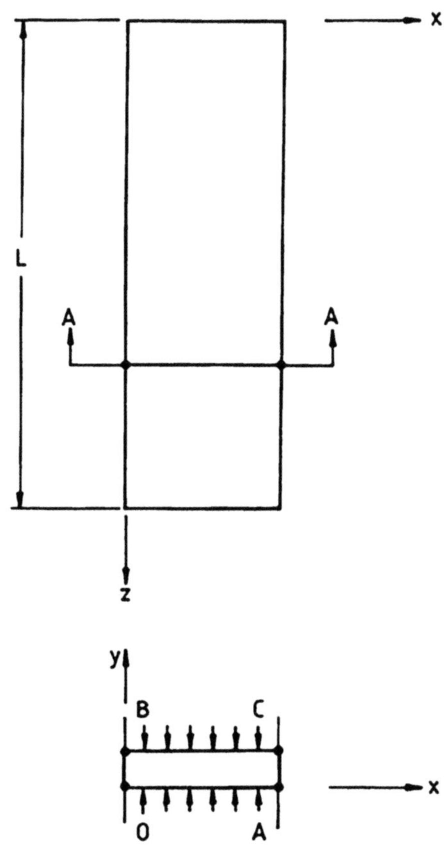

Figure 4.4 Plane strain configuration

Assuming that the applied loads are in the *xy*-plane, the resulting state of strain at such a section is two-dimensional, with $w = 0$ and $\frac{\partial}{\partial z} = 0$.

As a result,

$$\varepsilon_{zz} = \gamma_{xz} = \gamma_{yz} = 0 \tag{4.24}$$

and a state of plane strain is said to exist. In this case, the elasticity matrix [***D***] is reduced to

$$\begin{bmatrix} \lambda + 2G & \lambda & 0 \\ \lambda & \lambda + 2G & 0 \\ 0 & 0 & G \end{bmatrix} \tag{4.25}$$

4.3.4 Equilibrium Equations

The differential equations of equilibrium for the three coordinate directions can be expressed as

$$\frac{\partial \sigma_{xx}}{\partial x} + \frac{\partial \tau_{xy}}{\partial y} + \frac{\partial \tau_{xz}}{\partial z} + \overline{X} = 0 \tag{4.26}$$

$$\frac{\partial \sigma_{yy}}{\partial y} + \frac{\partial \tau_{xy}}{\partial x} + \frac{\partial \tau_{yz}}{\partial z} + \overline{Y} = 0 \tag{4.27}$$

$$\frac{\partial \sigma_{zz}}{\partial z} + \frac{\partial \tau_{xz}}{\partial x} + \frac{\partial \tau_{yz}}{\partial y} + \overline{Z} = 0 \tag{4.28}$$

where $\overline{X}$, $\overline{Y}$ and $\overline{Z}$ are the local components of the body forces per unit volume acting on the continuum along the coordinates directions. Both inertia and gravity forces are the most common causes of such forces.

4.3.5 Compatibility Equations

Strains must be compatible with each other. This effectively means that no discontinuities such as gaps or overlaps of material exist in a continuum. The mathematical formulation of this condition is as follows:

$$\frac{\partial^2 \varepsilon_{xx}}{\partial y^2} + \frac{\partial^2 \varepsilon_{yy}}{\partial x^2} = 2\frac{\partial^2 \varepsilon_{xy}}{\partial x \partial y} \tag{4.29}$$

$$\frac{\partial^2 \varepsilon_{yy}}{\partial z^2} + \frac{\partial^2 \varepsilon_{zz}}{\partial y^2} = 2\frac{\partial^2 \varepsilon_{yz}}{\partial y \partial z} \tag{4.30}$$

$$\frac{\partial^2 \varepsilon_{zz}}{\partial x^2} + \frac{\partial^2 \varepsilon_{xx}}{\partial z^2} = 2\frac{\partial^2 \varepsilon_{xz}}{\partial x \partial z} \tag{4.31}$$

$$\frac{\partial}{\partial x}\left(-\frac{\partial \varepsilon_{yz}}{\partial x} + \frac{\partial \varepsilon_{zx}}{\partial y} + \frac{\partial \varepsilon_{xy}}{\partial z}\right) = \frac{\partial^2 \varepsilon_{xx}}{\partial y \partial z} \tag{4.32}$$

$$\frac{\partial}{\partial y}\left(-\frac{\partial \varepsilon_{zx}}{\partial y} + \frac{\partial \varepsilon_{xy}}{\partial z} + \frac{\partial \varepsilon_{yz}}{\partial x}\right) = \frac{\partial^2 \varepsilon_{yy}}{\partial z \partial x} \tag{4.33}$$

$$\frac{\partial}{\partial z}\left(-\frac{\partial \varepsilon_{xy}}{\partial z} + \frac{\partial \varepsilon_{yz}}{\partial x} + \frac{\partial \varepsilon_{zx}}{\partial y}\right) = \frac{\partial^2 \varepsilon_{zz}}{\partial x \partial y} \tag{4.34}$$

where ε_{xy}, ε_{yz} and ε_{zx} are the tensor components of the shear strains.

4.4 Application of Finite Elements

4.4.1 Discretization of a Body

The discretization of a body into sub-regions is the first of a series of steps that must be performed when solving an engineering problem. This particular step does not have a theoretical basis. It is an art and depends on the use of engineering judgement. The application of poor or improper judgement will produce inaccurate results even though all the other steps are rigorously adhered to.

The discretization of a body involves the decision as to the number, size and shape of elements used to model the body. The general objective of such a discretization is to divide the body into elements sufficiently small so that the simple displacement functions can adequately approximate the solution. At the same time, one must remember that too fine a mesh will lead to extra computational effort. The judicious sub-division of a domain comes with experience. However, several general rules can be used to guide the inexperienced. These rules, and some general advice relative to the physical process of discretization, are discussed in the present section. However, let us begin by providing some of the common elements used in the discretization of engineering components.

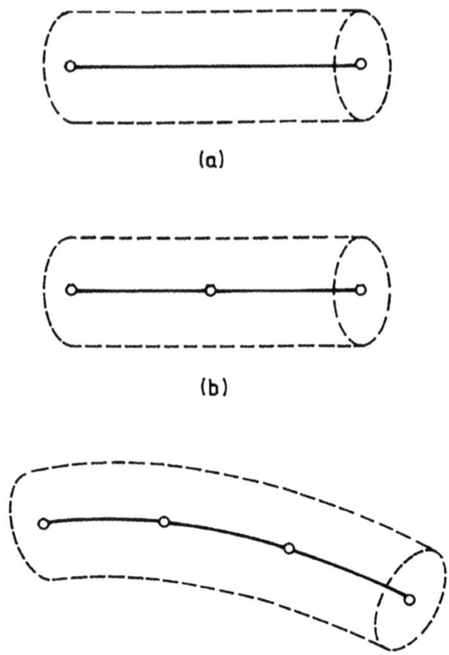

Figure 4.5 One-dimensional finite elements: (**a**) two-node (linear) element, (**b**) three-node (quadratic) element, and (**c**) four-node (cubic) element

4.4.2 Types of Elements

One-Dimensional Elements

When the geometry, material properties and such dependent variables as temperature, displacement or stress can be described in terms of only one independent spatial coordinates, a one-dimensional element can be used. The simplest one-dimensional element has two nodes, one at each end. Although this element has a cross-sectional area, it is generally shown schematically as a line segment, as shown in Figure 4.5a. The most common higher order elements are the three-node (quadratic) and four-node (cubic) elements, shown in Figure 4.5b, c. The most common usages of this element are in the one-dimensional heat transfer problems and truss and frame analyses.

Two-Dimensional Elements

In Section 4.3.3, we demonstrated that some problems in solid mechanics can be approximated by a two-dimensional formulation. Among these problems are plane stress, plane strain, and axisymmetric geometries. There are two general element families used to model the plane two-dimensional domain, the triangle and the quadrilateral of Figure 4.6a, b. In this case, the linear elements in each family have straight sides, but the higher order elements, quadratic and cubic, can have either straight or curved sides or both. The ability to model curved boundaries is made

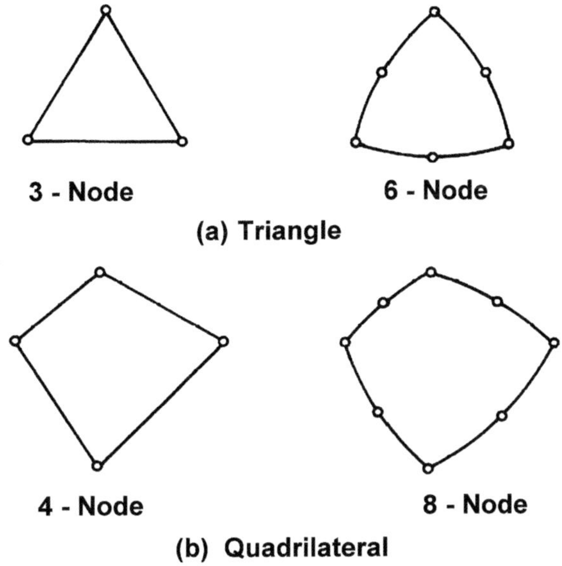

Figure 4.6 Examples of two-dimensional plane elements: (**a**) triangle element and (**b**) quadrilateral element

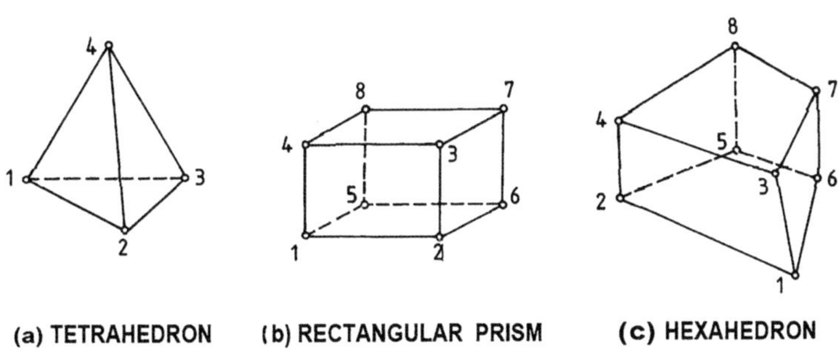

Figure 4.7 Examples of three-dimensional elements: (**a**) tetrahedron element, (**b**) hexahedron element and (**c**) rectangular prism element

possible by the addition of mid-side nodes. Both families of elements can be used within the same body provided that both have the same number of nodes along a given side in order to ensure topological compatibility between the nodes.

Three-Dimensional Solid Elements

The most common three-dimensional elements are variations of the two-dimensional elements, namely the tetrahedron and the hexahedron. A tetrahedron has four primary external nodes, while a general hexahedron and a rectangular prism have eight primary external nodes as depicted in Figure 4.7. In this case, the linear elements are restricted to plane surfaces, while the higher order elements can have curved surfaces for their sides.

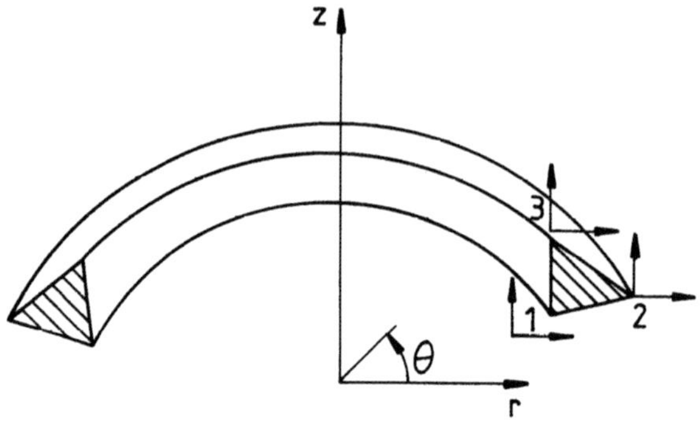

Figure 4.8 Axisymmetric toroidal element

Some problems which are actually three-dimensional can be approximated by two independent coordinates. Such problems can be approximated by using the axisymmetric element shown in Figure 4.8. The problems that possess axial-symmetry, like pistons, pressure vessels, nozzles, rotating discs, fall into this category. It is convenient to express these problems in terms of cylindrical coordinates. Because of symmetry, the stress components are independent of the tangential direction (θ). In this case, the elasticity matrix [***D***] is given by

$$[D] = \frac{E}{(1+v)(1-2v)}\begin{bmatrix} 1-v & v & v & 0 \\ v & 1-v & v & 0 \\ v & v & 1-v & 0 \\ 0 & 0 & 0 & \frac{1-2v}{2} \end{bmatrix} \tag{4.35}$$

4.4.3 Division of a Continuum into Elements

The actual discretization process can be divided into two parts: the first deals with the division of the body into elements and the second deals with the labeling of the elements and the numbering of the nodes; the latter sounds quite simple but is complicated by our desire to increase the computational efficiency.

Often the type of elements to be used will be evident from the physical characteristics of the problem itself. For example, one-dimensional elements are used if the problem involves the analysis of a truss structure or one-dimensional heat transfer or fluid flow problems or the dynamic response of some rotating machines. Similarly, in the case of stress analysis of engineering components, the modeling of the component can be done using three-dimensional solid elements, as shown in Figure 4.9.

In some cases, however, the type of element to be used for the division of the body may not be apparent. In this situation, the type of element is dictated by the

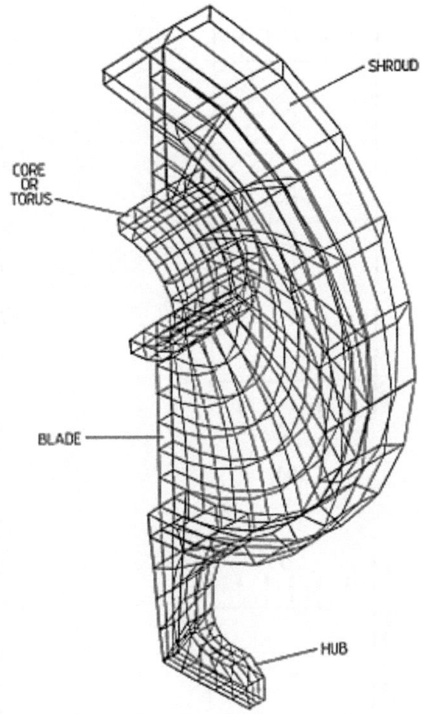

Figure 4.9 Three-dimensional finite element modeling of a sector of an impeller of a fluid coupling using 20-noded parabolic solid elements

number of degrees of freedom, accuracy required, computational power available and other related sensible parameters. In certain problems, the division of the body cannot be made using only one type of element, e.g., bending of a plate which is stiffened by edge beams as in aircraft wings. Here, two-dimensional elements supplemented by beam elements will be used.

The size of the element should be selected with care. If the size of the element is relatively small, then the final solution is expected to be more accurate. However, this is always associated with increased computational time and hence increased cost. In true engineering situations, most bodies have zones in which pronounced variations in the stresses and strains occur. The most familiar examples of such regions of steep gradients of the variables are locations where stress concentrations exist. In this case, a finer mesh is needed at and near the geometrical discontinuity in order to provide a satisfactory solution to the problem.

A typical example of this is a plate with a central hole under uniaxial tension in which the size of the elements near the hole should be very small compared with those farther away. In view of the symmetric nature of both geometry and the externally applied loads, only one quadrant can be used to model the situation. The symmetry conditions, however, must be incorporated in the solution procedure as depicted in Figure 4.10 by the condition imposed upon the u and v displacements.

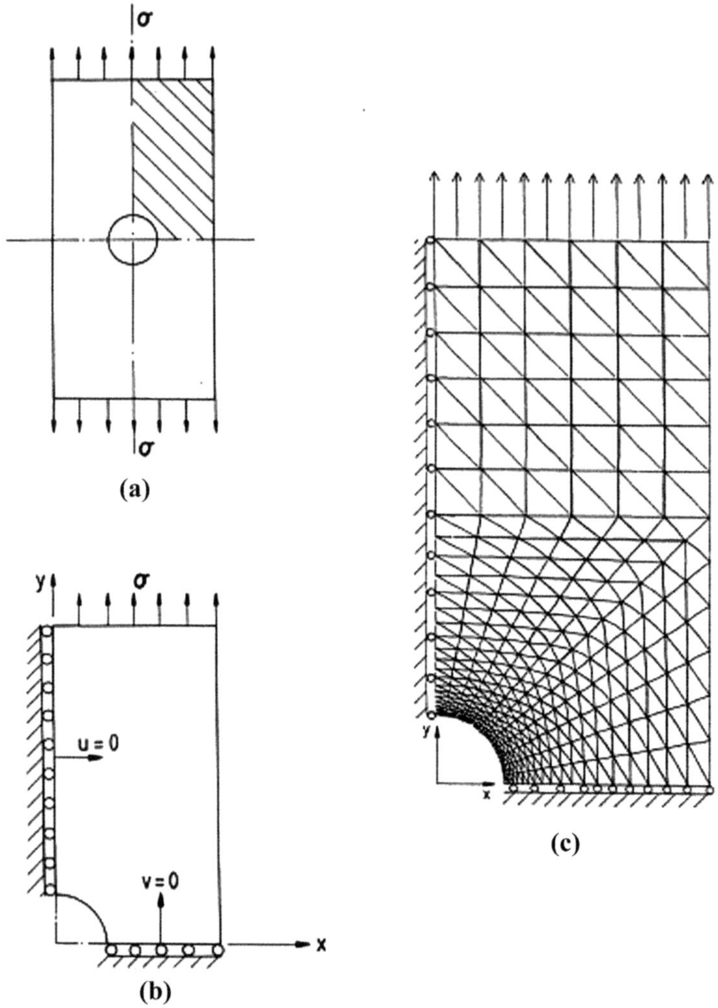

Figure 4.10 A plate with a central hole under uniaxial tension: (**a**) definition of the problem, (**b**) one quadrant of the plate with central hole, showing imposed restraints and applied load, and (**c**) discretized quadrant showing mesh used to capture stress gradients, applied load and imposed restraints

4.4.4 Labeling of Nodes

The labeling of the nodes (assigning a number) influences the computational efficiency associated with obtaining a solution. The set of linear equations, which arises when using the finite element method, has a large number of coefficients which are zero. A listing of the equations would show that all the non-zero values and some zero values fall between two narrow lines, which can be constructed parallel to the main diagonal of the global stiffness matrix, as shown below:

← Semi-Band Width →

$$\begin{bmatrix} a & a & a & 0 & a & 0 & 0 & 0 & 0 \\ a & 0 & a & a & a & a & 0 & 0 & 0 \\ a & a & a & 0 & a & a & a & 0 & 0 \\ a & a & a & a & 0 & a & a & a & 0 \\ a & 0 & a & 0 & a & a & a & 0 & a \\ 0 & a & a & a & a & 0 & a & a & a \\ 0 & 0 & a & a & a & a & a & a & a \\ 0 & 0 & 0 & a & a & a & 0 & a & a \\ 0 & 0 & 0 & 0 & a & a & a & a & a \end{bmatrix}$$

(a = non-zero value)

The distance from the diagonal to the two dotted lines is called the bandwidth. All coefficients outside of the bandwidth are zero, and they do not have to be stored.

The bandwidth *BW* is calculated using the following formula:

$$\mathrm{BW} = (\mathrm{N} + 1)\mathrm{D} \tag{4.36}$$

where *N* is the largest difference between the node numbers in a single element and *D* is the number of unknowns at each node (degrees of freedom). The minimization of *BW* depends upon the minimization of *N* which can be partially achieved by labeling the nodes across the shortest dimension of the continuum. Examples of the two different numbering systems are shown in Figure 4.11. The value of *BW* for the first numbering system is 9, whereas *BW* for the second system is only 6.

4.4.5 Effect of Mesh refinement and Higher Order Elements upon Accuracy of Solution

The accuracy of the finite element simulations can be improved by (i) refining the mesh and (ii) selecting higher order interpolation functions. In (i), the solution relies on an increased number of linear piecewise displacement models of simple form to represent a complex exact solution. In (ii), the solution depends upon the selection of a higher order interpolation function.

An increase in the number of elements generally means more accurate results. However, for any given problem, there will be a certain number of elements beyond which the accuracy of the solution cannot be improved significantly. Another characteristic of the discretization process that affects a finite element solution is

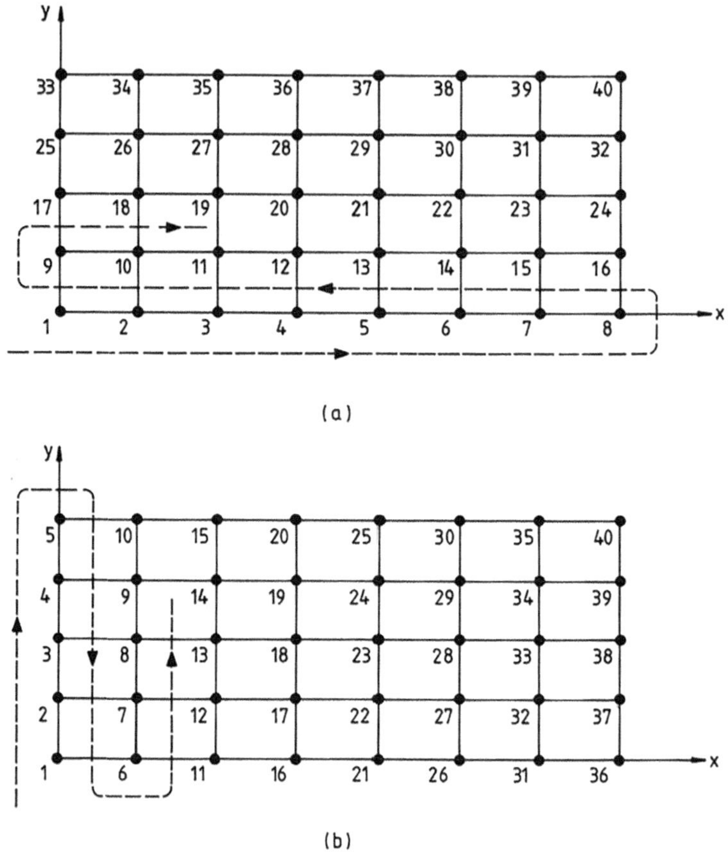

Figure 4.11 Numbering of nodes to reduce bandwidth from (**a**) BW = 9 to (**b**) BW = 6

the aspect ratio of the elements. The aspect ratio describes the shape of the element in the assemblage. For two-dimensional elements, this parameter is conveniently defined as the ratio of the largest dimension to the smallest dimension of the element. The optimum aspect ratio at any location within the discretized body depends largely upon the difference in the rate of change of the displacement field in the different directions. If the displacement varies at about the same rate in each direction, the closer the aspect ratio to unity the better the quality of the solution. This effectively means that we should avoid long narrow elements similar to those depicted in Figure 4.12.

The most popular form of element interpolation function is the polynomial series. The order of the polynomial depends upon the number of items known about the continuous interpolation function at each nodal point. Interpolation functions can be classified into three groups according to the order of the element interpolation function. These groupings are simplex, complex and multiplex. The simplex

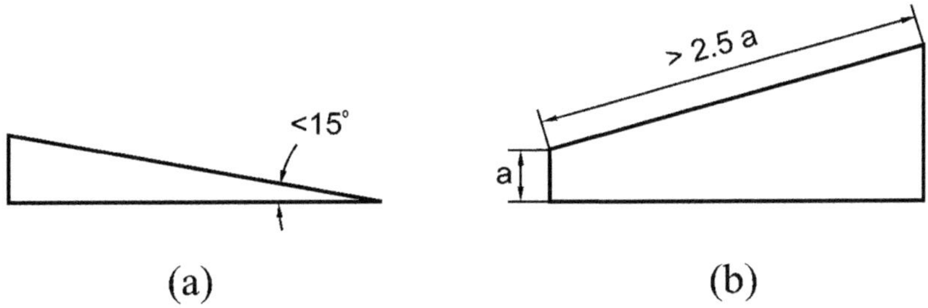

Figure 4.12 Poor element proportions for (**a**) triangular and (**b**) quadrilateral elements

elements have an approximating polynomial that consists of the constant term plus the linear terms. The number of coefficients in the polynomial is equal to the dimension of the coordinates space plus one. The polynomial

$$\varphi = a + bx + cy \tag{4.37}$$

is the simplex function for the two-dimensional triangular element. The polynomial is linear in x and y and contains three coefficients because the triangle has three sides.

The multiplex elements also use polynomials containing higher order terms, but the element boundaries must be parallel to the coordinate axes to achieve inter-element continuity. The element boundaries of the simplex and complex elements are not subjected to this restriction. An interpolating polynomial for a two-dimensional multiplex triangular element is given by,

$$\varnothing = a + bx + cx + dx_2 + exy + fy^2 \tag{4.38}$$

Experience indicates that fewer higher-order elements are needed to obtain the same degree of accuracy of linear elements. The use of higher order elements does not, however, always lead to a reduction in the total computation time. Numerical integration techniques are required to obtain the element matrices. These techniques can involve a large number of calculations, increasing the total computation time per element. Any additional expense for computation time, however, is probably offset by the cost decrease in the data preparation process.

The use of higher-order elements generally produces more accurate results in simulations where the gradient cannot be properly approximated by a set of linear values. The question that concerns engineers is which of the two methods is the most economical. Unfortunately, this question has not been answered satisfactorily. There are trade-offs in time and cost as well as accuracy that must be evaluated, if the question is to be answered correctly. In the final analysis, what should govern the selection of an approach is the degree of accuracy obtained per unit cost (analyst time and CPU).

4.5 Interpolating Polynomial Functions

From the above discussions, it is evident that the choice of suitable interpolation functions represents an important part of any finite element simulations. The appropriate choice of interpolation functions will lead to elements of high accuracy and with converging characteristics. Conversely, wrongly chosen interpolation functions may lead to poor and/or inaccurate results.

Generally, an interpolation function is either given as:

(i) a simple polynomial with undetermined coefficients which are subsequently transformed to become the relevant nodal displacements, or
(ii) directly in terms of shape functions, which are usually oscillatory polynomials having zero values at all other nodes of an element, but unit values for the displacement or its partial derivatives at the node under consideration.

Thus, the displacement field (u, υ) in a two-dimensional element can either be given as

$$\left.\begin{aligned} u(x, y) &= a_1 + a_2 x + a_3 y + \ldots \\ \upsilon(x, y) &= b_1 + b_2 x + b_3 y + \ldots \end{aligned}\right] \tag{4.39}$$

according to (i) above, or

$$\left.\begin{aligned} u(x, y) &= u_1 N_1 + u_2 N_2 + u_3 N_3 + \ldots \\ \upsilon(x, y) &= \upsilon_1 N_1 + \upsilon_2 N_2 + \upsilon_3 N_3 + \ldots \end{aligned}\right] \tag{4.40}$$

according to (ii). In the above expressions a_1, a_2, a_3 ... b_1, b_2, b_3, etc., are undetermined polynomial coefficients, N_1, N_2, N_3, etc., are shape functions and u_1, u_2, u_3, ..., and υ_1, υ_2, υ_3, etc., are the unknown nodal displacements.

The detailed analysis of displacement or interpolation functions can be found in any finite element text. However, we present here a summary of the pertinent points which must be taken into account in deciding upon the order of the polynomial in a polynomial type interpolation function. They are:

(i) the interpolation function must satisfy the convergence requirements,
(ii) the number of generalized coordinates should be equal to the number of the nodal degrees of freedom of the element, and
(iii) the pattern of variation of the field variable (stress, strain, temperature, etc.) should be independent of the local coordinates system. This property is known as geometric invariance or spatial isotropy. In order to achieve this, the polynomial should contain terms which do not violate symmetry.

At this stage, it is appropriate to become familiar with concepts used in the construction of shape/interpolation functions.

4.5.1 Pascal Triangle

For displacement functions of two- and three-dimensional elements given in terms of simple polynomials, the Pascal triangle of Figure 4.13 is a useful aid for determining the combination of terms which should be used. Thus, in the case of a two-dimensional simplex element, the polynomial should include a constant term plus terms containing x and y and not only one of them. Table 4.1 provides a few examples to demonstrate the use of Pascal triangle in the development of the appropriate shape functions for two-dimensional elements.

4.5.2 Lagrange Polynomials

These are often used for the construction of shape functions for elements in which only function values but not derivatives are specified at the nodes. The basic form of Lagrange's polynomial in a single coordinate system with n nodes is

$$u(x) = \sum_{i=0}^{n} L_i^n(x) u_i \tag{4.41}$$

where $L_i^n(x)$ is known as Lagrange multiplier and is given by

$$L_i^n(x) = \frac{(x - x_0)(x - x_1)\ldots(x - x_{i-1})(x - x_{i+1})\ldots(x - x_n)}{(x_i - x_0)(x_i - x_0)\ldots(x_i - x_{i-1})(x_i - x_{i+1})\ldots(x_i - x_n)} \tag{4.42}$$

It can be seen that $L_i^n(x)$ is an nth degree polynomial as it is given by the product of n linear factors. It is also obvious that $L_i^n(x)$ possesses the properties

$$L_i^n(x_k) = \begin{cases} 0 & k \neq i \\ 1 & k = i \end{cases} \tag{4.43}$$

Figure 4.13 Pascal triangle

$$\begin{array}{ccccccccccc}
 & & & & & 1 & & & & & \\
 & & & & x & & y & & & & \\
 & & & x^2 & & xy & & y^2 & & & \\
 & & x^3 & & x^2y & & xy^2 & & y^3 & & \\
 & x^4 & & x^3y & & x^2y^2 & & xy^3 & & y^4 & \\
x^5 & & x^4y & & x^3y^2 & & x^2y^3 & & xy^4 & & y^5
\end{array}$$

Table 4.1 Typical two-dimensional finite elements using pascal triangle

Pascal triangle	Element type	Comments
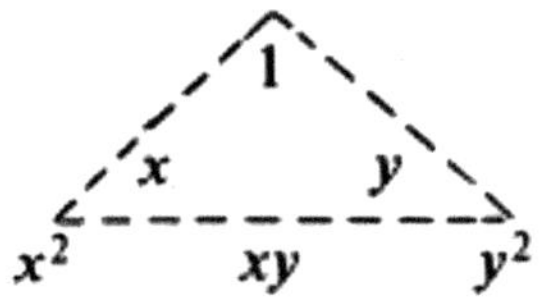	Constant strain triangle	Linear variation of u and υ $u \rightarrow 3$ DOF $\upsilon \rightarrow 3$ DOF
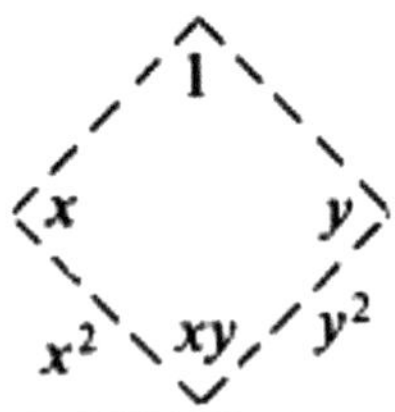	Plane stress rectangle	Linear variation of u and υ along edges of element. xy term degenerates to either x or y along the appropriate edge $u \rightarrow 4$ DOF $\upsilon \rightarrow 4$ DOF
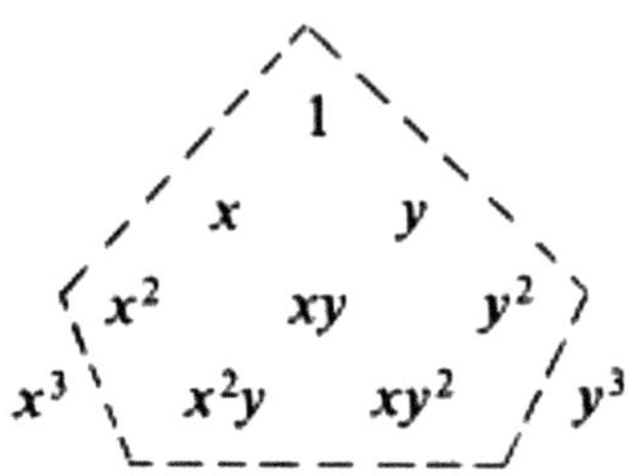	Parabolic plane stress rectangle with corner and mid-side nodes	Parabolic variation of u and υ along edges. x^3 and y^3 are excluded $u \rightarrow 8$ DOF $\upsilon \rightarrow 8$ DOF

and thus fits in with the definition of a shape function. It is obviously possible to use Lagrange's formulation to shape functions involving two and three coordinates, e.g. the shape functions for a two-dimensional problem would be,

$$u(x, y) = \sum_{i=0}^{n} \sum_{j=0}^{m} L_i^n(x) L_j^m(y) u_{ij} \tag{4.44}$$

where n and m are the number of subdivisions in the x and y directions, respectively. In fact, the shape functions given by (4.44) form the basis of a family of plane stress elements.

As an extension to Lagrange polynomials, the Hermitian polynomials were developed not only to agree in value with a given function at specified locations, but their derivatives will also match with the derivatives of the given functions up to any given order at any particular location. These functions are commonly used in the analysis of bending of finite strips and plates.

It is worth noting that all the previous discussions make use of the Cartesian coordinates (x, y). In finite element other than the linear bar, truss and beam elements, the element stiffness matrix must be integrated numerically. For this, we use Gaussian Quadrature. This leads to the introduction of intrinsic coordinates (ξ, η) to facilitate this integration using a parent element. A detailed discussion of this will be provided in the following section.

4.6 Isoparametric Finite Element Representation

4.6.1 Intrinsic coordinates System

A local coordinates system (ξ, η) is one which is defined for a particular element and not necessarily for the entire body or structure. The coordinate system for the entire body is known as the global system (x, y). An intrinsic (natural) coordinates system is a local system which permits the identification of a point within the element by a set of dimensionless numbers whose magnitudes either never exceed unity (Figure 4.14) or from $(-1$ to $+1)$. These systems are usually so arranged that some of the intrinsic coordinates have unit magnitudes at specific nodal points. Furthermore, such a structure simplifies the implementation and storage of finite element codes and as stated above facilitates the numerical integration necessary for the determination of element stiffness matrices.

4.6.2 Isoparametric Elements

The displacement field (u, v) for a two-dimensional element of general shape can be given by,

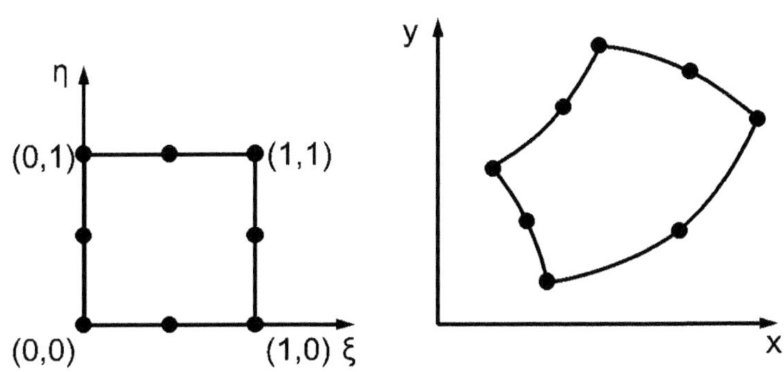

Figure 4.14 Isoparametric mapping: (**a**) parent element and (**b**) eight node element

$$\left.\begin{aligned} u(x,y) &= \sum_{i=1}^{n} u_i N_i \\ &= u_1 N_1 + u_2 N_2 + u_3 N_3 \ldots u_n N_n \end{aligned}\right\} \quad (4.45)$$

and

$$\left.\begin{aligned} \upsilon(x,y) &= \sum_{i=1}^{n} \upsilon_i N_i \\ &= \upsilon_1 N_1 + \upsilon_2 N_2 + \upsilon_3 N_3 \ldots \upsilon_n N_n \end{aligned}\right\} \quad (4.46)$$

where N_i is a function of the intrinsic coordinates (ξ, η) and n is the number of nodes. The coordinates (x, y) inside the element domain can also be described in a similar manner by the following expressions:

$$\left.\begin{aligned} x &= \sum_{i=1}^{n} x_i M_i \\ &= x_1 M_1 + x_2 M_2 + x_3 M_3 \ldots x_n M_n \end{aligned}\right\} \quad (4.47)$$

and

$$\left.\begin{aligned} x &= \sum_{i=1}^{n} x_i M_i \\ &= x_1 M_1 + x_2 M_2 + x_3 M_3 \ldots x_n M_n \end{aligned}\right\} \quad (4.48)$$

in which M_i is also a function of ξ and η. For the particular case in which N_i and M_i are identical, i.e. the shape functions defining the displacement fields and geometry are the same, the element is said to be isoparametric.

4.7 General Expression for Element Stiffness Matrix

One of the most important items in finite element analysis is the development of general element stiffness matrices for two- and three-dimensional elasticity problems. In the present work, we adopt the principle of minimum potential energy for deriving the equilibrium equations. Since the nodal degrees of freedom are treated as unknown in the present displacement formulation, the total potential energy U must be expressed in terms of nodal degrees of freedom. The necessary equilibrium can be obtained by setting the first partial derivatives of U with respect to each of the nodal degrees of freedom equal to zero. The following is an outline of the required steps for the derivation of the element stiffness matrix for three-dimensional elasticity problems of the simplex type. This approach is not directly applicable to the isoparametric elements discussed earlier.

(i) Define the displacement and force vectors for a general element as follows:

$$\{\delta\}_{3n\times 1} = \{u_1 \upsilon_1 w_1 u_2 \upsilon_2 w_2 \ldots u_n \upsilon_n w_n\}^T \quad (4.49)$$

$$\{F\}_{3n\times 1} = \{F_{x1} F_{y1} F_{z1} F_{x2} F_{y2} F_{z2} \ldots F_{xn} F_{yn} F_{zn}\}^T \quad (4.50)$$

where n is the number of nodes in the desired element.

(ii) Express the displacement at any point within the element in terms of the nodal displacements, i.e.

$$\left.\begin{aligned} u(x,y,z) &= \textstyle\sum_{i=1}^{n} u_i N_i \\ \upsilon(x,y,z) &= \textstyle\sum_{i=1}^{n} \upsilon_i N_i \\ w(x,y,z) &= \textstyle\sum_{i=1}^{n} w_i N_i \end{aligned}\right\} \quad (4.51)$$

where u_i, υ_i and w_i are the nodal displacements of the ith node and N_i are the corresponding element shape functions.

(iii) Express the strain components at any point within the element in terms of the nodal displacements:

$$\{\varepsilon\} = \begin{Bmatrix} \varepsilon_{xx} \\ \varepsilon_{yy} \\ \varepsilon_{zz} \\ \gamma_{xy} \\ \gamma_{yz} \\ \gamma_{zx} \end{Bmatrix} = \begin{Bmatrix} \dfrac{\partial u}{\partial x} \\ \dfrac{\partial \upsilon}{\partial y} \\ \dfrac{\partial w}{\partial z} \\ \dfrac{\partial u}{\partial y} + \dfrac{\partial \upsilon}{\partial x} \\ \dfrac{\partial \upsilon}{\partial z} + \dfrac{\partial w}{\partial y} \\ \dfrac{\partial w}{\partial x} + \dfrac{\partial u}{\partial z} \end{Bmatrix}$$

$$= \begin{bmatrix} \frac{\partial N_1}{\partial x} & 0 & 0 & \frac{\partial N_2}{\partial x} & 0 & 0 & \cdots & \frac{\partial N_n}{\partial x} & 0 & 0 \\ 0 & \frac{\partial N_1}{\partial y} & 0 & 0 & \frac{\partial N_2}{\partial y} & 0 & \cdots & 0 & \frac{\partial N_n}{\partial y} & 0 \\ 0 & 0 & \frac{\partial N_1}{\partial z} & 0 & 0 & \frac{\partial N_2}{\partial z} & \cdots & 0 & 0 & \frac{\partial N_n}{\partial z} \\ \frac{\partial N_1}{\partial y} & \frac{\partial N_1}{\partial x} & 0 & \frac{\partial N_2}{\partial y} & \frac{\partial N_2}{\partial x} & 0 & \cdots & \frac{\partial N_n}{\partial y} & \frac{\partial N_n}{\partial x} & 0 \\ 0 & \frac{\partial N_1}{\partial z} & \frac{\partial N_1}{\partial y} & 0 & \frac{\partial N_2}{\partial z} & \frac{\partial N_2}{\partial y} & \cdots & 0 & \frac{\partial N_n}{\partial z} & \frac{\partial N_n}{\partial y} \\ \frac{\partial N_1}{\partial z} & 0 & \frac{\partial N_1}{\partial x} & \frac{\partial N_2}{\partial z} & 0 & \frac{\partial N_2}{\partial x} & \cdots & \frac{\partial N_n}{\partial z} & 0 & \frac{\partial N_n}{\partial x} \end{bmatrix}$$

$$= \begin{Bmatrix} u_1 \\ \upsilon_1 \\ w_1 \\ u_2 \\ \upsilon_2 \\ w_2 \\ - \\ - \\ - \\ - \\ u_n \\ \upsilon_n \\ w_n \end{Bmatrix} \tag{4.52}$$

i.e.,

$$\{\varepsilon\}_{6\times1} = [B]_{6\times3n}\,\{\delta\}_{3n\times1} \tag{4.53}$$

(iv) Express the stress components at a general point in the element in terms of the nodal displacements. Assuming the material is perfectly elastic, homogeneous and isotropic, then,

$$\{\sigma\}_{6\times1} = [D]_{6\times6}\{\varepsilon\}_{6\times1} \tag{4.54}$$

or alternatively

$$\{\sigma\} = [D][B]\{\delta\} \tag{4.55}$$

where [***D***] is the elasticity matrix, and is defined by Equation (4.15).

(v) The element stiffness matrices are to be derived from the application of the principle of minimum potential energy to the above system. For this, the potential energy *U* is obtained by considering the strain energy and the work done by point loads; as follows:

$$\begin{aligned} U &= \left(\frac{1}{2}\right) \int_{V^{(e)}} \{\varepsilon\}^T\{\sigma\}dV - \{\delta\}^T\{F\} \\ &= (1/2) \int_{V^{(e)}} \{\delta\}^T[B]^T[D][B]\{\delta\}dV - \{\delta\}^T\{F\} \end{aligned} \tag{4.56}$$

The minimization of the above equation with respect to the displacement vector $\{\boldsymbol{\delta}\}$ will yield the following expression for the element stiffness matrix for general three-dimensional elastic problems:

$$\left(\int_{V^{(e)}} [B]^T [D][B] dV\right)\{\delta\} = \{F\} \tag{4.57}$$

or

$$\left[K^{(e)}\right]_{3n\times 3m}\{\delta\}_{3n\times 1} = \{F\}_{3n\times 1} \tag{4.58}$$

where $[\boldsymbol{K}^{(e)}]$ is known as the element stiffness matrix and is given by

$$\left[K^{(e)}\right] = \int_{V^{(e)}} [B]^T [D][B] dV \tag{4.59}$$

for three-dimensional problems, and

$$\left[K^{(e)}\right] = t\int_{A^{(e)}} [B]^T [D][B] dA \tag{4.60}$$

for two-dimensional problems.

4.7.1 *Role of Isoparametric Elements*

The $[\boldsymbol{B}]$ matrix given in Equation (4.60) for two-dimensional problems expresses the strains in terms of the nodal displacements and therefore contains first derivatives of the shape functions with respect to the global x and y axes. However, the shape functions for isoparametric elements are defined in terms of the intrinsic coordinates ξ and η, and therefore cannot be differentiated directly with respect to the global x and y axes. (Typical shape functions are provided in Appendix).

In order to overcome this difficulty, it is necessary to obtain a relationship between the derivatives of the two sets of coordinates; as follows:

$$\left.\begin{aligned} \frac{\partial N}{\partial \xi} &= \frac{\partial N}{\partial x}\frac{\partial x}{\partial \xi} + \frac{\partial N}{\partial y}\frac{\partial y}{\partial \xi} \\ \frac{\partial N}{\partial \eta} &= \frac{\partial N}{\partial x}\frac{\partial x}{\partial \eta} + \frac{\partial N}{\partial y}\frac{\partial y}{\partial \eta} \end{aligned}\right\} \tag{4.61}$$

which in a matrix form can be written as:

$$\left\{ \begin{array}{c} \frac{\partial N}{\partial \xi} \\ \frac{\partial N}{\partial \eta} \end{array} \right\} = \left[\begin{array}{cc} \frac{\partial x}{\partial \xi} & \frac{\partial y}{\partial \xi} \\ \frac{\partial x}{\partial \eta} & \frac{\partial y}{\partial \eta} \end{array} \right] \left\{ \begin{array}{c} \frac{\partial N}{\partial x} \\ \frac{\partial N}{\partial y} \end{array} \right\} = [J] \left\{ \begin{array}{c} \frac{\partial N}{\partial x} \\ \frac{\partial N}{\partial y} \end{array} \right\} \tag{4.62}$$

The matrix [***J***] relating the derivatives of the two systems is known as the Jacobian matrix and its coefficients can be obtained by differentiating expressions (4.47) and (4.48) with respect to the ξ and η coordinates. To complete the transformation between the two systems, it is also necessary to express the elemental area dA of Equation (4.60) in terms of ξ and η, as follows:

$$dA = |J| d\xi d\eta \tag{4.63}$$

It follows that the stiffness matrix for two-dimensional isoparametric elements can be written as:

$$\left[K^{(e)}\right] = t \int_0^1 \int_0^1 [B]^T [D][B] |J| d\xi d\eta \tag{4.64}$$

in which $[\boldsymbol{B}]^T \cdot [\boldsymbol{D}] \cdot [\boldsymbol{B}]$ is only a function of the intrinsic coordinates ξ and η. It is worth noting that in order to obtain $\partial N/\partial x$ and $\partial N/\partial y$, it is necessary to invert [***J***]. This inverse exists, if there is no excessive distortion of the element.

The above integrals are commonly evaluated numerically using Gaussian quadrature. The details can be found in the appropriate literature on finite elements and numerical analysis text books. It is worth remembering that in order to effect a solution to the entire system, one must apply (4.64) to that entire system. This will enable the determination of the equilibrium equation for the body or structure.

4.8 Computer Implementation of Finite Element

One of the advantages of the finite element method is that many of the solution steps are common to most application areas. The procedure for solving problems in heat transfer and fluid flow, involve many of the same steps found in structural stress and dynamic response analyses. Figure 4.15 shows a simplified computer flow diagram for an elastic finite element code, the major steps of which will be discussed in general terms rather than with respect to any specific example.

Before any analysis can be made, it is necessary for the analyst to make two important decisions: (i) the type of element and (ii) the mesh configuration to be used. Both of these decisions are dictated by the geometrical features of the component, the applied loads, the boundary conditions, stress gradients and accuracy in relation to cost.

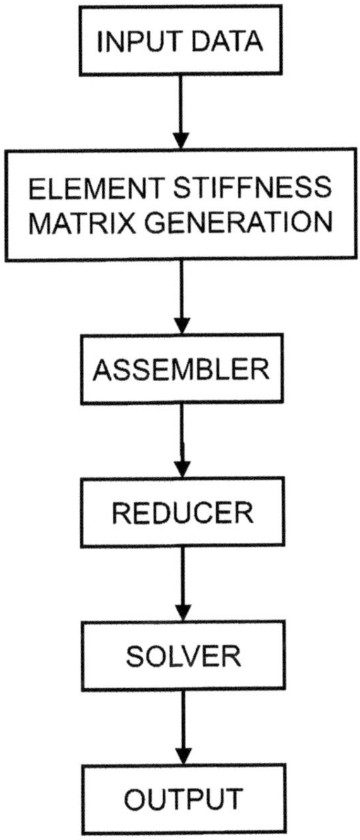

Figure 4.15 A general computer flow diagram for a finite element program

For a number of elements, and in particular the constant strain triangular elements, the accuracy of the results deteriorates rapidly with increased aspect ratio and, in general, long and narrow elements should be avoided (Figure 4.12). It is also worth noting that there are no specific rules for establishing the optimum mesh for a new design, except that:

(i) common boundaries must have the same number of divisions and ratios to ensure compatibility, and
(ii) geometry must be modeled into groups of simpler geometric elements with known shape functions.

In these situations however, experience plays an important role in the decision making process. One common approach relies on the use of successfully refined meshes until acceptable convergence has been achieved. This is known as adaptive convergence.

4.8.1 Automatic Mesh Generation

Most practical engineering problems involve a large number of elements and nodes and the task of preparing the input data becomes extremely lengthy, time consuming and tedious. Furthermore, during the preparation of data some human errors may be introduced and remain undetected in spite of the several checks which are usually made. The presence of such errors will inevitably bring about incorrect results. It is therefore important to eliminate such data errors. This can be achieved, to a large extent, by automating the discretization process. In this case, nodal numbers and their associated coordinates together with element numbers and their definitions are prepared automatically by the computer, using as input data the minimal amount of information necessary to describe the component and the desired mesh divisions.

In the past few years, considerable efforts have been expended in developing mesh generation algorithms and thus improving the efficiency of data preparation and minimizing data errors (these algorithms can be found in the appropriate literature). This has also been supplemented by the development of interactive mesh generation facilities.

4.8.2 Input Data

After having decided on element type and mesh division, it is now possible to prepare the data for input to the program. Such data includes all information concerning component geometry, element types, node numbering and coordinates, topology, material mechanical and physical properties, applied loads and imposed restraints.

In most finite element programs data echo is included and the input data are printed as a matter of course. The inclusion of such data echo will facilitate the checking of input data.

4.8.3 Assembly of Element Stiffness Matrices

Traditional finite elements computer programs require information regarding the number of equations, the number of elements and the bandwidth. Information about the number of equations is needed so that the global (overall) stiffness matrix and the global force vector can be initialized to zero. Initialization is necessary because the matrices are built by a summation process. The initialization process is followed by a 'DO' loop through the element. At this instant, the element data (described earlier) are read and the appropriate element matrices are generated using the relevant parent elements and then inserted in their appropriate position in the global matrix (the structure). Obviously, the computation of element stiffness matrices requires information about the material mechanical and physical properties. These properties include Young's modulus, Poisson's ratio, coefficient of thermal conductivity, convection heat transfer coefficient, density, etc. The evaluation of the Jacobian matrix along with its inverse and determinant are also needed. The partial derivatives needed to evaluate the Jacobian matrix for the appropriate elements are computed in a subroutine which has equations for both the shape functions and their derivatives.

It must be pointed out however that in a typical finite element using numerically integrated isoparametric elements and an advanced solution subroutine such as "the front solver," the computer time spent in calculating the element stiffness can vary from 25% to 85% of the total time required for a complete analysis. Therefore, the algorithm used to compute the element stiffness has a significant effect on the overall efficiency of the program. With the algorithms in common use, the computer time required to calculate an element stiffness matrix varies by a factor of 10 between the best and the least efficient.

The stiffness relationship for an element has been given by Equation (4.58). For a 4-noded rectangular element, the above equation can be written as:

$$\begin{bmatrix} K_{ii} & K_{ij} & K_{il} & K_{im} \\ K_{ji} & K_{jj} & K_{jl} & K_{jm} \\ K_{li} & K_{lj} & K_{ll} & K_{lm} \\ K_{mi} & K_{mj} & K_{ml} & K_{mm} \end{bmatrix} \begin{Bmatrix} \delta_i \\ \delta_j \\ \delta_l \\ \delta_m \end{Bmatrix} = \begin{Bmatrix} P_i \\ P_j \\ P_l \\ P_m \end{Bmatrix} \tag{4.65}$$

in which all the terms are in fact submatrices.

Since a component is made up of many elements, it follows that the component global stiffness matrix is also made up of a corresponding number of element stiffness matrices, thus

$$[K^g]\{\delta^g\} = \{F^g\} \tag{4.66}$$

where $[K^g]$ is the global stiffness matrix

$\left(= \sum_{e=1}^{n} \left[K^{(e)}\right], \text{where n is the number of elements}\right)$,
$\{\delta^g\}$ is the global vector of nodal displacements, and
$\{F^g\}$ is the global force vector with a number of load cases.

It is worth noting that the nodal force vector consists of quantities that are associated with a specific node rather than a specific element. In general, most of these quantities are associated with boundary nodes such as concentrated or distributed forces in solid mechanics and heat losses in heat transfer problems.

4.8.4 Solution of Simultaneous Equations

The above global system of equations which are characterized by the elastic spring formulations and obtained by the summation of all the elements, must be modified whenever some of the values in the global displacement vector $\{\delta^g\}$ are known. This is more the rule rather than the exception.

Most field problems have some of the boundary values specified, and all elasticity problems must have some of the displacements specified in order to eliminate rigid body degrees of freedom. As a matter of fact, the global stiffness matrix $[K^g]$ discussed earlier is singular until some of the displacement boundary restrictions have been applied, leading to a reduced system of equations.

The modification of the system of equations is followed by the solution of these equations for the nodal unknowns. There are several procedures for accomplishing this solution. These include the direct solvers using Gaussian elimination, Gauss-Jordan elimination or Choleski's factorization and the iterative solvers using the relaxation approach, Gauss-Seidel or the conjugate-gradient methods. The techniques adopted by these methods as well as merits and limitations can be further investigated in general numerical analysis books.

The resulting system of equations is special because it is banded and the diagonal terms are usually positive and dominate their respective row and column. This allows many of the above mentioned solvers to be modified for increased efficiency and compactness.

The solution of the system of equations is followed by output of the nodal displacement. Once the displacements have been obtained, it is possible to compute the stresses of each element in turn by using Equation (4.55), viz.

$$\{\sigma\} = [D][B]\{\delta\}$$

The above description assumes the use of programming languages such as C++. However, it is more efficient to use MATLAB in finite element implementation.

4.8.5 Adaptive Convergence

Adaptive convergence is a technique applied to identify the mesh in regions with large gradients and/or large discretization errors and to refine the mesh in these regions. As a consequence, computational efficiency can be obtained without sacrificing accuracy. Adaptive convergence is really important because the selection of elements and the size of these element govern the accuracy of FE simulations.

To demonstrate the importance of adaptive meshing, we conducted a simple analysis of a plate 20 mm thick and 1 m long containing a hole at one end. The plate, which is treated as a plane stress problem, is subjected to an internal pressure of P = 1 MPa and a point load F = 1 kN, as depicted in Figure 4.16. The material of

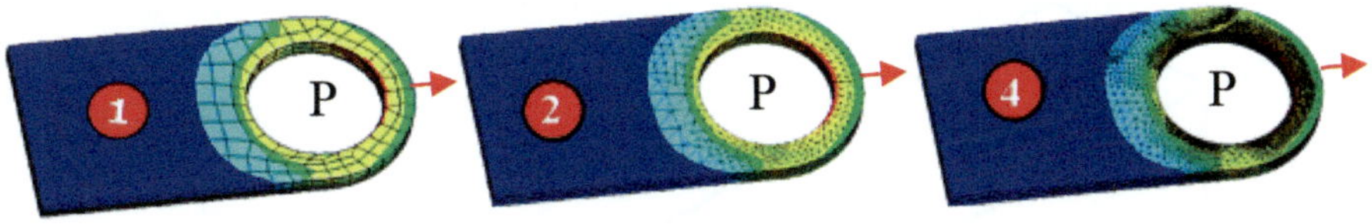

Figure 4.16 Adaptive meshing of a plate containing a hole at one end using commercial code ANSYS Workbench. Plate subjected to internal pressure and point load. Fourth run shows a much reduced element size compared to first run

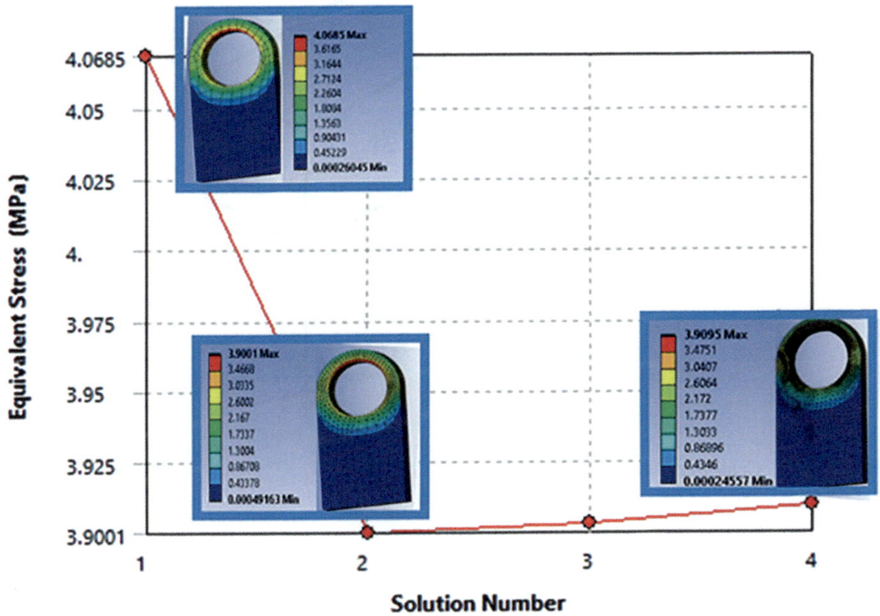

Figure 4.17 Example of adaptive convergence using ANSYS Workbench

the plate was assumed to be structural steel and the simulations was conducted using ANSYS Workbench. The equivalent stress was deemed to be a critical parameter for the current system, because of the presence of stress concentration. As can be seen from Figure 4.17, increasing the mesh at the critical areas, results in improved convergence. For example, in the first run (course elements), the equivalent stress was 4.1 MPa, while in the fourth trial (refined elements) it was 3.9 MPa.

4.8.6 Vsalidation of FE Predictions

The finite element method is an approximate numerical technique that must be validated experimentally. Of particular interest to this heading is the elastic and elasto-plastic behavior of a cantilever beam containing rectangular, circular or

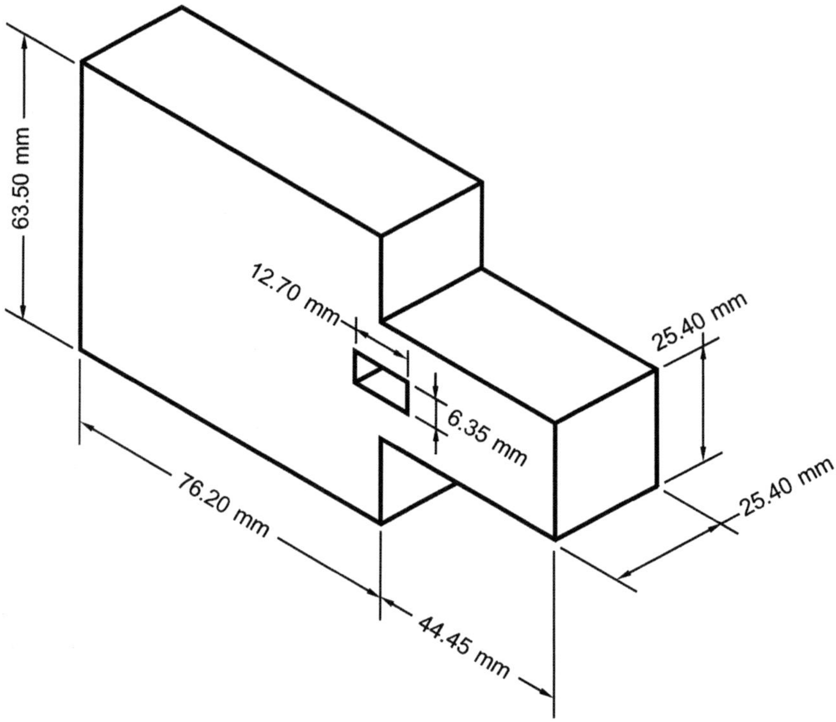

Figure 4.18 Cantilever beam with cut out

diamond cut outs under plane strain conditions conducted recently by the author. Three aspects of this work were are examined; see Ref. [4.3] for details. The first is concerned the determination of the stress concentration resulting from the aforementioned cut out features under elastic loading conditions of the beam. The second with the Elasto-plastic behavior of a beam containing the three varied cut out geometries, with the focus being the determination of the effective plastic strains around the cut out features. The third with the remarkable agreement of the predicted plastic strains with chemical etching results of a comparable cantilever beam.

Specifically, the situation envisaged is that of the cantilever beam shown in Figure 4.18, which contains cut out of different geometries: rectangular, circular or diamond-shaped, as shown in Figure 4.19. The cantilever beam, which contains the cut out placed along its center line, is fixed at one end and a uniform quasistatic shear load is applied at the other end. The selection of these specific dimensions was motivated by our desire to validate our plane strain finite element predictions with existing plane strain etching results of a cantilever beam containing cut outs of a similar beam intended to detect slipline fields. Figure 4.20a shows a schematic of a beam containing rectangular cut out, a constraint, an externally applied load, and a

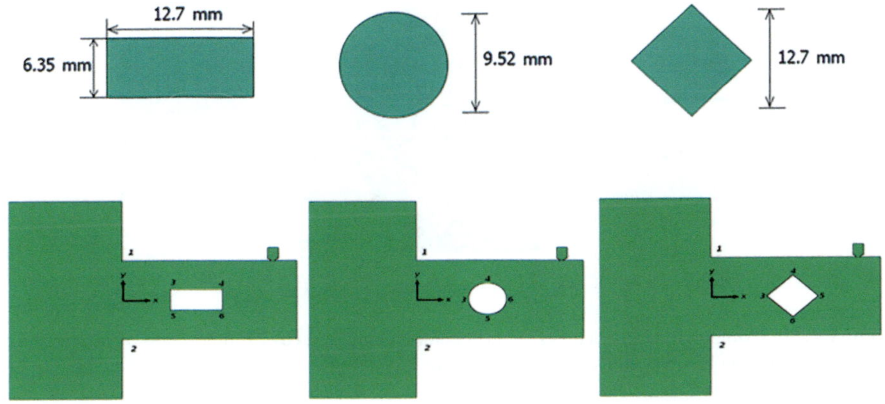

(a) (b) (c)

Figure 4.19 Three different cut outs in three cantilever beams: (**a**) ractangular, (**b**) circular and (**c**) diamond-shaped

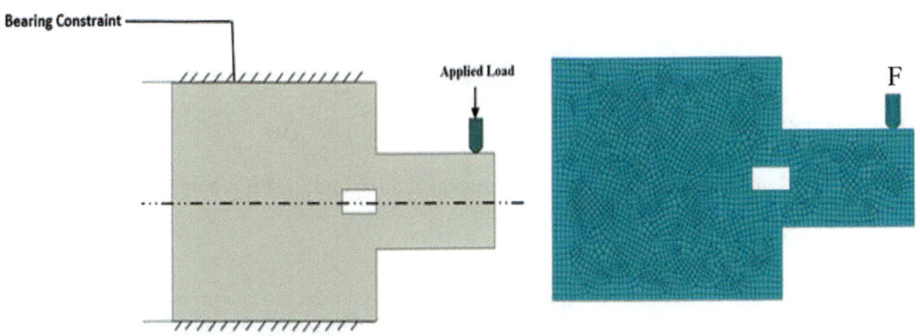

(a) **(b)**

Figure 4.20 Cantilever beam: (**a**) free body diagram of beam and (**b**) discretized beam containing cut out

typical mesh of the discretized finite element model is shown in Figure 4.20b. The beam is meshed using 2D plane strain element (CPS4R) of ANSYS. The indenter is discretized using rigid elements with very high stiffness values. A uniform quasistatic pressure load is applied using a rigid indenter. The beam is made of steel with Young's modulus of 200 GPa and Poisson's ratio of 0.3. For simplicity, the steel is assumed to behave in an elastic perfectly-plastic fashion with the yield stress being 220 MPa, again to facilitate the comparison with exiting experimental work. A surface-to-surface contact is defined between the indenter and the upper surface of the beam, where the pressure is applied.

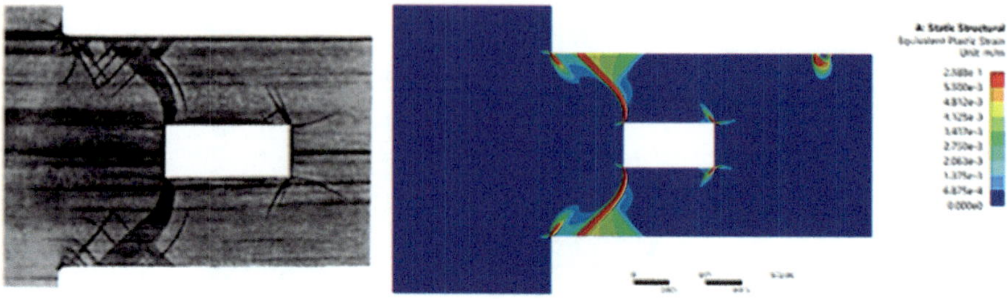

(a) (b)

Figure 4.21 Comparison between (**a**) experimental etching and (**b**) FE prediction of plastic strain distribution for the beam with a rectangular cut-out

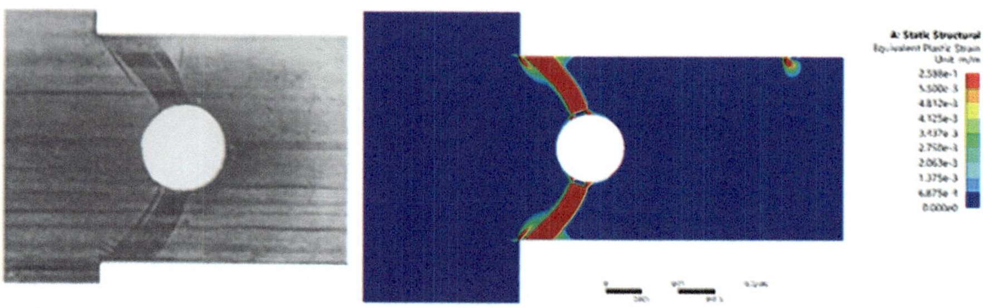

(a) (b)

Figure 4.22 Comparison between (**a**) experimental etching and (**b**) FE prediction of plastic strain distribution for the beam with a circular cut-out

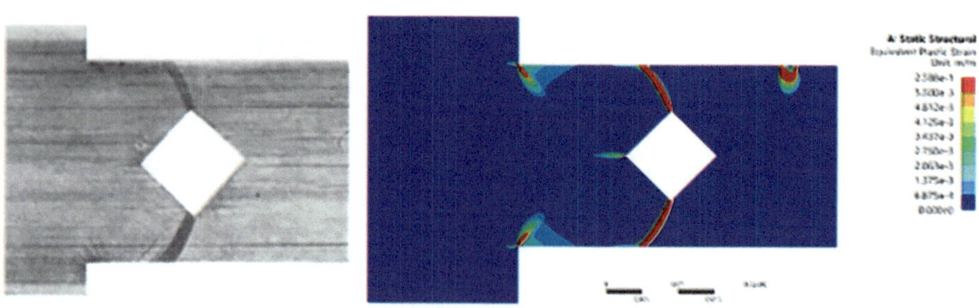

(a) (b)

Figure 4.23 Comparison between (**a**) experimental etching and (**b**) FE prediction of plastic strain distribution for the beam with a diamond-shaped cut-out

The plastic zone developed during the loading process is simulated using elasto-plastic finite element simulations. Finite element predictions depicted in Figures 4.21b, 4.22b and 4.23b show truly remarkable agreements with the plastic zone developed experimentally using etching techniques in Figures 4.21a, 4.22a and 4.23a.

Appendix: Shape Functions for a Few Common Elements in Terms of Intrinsic Coordinates

(i) *Two-noded linear bar elements*

$$N_1(\xi) = 1\text{-}\xi$$
$$N_2(\xi) = \xi$$

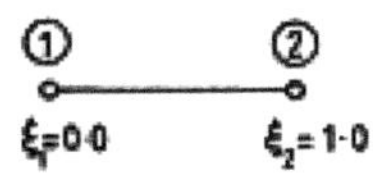

(ii) *Three-node parabolic bar elements*

$$N_1(\xi) = (1\text{-}\xi)(1\text{-}2\xi)$$
$$N_2(\xi) = 4\xi(1\text{-}\xi)$$
$$N_3(\xi) = \text{-}\xi(1\text{-}2\xi)$$

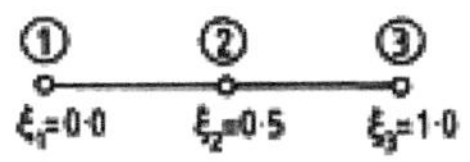

(iii) *Four-node linear plane elements*

$$N_1(\xi,\eta) = (1\text{-}\xi)(1\text{-}\eta)$$
$$N_2(\xi,\eta) = \xi(1\text{-}\eta)$$
$$N_3(\xi,\eta) = \xi\eta$$
$$N_4(\xi,\eta) = (1\text{-}\xi)\eta$$

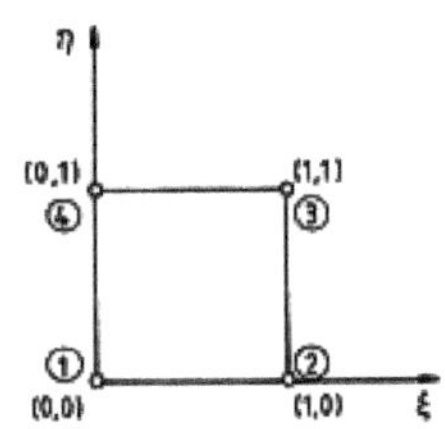

(iv) *Eight-node quadratic elements*

$$N_1(\xi,\eta) = (1\text{-}\xi)(1\text{-}\eta)(1\text{-}2\xi\text{-}2\eta)$$
$$N_2(\xi,\eta) = 4\xi(1\text{-}\xi)(1\text{-}\xi)$$
$$N_3(\xi,\eta) = \xi(1\text{-}\eta)(\text{-}1 + 2\xi\text{-}2\eta)$$
$$N_4(\xi,\eta) = 4\xi\eta(1\text{-}\eta)$$
$$N_5(\xi,\eta) = \xi\eta(\text{-}3 + 2\xi + 2\eta)$$
$$N_6(\xi,\eta) = 4\xi(1\text{-}\xi)\eta$$
$$N_7(\xi,\eta) = (1\text{-}\xi)\eta(\text{-}1\text{-}2\xi + 2\eta)$$
$$N_8(\xi,\eta) = 4(1\text{-}\xi)\eta(1\text{-}\eta)$$

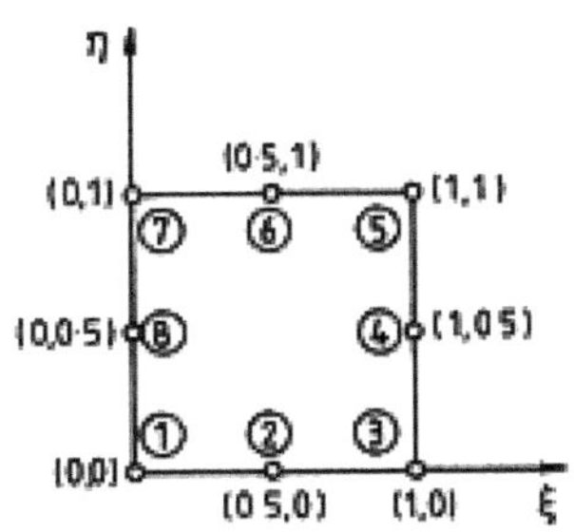

(v) *Eight-node solid brick elements*

$$N_1(\xi,\eta,\zeta) = (1\text{-}\xi)(1\text{-}\eta)(1\text{-}\zeta)$$
$$N_2(\xi,\eta,\zeta) = \xi(1\text{-}\zeta)(1\text{-}\eta)$$
$$N_3(\xi,\eta,\zeta) = \xi\eta(1\text{-}\zeta)$$
$$N_4(\xi,\eta,\zeta) = (1\text{-}\xi)\eta(1\text{-}\zeta)$$
$$N_5(\xi,\eta,\zeta) = (1\text{-}\xi)(1\text{-}\eta)\zeta$$
$$N_6(\xi,\eta,\zeta) = \xi(1\text{-}\eta)\zeta$$
$$N_7(\xi,\eta,\zeta) = \xi\eta\zeta$$
$$N_8(\xi,\eta,\zeta) = (1\text{-}\xi)\eta\zeta$$

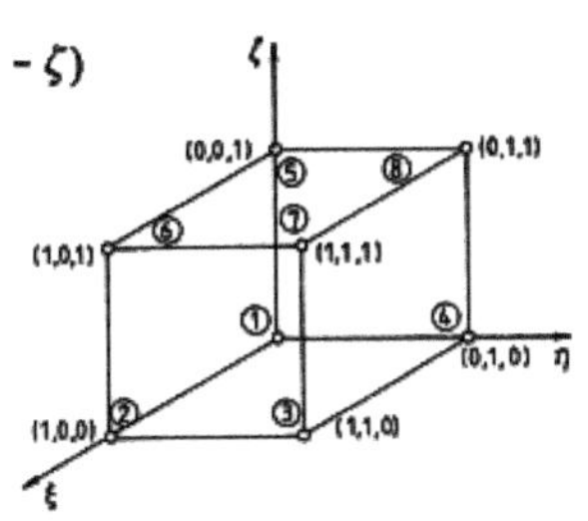

References

4.1 D.L. Logan, *A First Course in the Finite Element Method* (Cengage Learning, Boston, 2016)
4.2 T.R. Chandrupatla, A.D. Belegundu, *Introduction to Finite Elements in Engineering, Pearson*, 4th edn. (Pearson, Upper Saddle River, 2012)
4.3 S.A. Meguid, Elasto-plastic behaviour of cantilever beams containing varied stress concentration cut out features. Int. J. Mech. Mater. Des. **17**, 453–462 (2021)

Chapter 5
Interatomic Bonds and Defects in Solids

Summary Defects in solids are imperfections in the regular geometrical arrangements of the atoms. They are irregularities in the arrangement of the constituent particles in the form of point (impurity atoms and vacancies), line (dislocations) and planar defects (interfaces and grain boundaries). Defects play a major role in dictating the properties of crystalline solids not only by themselves, but also through interaction with other defect types. In fact, at times, imperfections are deliberately introduced into a solid to enhance its mechanical, thermal or electrical properties. In this chapter, we address the varied interatomic bonds and the role played by defects in determining the mechanical properties of solids.

5.1 Electrons and Bonding

Interatomic bonds govern the mechanical strength of a given material. These bonds are determined by the way the atoms are arranged in a given material. For example, a knowledge of the characteristics (energy and location) of the negatively charged electrons with respect to the positively charged ions determines the strength of that interatomic bond. Actually, the strength of the bond is controlled by the way the electrons are locally arranged within the atom [5.1–5.4].

Knowing that atoms are very small particles, can we treat those using Newton's laws of motion? The answer to that question is not straightforward. According to quantum mechanics, very small objects do not always behave like large objects, such as tennis or billiard balls. In fact, two gifted scientists, Newton and Maxwell, contradicted each other in their definitions. According to Newton, light is described in terms of particle theory. However, according to Maxwell, light is described in terms of wave theory. In 1905, Einstein did not accept the wave theory and instead proposed that a beam of light can be thought of as a stream of particles, which he called *photons*. Einstein stipulated that the number of photons is proportional to the intensity of the light, and the energy E of each photon is proportional to its frequency f. In 1920, de Broglie stipulated that light is characterized by *wave-particle duality*. This effectively meant that in certain situations, electrons behave like particles and in

S. A. Meguid, *Atomistic and Continuum Fracture Mechanics of Solids*,
https://doi.org/10.1007/978-3-031-56085-9_5

other situations behave like waves. The wave-particle duality was later confirmed using the well-known double slit experiment.

Electrons are considered quantum mechanical particles, and as such must obey the rules of quantum mechanics. In the following, we will only consider a summary of the laws relevant to atomistic bonding of solids. The first rule states that an electron can only have quantified energies and it cannot have energies between these quantified energies. Accordingly, a change in the electron energy is associated with a quantum jump to another allowed energy level. In particular, an electron can jump to a higher energy level by absorbing a radiant energy or to a lower energy level by emitting a radiant energy. The change in Energy ΔE is given by Einstein equation [5.1],

$$\Delta E = E_2 - E_1 = hf \tag{5.1}$$

Where E_1 is the energy prior to the jump, E_2 is the energy post the jump, f is the frequency of the energy emitted or absorbed and h is Planck's constant (6.626×10^{-34} m^2kg/s); a very small number, but an important one. It is worth noting that a positive ΔE implies energy is absorbed and a negative ΔE implies energy is emitted.

The second quantum mechanics rule is "Pauli's Exclusion Principle," which states that only two electrons can occupy the same energy level and that these two electrons must have opposite spins (Figure 5.1). It is important to recognize that the quantization of electron energies together with the Pauli Exclusion Principle will dictate the different bonding states of solids.

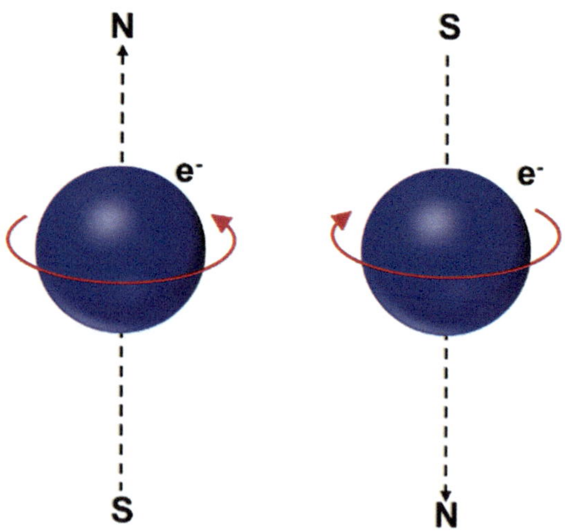

Figure 5.1 A schematic of the Pauli's Exclusion Principle showing electron with different spin

5.2 Crystal Structure

Solids exist in three forms: single crystalline, polycrystalline and amorphous. They hold finite geometric shapes and are mostly incompressible. Figure 5.2 shows a schematic of the three different types. A lattice is a regular arrangement of atoms. A single crystal is one that has repeated and orderly atomic structure forming a lattice that extends in the three spatial directions. Examples of single crystals are solid state transistors, light emitting diode, longitudinally solidified single crystal turbine blades. Polycrystal, on the other hand, has many crystals or grains separated interfaces known as grain boundaries (Figure 5.2b). Each grain can be considered as a single crystal. Metals are typically polycrystal in nature. Since these crystals do not line up, an interface in the form of a grain boundary develops. Examples include iron, copper aluminium and zinc. Unlike single and polycrystals, amorphous solids do not have repeated and ordered atomic arrangement, as depicted in Figure 5.2c. Typical examples include glass and polymers [5.2–5.5].

Let us devote our attention to crystalline solids. They are typically formed from unit cells. These unit cells are repeated in all three spatial directions to form the different solids, known as, "Lattice structures." Figure 5.3 shows a schematic of a generalized unit cell. Typical unit cells include cubical, tetragonal, orthorhombic, hexagonal, rhombohedral, monoclinic and triclinic crystal lattice systems. The symmetry of each system is described by the relationship between the lattice sides a, b, and c and angles α, β and γ described in Figure 5.4. Table 5.1 gives the six parameters, which define the seven lattice structures [5.3, 5.4].

5.3 Interatomic Bonding in Solids

There are four types of bonds in solids and they can be reclassified as primary bonds and secondary bonds. All four types involve packing discrete atoms or molecules into the previously explained three types of lattice structures. The primary bonds are covalent, ionic, and metallic, and the secondary bonds are van der Waal bonds and

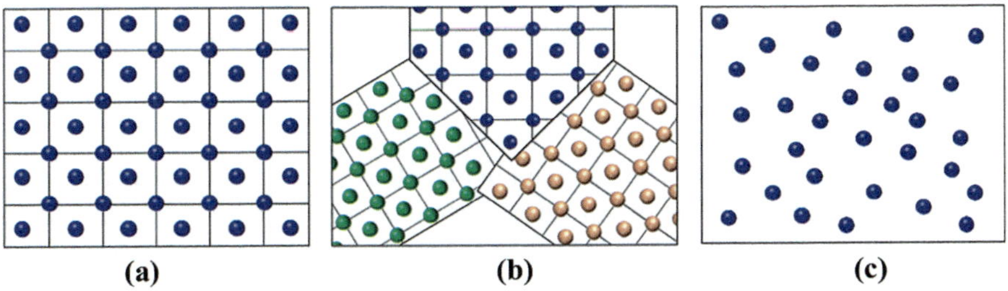

Figure 5.2 A two-dimensional projection representation of different solids: (**a**) single crystal, (**b**) polycrystal, and (**c**) amorphous

Figure 5.3 Generalized unit cell defined by three lengths a, b and c along x, y and z axes and three angles α, β and γ

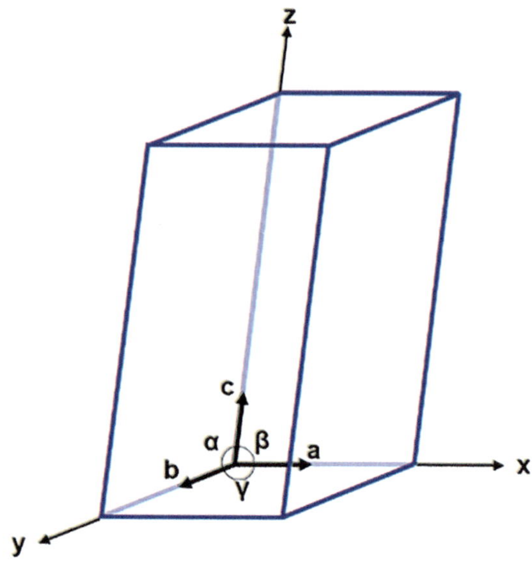

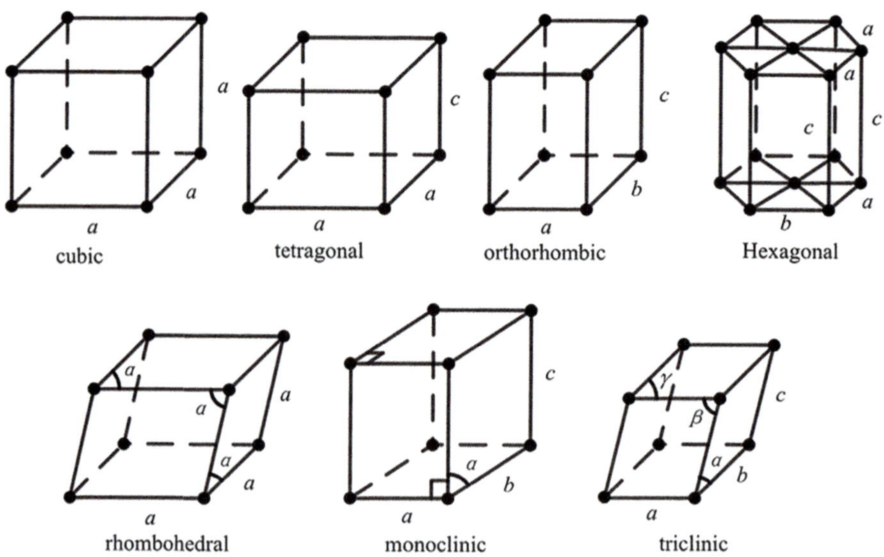

Figure 5.4 Typical lattice structures

intermolecular interactions. The variation in the relative strengths of these five types of interactions determine the atomic arrangement of solids (short or long range) as well as dictates the mechanical, optical, magnetic and electronic properties of that solid. For example, the ability of valence free electrons to travel throughout a solid explains both the high electrical conductivity and thermal conductivity of metals [5.2, 5.4, 5.5].

Table 5.1 Relationship between lattice parameters

Type	Linear relationships	Angular relationships
Cubic	$\mathbf{a = b = c}$	$\alpha = \beta = \gamma = 90°$
Tetragonal	$\mathbf{a = b \neq c}$	$\alpha = \beta = \gamma = 90°$
Orthorhombic	$\mathbf{a \neq b \neq c}$	$\alpha = \beta = \gamma = 90°$
Hexagonal	$\mathbf{a = b \neq c}$	$\alpha = \beta = 90°$ and $\gamma = 120°$
Rhombohedral	$\mathbf{a = b = c}$	$\alpha = \beta = \gamma \neq 90°$
Monoclinic	$\mathbf{a = b \neq c}$	$\alpha = \beta = 90°$ and $\gamma \neq 90°$
Triclinic	$\mathbf{a \neq b \neq c}$	$\alpha \neq \beta \neq \gamma \neq 90°$

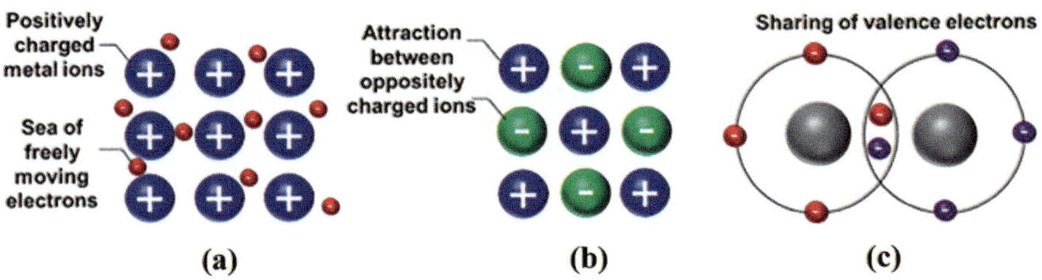

Figure 5.5 Bonding in solids: (**a**) metallic, (**b**) ionic, and (**c**) covalent

5.3.1 *Metallic Bonding*

Atoms in metals lose electrons to form cations. Delocalized electrons surround the ions in a cloud. Metallic bonding is generated as a result of the electrostatic interactions between the ions and the electron cloud. This is due to a tightly packed crystal lattice of the metallic structure, as depicted in Figure 5.5a. This kind of bonding is uniform in all directions, so that metallic structures are dense and closely-packed. The electron mobility renders metals both malleable and ductile, which are vital properties for producing thin sheets and for the fabrication and stretching of metals. Examples of metallic bonds exist in iron, cobalt, silver, gold, barium, platinum, chromium, copper, zinc, and lithium.

5.3.2 *Ionic Bonding*

The ability of valence electrons to travel due to their affinity to another compound leads to the creation of negative and positive ions. A negative ion is an ion that gains an electron and a positive ion is one that loses an electron. Most minerals are ionically bonded compounds formed by the transfer of one or more electrons from one atom to another. This creates two ions with opposite charges, which are attracted to each other to form an ionic bond; depicted in Figure 5.5b. For example, sodium chloride is an ionic compound in which the positive sodium ions and negative chloride ions form an extended array of regular atoms [5.4–5.6].

5.3.3 Covalent Bonding

While ionic bonds join metals to non-metals and metallic bonds form the bonding between metals, covalent bonds are responsible for bonding non-metals to non-metals. Covalent bonding is formed when two atoms share two electrons through intersection of two electron orbitals; see Figure 5.5c. Unlike metallic and ionic bonds, covalent bonding is directional and tends to have more complex structures than those of ionic or metallic structures. Diamond and graphite are examples of covalently bonded solids. In this case, each carbon atom is covalently bonded to four other carbon atoms, arranged at the corners of a tetrahedron, while graphite atoms are covalently bonded to three neighbors in the same plane, producing a strong sheet of carbon atoms. It is important to note that each carbon atom has one extra electron available for bonding that form a very weak bond that keeps the carbon sheets together.

5.3.4 Secondary Bonds

Secondary bonds are not bonds. They do not involve valence electron being shared or donated. They are usually formed because an uneven charge distribution exists between atoms creating what is known as "dipoles." Effectively, the total atomic charge is zero, but there is slightly unbalanced charge; i.e., more positive or negative charge on one end of the atom than on the other, resulting in this secondary bond. These dipoles can be produced by a random fluctuation of the electrons around what is normally an electrically symmetric field in the atom. It is worth noting that once a dipole is formed in one atom, an induced dipole will be formed in the adjacent atom. This type of bonding is present in N2 molecules, and is known as Van der Waals Bonding.

5.4 Defects in Crystal Structure

An ideal solid with its atoms/crystal structure perfectly arranged and positioned does not exist. All solids have imperfections and these imperfections (also known as microdefects) dictate the mechanical properties of these solids. In fact, at times, imperfections are deliberately introduced into a solid to enhance its mechanical, thermal or electrical properties; e.g., by intentionally adding specific alloying elements. Thus, not all imperfections are undesirable. Nevertheless, since we are concerned with fracture, our attention in this chapter will be devoted to the different types of imperfections that exist in solids.

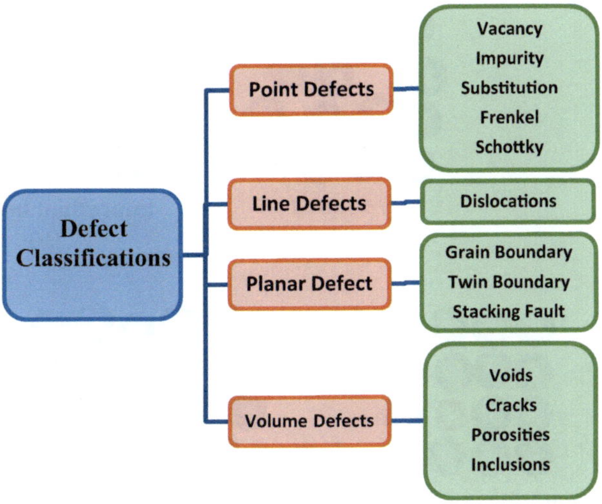

Figure 5.6 Defect classifications in solids

There are three types of defects in crystal structures: Point defects (Section 5.4.1), Line defect (Section 5.4.2) and Planner defects (Section 5.4.3). Figure 5.6 summarizes these three types in terms of dimensions. Noteworthy is the fact that every defect has a stress field associated with that defect [5.2–5.6].

5.4.1 Point Defects

As depicted in Figure 5.7, point defects can take one of the following forms:

- **Vacancy defect:** an atom is missing from its regular atomic position. This is due to imperfect packing during recrystallization. Vacancies lead to reduced density and strength but increased ductility (Figure 5.7a).
- **Interstitial or impurity defect:** an atom (or atoms) occupy a site in the lattice structure that they are not supposed to occupy. In this case, the number of atoms is greater than the number of lattice sites. The impurity atoms increase the density of the crystal structure and thus lead to atomic distortion (Figure 5.7b).
- **Substitution defect:** an atom replaces the host atom in the crystal lattice. The presence of substitution atoms either can be a natural defect or deliberately introduced to tailor the properties of the lattice (Figure 5.7c).
- **Frenkel defect:** also known as Frenkel pair or disorder**.** In this case, an extra self-interstitial atom relocates itself in a nearby site leaving behind a vacancy (Figure 5.7d).
- **Schottky defect:** takes place in ionic and non-ionic crystal structures. In this case, cation (positively charged) vacancy is associated with anion (negatively charged) vacancy in ionic structures. Typically, it occurs in those ionic crystals where the difference in size between cation and anion is small (Figure 5.7e).

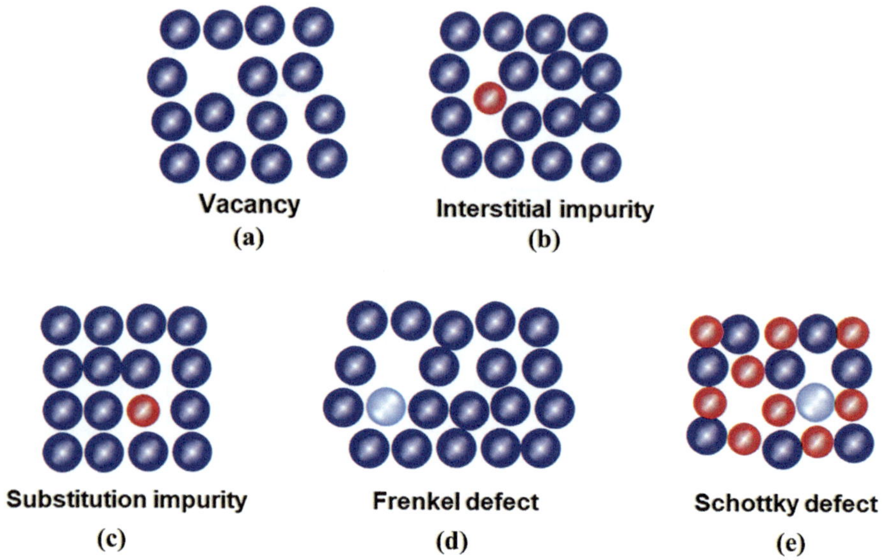

Figure 5.7 Different types of point defects: (**a**) vacancy, (**b**) interstitial impurity, (**c**) substitution impurity, (**d**) Frenkel defect and (**e**) Schottky defect

5.4.2 Line Defect

Unlike point defects, line defects, or dislocations are lines along which entire rows of atoms in a solid are irregularly arranged. Line defects may run the length of the crystal. Two types of dislocations exist. In a lattice, an edge dislocation will have an extra half-line of atoms, with the atoms being compacted at top of the missing line and sparse in the missing line of atoms. Screw dislocation, on the other hand, is the result of the lattice being sheared. Figure 5.8 shows schematics of edge and screw dislocations. Dislocations are extensively used to characterize plastic deformation and fracture mechanics problems.

Dislocation movement leads to plastic deformation. In Figure 5.9a, we have a perfect lattice structure, while in Figure 5.9b we have a missing atom. The former is unrealistic and does not exist. Solids typically contain defects during manufacture and processing.

Let us consider the force system shown in Figure 5.9 by the red arrows that is applied to the left of the perfect and dislocated lattice structure. For the perfect lattice in Figure 5.9a, unless the applied force is excessive, the bonds will not break. For the lattice in Figure 5.9b there is a missing atom, and the planes marked A will not line up because of that missing atom. As the load continues to increase, bonds will begin to stretch and ultimately break at end of line B. The atomic bonds will naturally reform, but we now observe a new edge dislocation moved to the right of the original line defect at C in Figure 5.9c and the process continues from D (Figure 5.9d) to E

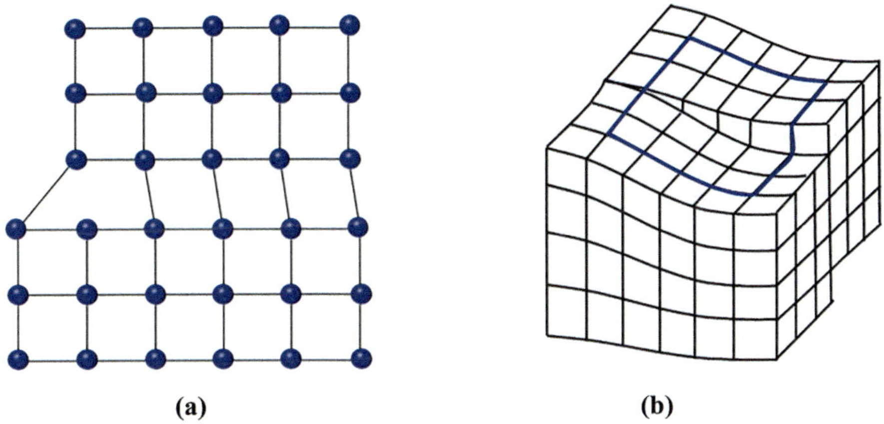

Figure 5.8 Schematics of (**a**) an edge and (**b**) a screw dislocations

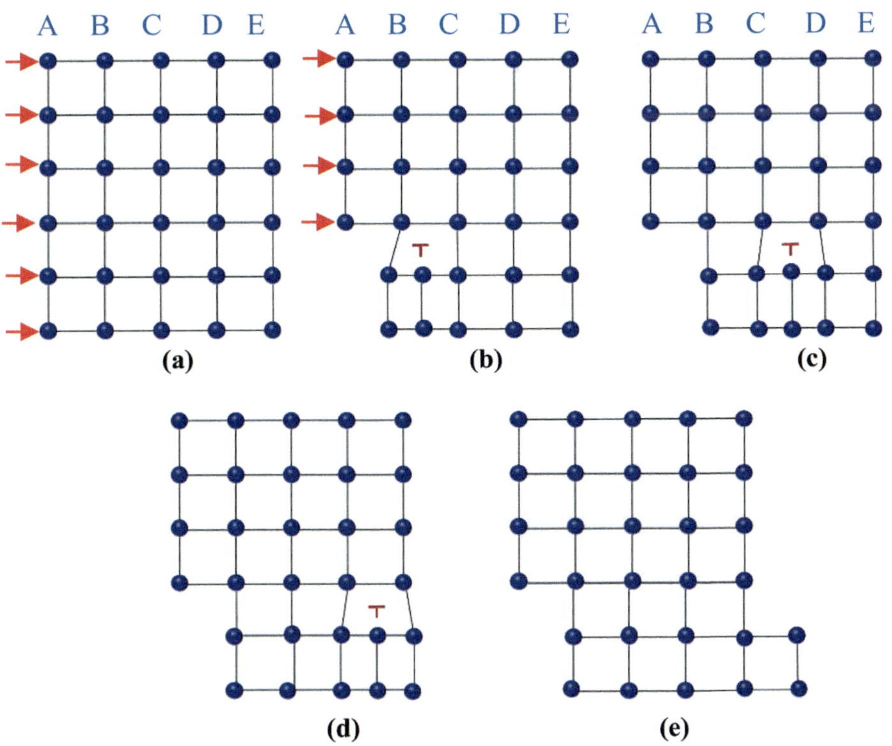

Figure 5.9 Movement of a single edge dislocation in a lattice: (**a**) a perfect lattice, (**b**) lattice missing an atom, (**c**–**e**) sequential dislocation movement in the lattice along slip or glide planes

(Figure 5.9e), resulting in a defect moving plastically to the edge of the lattice along a sliding or a gliding plane, as a result of the highest resolved shear stress. A detailed account of this topic can be viewed in [5.7].

In dislocation dynamics, the Burger vector is used to describe the character of the dislocation. Effectively, it is the displacement vector that closes the loop when traversing an equal number of lattice steps around a line defect. Further details concerning this topic can be obtained from numerous material science textbooks that exist in literature that are devoted wholly to this topic.

5.4.3 Planar Defects

Three types of planar defects exist: grain boundary, twin boundary and stacking faults.

Grain Boundary defined as the planar interface between adjacent grains (single crystal) to accommodate for the misorientation of the different grains in polycrystalline material. It is formed during melt solidification in metals. Figure 5.10 illustrates the sequential development of the grain boundaries. The process is divided into three stages in which the first stage is characterized by the formation of nuclei (Figure 5.10a). This is followed by growth of single crystals forming grains of different orientations, Figure 5.10b. To accommodate for orientation incompatibility of the grains, an interface develops. This interface is the grain boundary (Figure 5.10c). Grain boundaries influence the mechanical, thermal and electrical properties of the metal. For example, grain boundaries disrupt the movement of dislocations thus reducing glide and slip resulting in increased strength. Under certain conditions, the fracture behavior of a metal, aided by the precipitates and second phase particles that weakens the grain boundaries, can take place intergrannularly (along grain boundaries as in creep rupture at high temperatures).

Fracture can also occur transgrannularly (across the grain boundaries). Figure 5.11 shows the effect of temperature on the strength of the grain boundaries. At relatively low temperatures, the grain boundaries are stronger and the solid is likely to fracture in a transgrannular fashion. At higher temperatures, the grain boundaries are weaker and the fracture is likely to be ensued in an intergrannular fashion. It follows that a fine grained structure will offer higher strength and hardness than a coarse grain one at room temperature, but at a higher temperature, a coarse grain structure would be preferred. These remarks apply to pure metals and certain alloys, but do not apply to metals and alloys with grain boundary impurities. In fact, grain boundaries play an important role in steam turbines as a result of metal creep.

Twin Boundary A second type of planar defect is the twin boundary defect. Crystals grow by the progressive addition of atoms onto a surface. Faulty stacking of new planes with respect to previously formed planes are common during the crystal's growth. Figure 5.11 shows a boundary that separates two domains of a

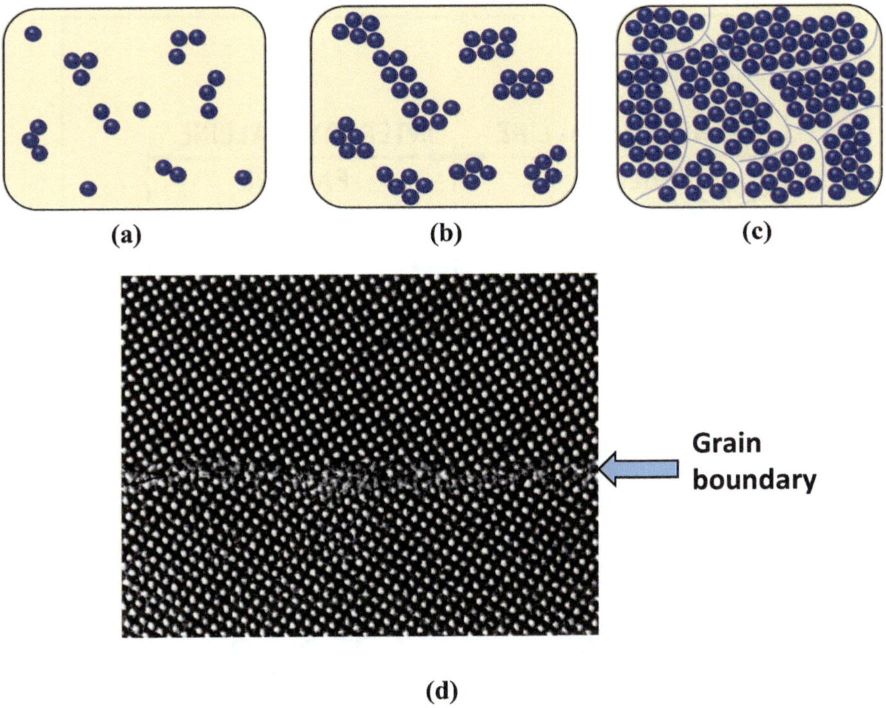

Figure 5.10 Formation of grain boundaries during melt solidification: (**a**) formation of nuclei, (**b**) growth of single crystals at varied positions, and (**c**) grains separated at grain boundaries and (**d**) typical high resolution image of a grain boundary at the atomic scale [5.8]

crystal that are mirror images of each other. The result is called a twinned crystal or twinning. Twin boundaries are types of planar surface defects. As can be seen from Figure 5.12, the atomic arrangement on one side of a twin boundary is a mirror reflection of the arrangement on the other side.

Stacking Fault A third type of planar defect is a stacking fault. They can be formed during crystal growth or during plastic deformation as partial dislocations movement as a result of dissociation of a perfect dislocation. They can also form by condensation of point defects during high-rate of plastic deformation. Stacking faults occur in a number of crystal structures, but they are commonly observed in close packed structures.

5.4.4 Volume Defects

Bulk or volume defects result from aggregation and agglomeration from point and line defects. They include pores, cracks, and foreign inclusions in crystalline

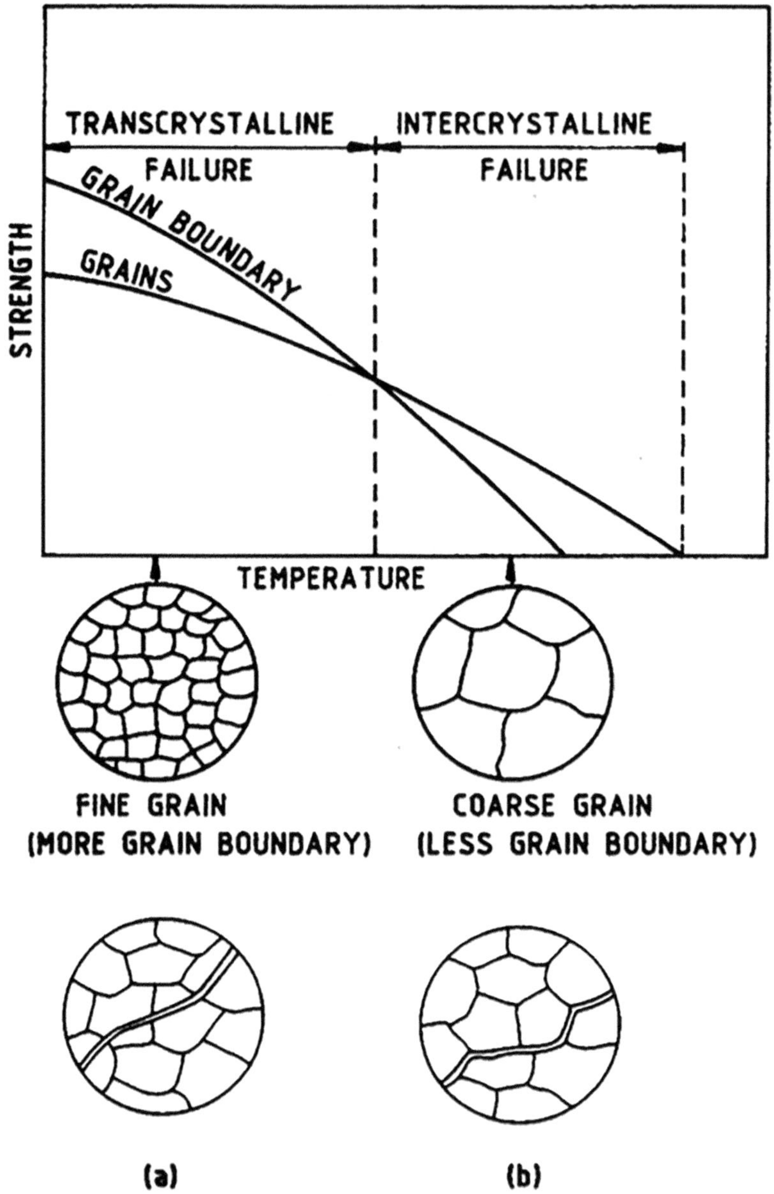

Figure 5.11 Variation of the strength of the grains and the grain boundaries with temperature showing (**a**) transgranular and (**b**) intergranular fractures

materials. For examples, vacancies can occur when a cluster of point defects combine together to form a three-dimensional imperfection. Volume defects also include microcracks, porosities and inclusions. These defects play a significant role on the performance of engineering materials.

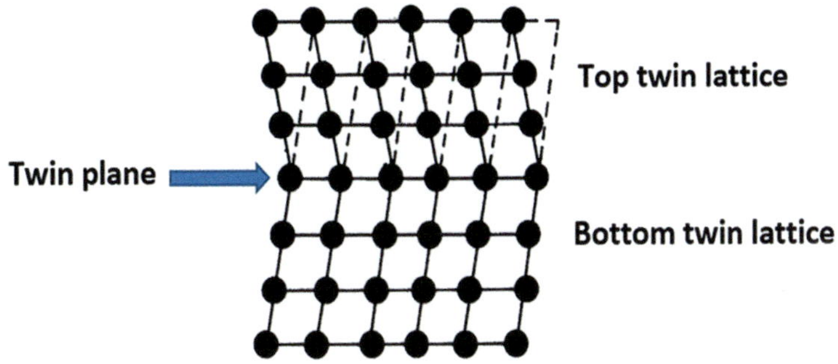

Figure 5.12 A schematic of twin boundary and twinned crystal

References

5.1 F. Wang, *Bonding Theory for Metals and Alloys*, 2nd edn. (Elsevier, San Diego, 2018)
5.2 A. Kelly, K.M. Knowles, *Crystallography and Crystal Defects*, 3rd edn. (Wiley, Hoboken, 2020)
5.3 M. Ashby, H. Shercliff, D. Cebon, *Materials – Engineering, Science, Processing and Design*, 4th edn. (Elsevier, Oxford, 2018)
5.4 M. de Graef, M.E. McHenry, *Structure of Materials—An Introduction to Crystallography, Diffraction and Symmetry*, 2nd edn. (Cambridge University Press, Cambridge, 2012)
5.5 T.-T. Fang, *Elements of Structures and Defects of Crystalline Materials*, 1st edn. (Elsevier, 2018). https://doi.org/10.1016/C2017-0-00834-0
5.6 H.V. Keer, *Principles of the Solid State* (New Age International Publishers, New Delhi, 2005)
5.7 T. Suzuki, S. Takeuchi, H. Yoshinaga, *Dislocation Dynamics and Plasticity* (Springer, New York, 2009)
5.8 B. N. Berkels, A. Rätz, M. Rumpf and A. Voigt, Identification of grain boundary contours at atomic scale. International conference on scale space and variational methods in computer vision, pp. 765–776, 2007. doi:https://doi.org/10.1007/978-3-540-72823-8_66

Chapter 6
Modes of Failure

Summary This chapter identifies the different modes of failure under the varied service loading conditions. This will undoubtedly enhance the readers' knowledge of the causes of failures and the types of remedies needed to mitigate them.

6.1 Introduction

The current trends in performance, reliability and safety of aeroengines, nuclear power plants, off-shore structures and other critical systems have resulted in the need for better materials and an improved understanding of their behavior under extreme thermo-mechanical loading conditions. Since current operating conditions are becoming more severe, it is important that the design engineer should be able to understand the different possible modes of failure of the available materials under service loading conditions. This will undoubtedly enable the designer to produce efficient, imaginative, economical and reliable yet robust and safe designs.

The purpose of this chapter (and the next) is to provide the readers with a detailed account of most common modes of failure experienced by metallic alloys. The principal modes of failure are:

(i) General yielding and plastic deformation,
(ii) Brittle fracture (in view of its importance, it is covered in the following chapter),
(iii) Plastic instability,
(iv) Buckling and bifurcation,
(v) Fatigue of metals,
(vi) Corrosion,
(vii) Stress-corrosion cracking,
(viii) Creep, and
(ix) Creep-Fatigue interaction.

From the outset, let us differentiate between failure and fracture, since the two are not synonymous. Collapse due to buckling of a strut may not involve any fracture, but still renders the strut useless. Similarly, the presence of a fatigue crack in a

S. A. Meguid, *Atomistic and Continuum Fracture Mechanics of Solids*,
https://doi.org/10.1007/978-3-031-56085-9_6

turbine disc or a compressor blade may condemn these components, even though complete fracture has not occurred. As a matter of fact, this raises another important issue, that frequently two or more modes of failure may be involved before final separation occurs. This is particularly so in cases involving fatigue, creep or corrosion initiated failures in which the final fracture may be due to brittle fracture. These interactions will be reviewed briefly near the end of this chapter.

6.2 Ductile Failure

6.2.1 Plastic Behavior of Crystals

The resistance of a crystalline body to elastic or plastic deformation is dependent upon the cohesive forces between the atoms. The most important mechanism of crystal deformation is slip [6.1, 6.2]. Slip is defined as the gliding of planes of atoms over one another, which occurs by the motion of dislocations over specific crystallographic planes and in definite crystallographic directions, as depicted in Figure 6.1. Generally, the crystallographic planes are the planes which possess the highest atomic density, i.e. the largest number of atoms per unit area. Accordingly, slip occurs in bcc crystals in [110] planes, in fcc on [111] and in hcp on [0001]. Typical of the bcc group are α-iron, molybdenum and tungsten, while the fcc group includes copper, aluminium and nickel and the hcp group includes zinc, titanium and beryllium.

Since slip in crystals is limited to certain planes and directions, the applied stress that is required to initiate plastic flow depends on the orientation of the stress relative to the crystallographic axis of the crystal.

Figure 6.1 Slip in a crystal

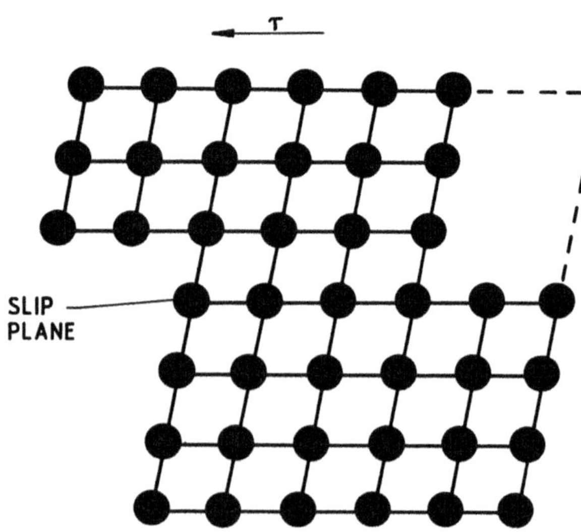

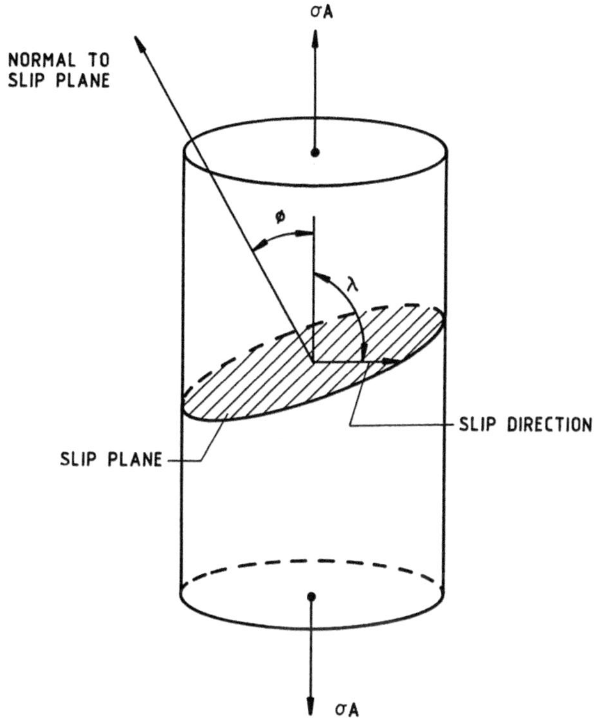

Figure 6.2 Relationship between slip plane, slip direction and the axis of the tensile load in a single crystal

In the plastic deformation of the single crystal depicted in Figure 6.2, the applied tensile stress is resolved into two components: a shear stress component in the slip direction on the slip plane, and a normal stress component perpendicular to the slip plane. The normal component is of little importance in initiating plastic deformation. The value of the shear stress required for the onset of the plastic deformation is called the critical resolved shear stress τ_c and is simply given by

$$\tau_c = \sigma_0 \cos\phi \cos\lambda \tag{6.1}$$

where σ_0 is the tensile yield stress of the material, ϕ is the angle between the tensile axis and the normal to the slip plane and λ is the angle between the tensile axis and the slip direction, as depicted in Figure 6.2. The term $\cos\phi \cos\lambda$ is known as the Schmid factor; τ_c is the single-crystal equivalent of the yield stress of an ordinary stress-strain curve and depends chiefly on composition and temperature.

Slip in single crystals does not begin simultaneously on all slip planes of a system, but only along some of them. Figure 6.3 shows the associated slip bands, each consisting of many slip lines, which are observed on the surface of the deformed crystals during plastic deformation. This indicates that plastic deformation is inhomogeneous, with extensive slip occurring on certain planes, while the crystal

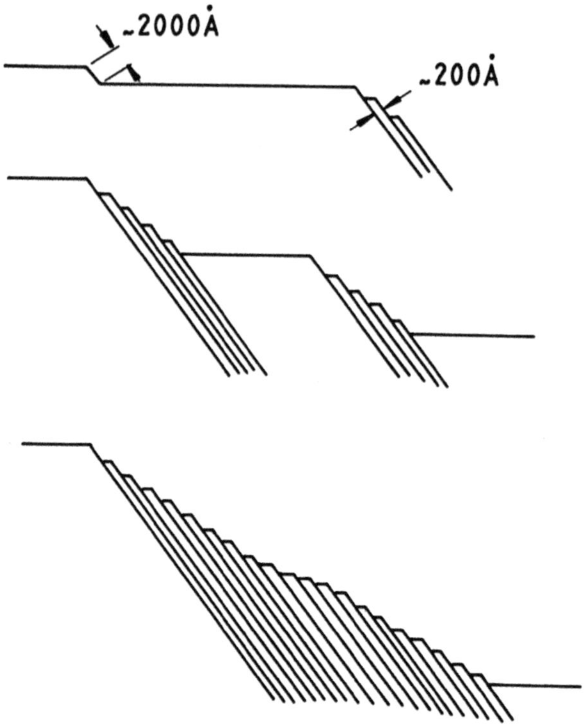

Figure 6.3 Profiles of slip bands and formulation of slip steps

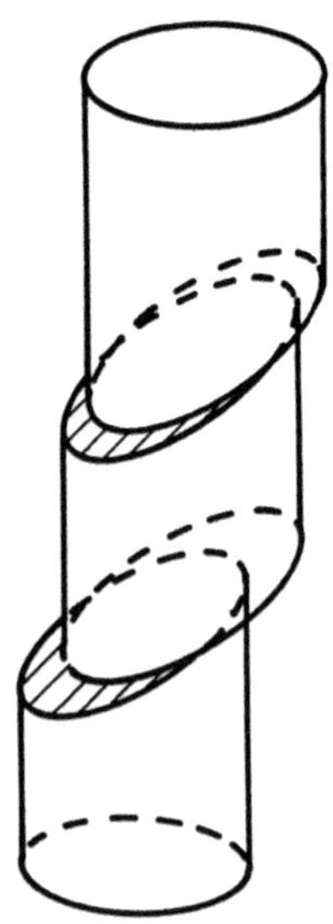

Figure 6.4 The slip process in a constrained single crystal

Figure 6.5 Twinning in a crystal

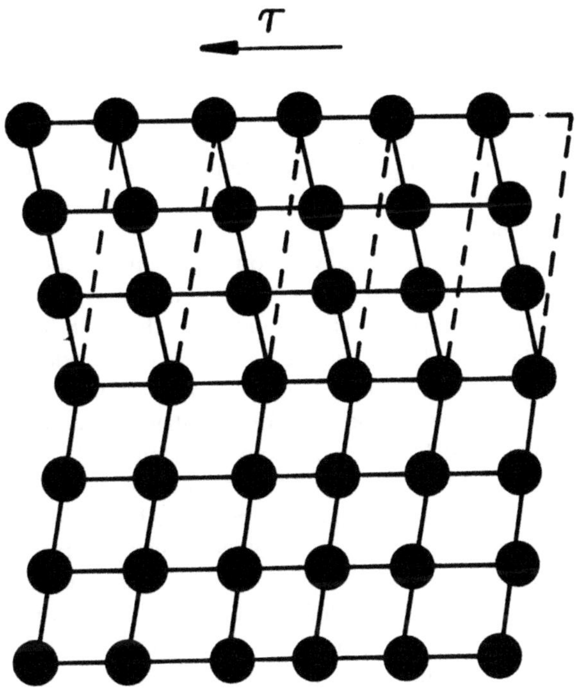

planes between them remain practically undeformed, as shown in Figure 6.4 by the constrained slip process of a single crystal. It must be noted that cessation of slip on the plane is caused by work hardening; slip is then transferred to the next plane, and so on.

If slip is constrained, plastic deformation takes place by twinning. Twinning is the conversion of the crystal into a mirror image of the parent lattice by homogeneous shearing of successive lattice planes, as depicted in Figure 6.5. The process of twinning takes place instantly and this is reflected in the shape of the stress-strain curve, which shows extensive jagging whenever twinning occurs. This type of plastic behavior is common in some steels deformed at very high strain rates or at low temperatures (Neumann's bands). The stress required to produce twinning tends to be higher than that necessary for normal slip.

6.2.2 Plastic Deformation of Polycrystalline Materials

In the above sub-section, we outlined the micro-mechanisms by which plastic deformation occurs in single crystal, where geometric constraints to dislocation motion are limited to those present within the crystal. Grain boundaries in polycrystalline materials introduce new constraints to dislocation movement. If integrity of the grain boundary is to be maintained, then each grain must deform in a manner compatible with the deformation of all its immediate neighbors. Since the constraints differ from one grain to another, the observed microscopic behavior of a polycrystalline specimen is the result of a compromised deformation pattern from a large number of grains. Accordingly, the macroscopic behavior of a polycrystalline metal can only be qualitatively associated with the behavior of simple crystals of the same metal.

Grain boundaries may be formed in a metal after solidification or after cold work and recrystallization. Since the orientation of the atoms in each grain is different, it is obvious that the atoms of the metal at the grain boundary cannot be arranged on a regular space lattice. Figure 6.6 shows that at relatively low temperatures the grain boundaries are stronger than the grains, whereas at elevated temperatures the grain boundaries are weaker. It follows that a fine-grained structure will give higher strength and hardness at room temperature, but at higher temperatures a coarse grain would be preferred. These remarks apply to pure metals and solid-solution alloys and may not apply to certain alloys with grain boundary impurities. The diagram also explains why, other influencing factors being the same, the type of fracture at room temperature is usually transgrannular; i.e., across the grain boundaries, whereas at elevated temperatures fractures may become intergrannular [6.1]; i.e., along the grain boundaries, as depicted in Figure 6.6.

In metals, the strength of grain boundaries is affected by alloying elements and impurities, both of which tend to segregate there. Such segregation can embrittle the material mostly at low temperatures. Segregation at grain boundaries may also accelerate local corrosion; e.g., sensitization of stainless steels.

6.2.3 *The Tensile Test*

The simplest and most common test, as well as the most important, is the standard tensile test. An appropriate test specimen is inserted into a tensile test machine, the load is increased, and the readings of the load and the corresponding extension of the gauge length are recorded and converted into either nominal or true stresses and strains. The results are depicted in Figure 6.7a. The expressions for the nominal stresses and strains are referred to the original geometrical dimensions of the specimen, while the true stresses and strains are determined using the instantaneous geometrical dimensions.

Initially, the relation between stress and strain is essentially linear. This linear part of the curve extends up to point A, which is known as the proportional limit. It is in this range that the linear theory of elasticity applies. Upon further increase of the load, the strain no longer increases linearly with the stress, but the material still remains elastic. This condition will prevail until point B, called the elastic limit or yield point, is reached. In most materials, the difference between A and B is not noticeable. For some materials, such as aluminium alloys and copper, the yield point

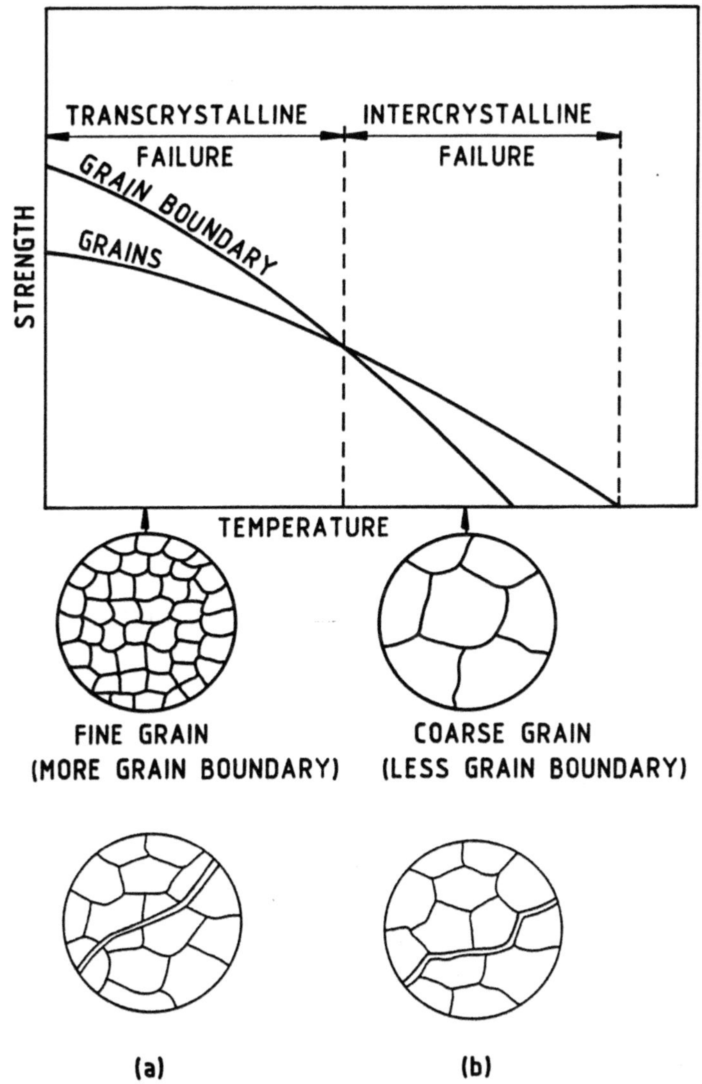

Figure 6.6 Variation of the strength of the grains and the grain boundaries with temperature showing (**a**) transgranular and (**b**) intergranular fractures

is so poorly defined that it is arbitrarily taken to be at some fixed value of permanent strain, such as 0.1–0.2%, as shown in Figure 6.7b. The stress at this point is usually called the offset yield stress or the proof strength. Beyond the elastic limit, permanent or plastic deformation takes place.

The strain at the elastic limit is of the order of magnitude of 0.001–0.002. As the load increases beyond the elastic limit, the strain increases at a greater rate. However,

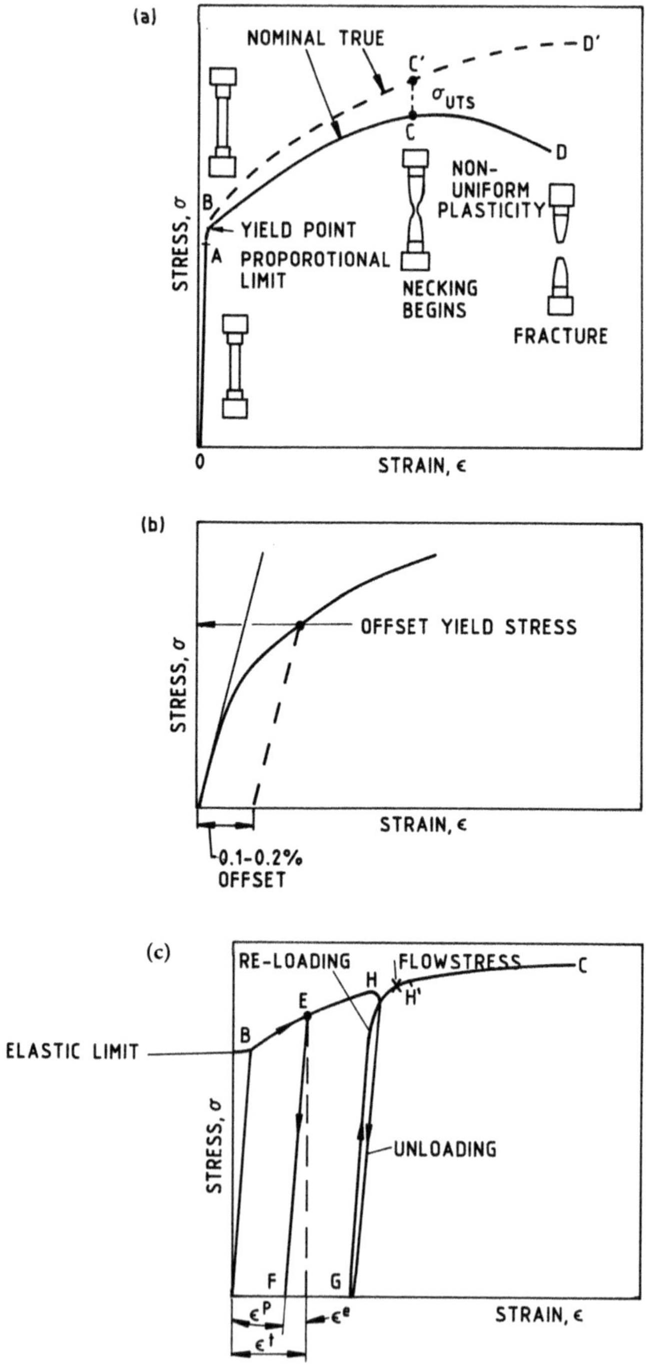

Figure 6.7 Tensile stress-strain curves: (**a**) metallic material with defined yield point; (**b**) a metallic material with poorly defined yield point; (**c**) unloading and reloading of a material in the plastic range

the specimen will not deform further unless the load is increased. This condition is called work hardening or strain hardening and the stress required for further plastic flow is known as the flow stress. Point C represents the point of instability where the specimen begins to neck down. Beyond C the specimen suffers from non-uniform plastic deformation and fractures at D. Beyond C, a triaxial state of stress exists in the necked region. The stress at the maximum load point C is called the nominal ultimate tensile strength σ_{UTS}, while at D it is called the nominal fracture stress σ_f.

If at any point between the elastic limit B and the maximum load point C the load is removed, unloading will take place along a line parallel to the elastic line, as shown in Figure 6.7c by line EF. Part of the strain is thus recovered and part remains permanently in the specimen, i.e.

$$\varepsilon^t = \varepsilon^e + \varepsilon^p \tag{6.2}$$

where ε^e is the recoverable elastic component and ε^p is the non-recoverable plastic component. Upon reloading, the unloading line is generally retraced with very minor deviations. Actually a very thin hysteresis loop is formed, as shown by the reloading state GH. Plastic flow does not begin again until point H′ is reached. With further loading, the stress-strain curve is continued along H′C as if no unloading had occurred.

A few materials, such as annealed steel, exhibit a sharp drop in yield after the upper point B is reached, as shown in Figure 6.8a. The specimen will then extend at approximately constant load to a strain of about 10 times the initial yield before the load will start increasing again as the material begins to work harden. The flat portion of the curve, known as the lower yield, actually represents a series of unstable jumps between the upper and lower yield points caused by the propagation of the Lüders bands across the specimen.

The stress-strain curves discussed in Figure 6.7 above refer to quasi-static tests carried out at a very low strain rate, typically of the order 10^{-6} s^{-1}. The effect of increasing the strain rate will generally lead to an increase in the yield stress of the material. Figure 6.8a shows the result of a constant strain rate test, while Figure 6.8b shows the rate-jump test results for mild steel performed by the author. In the jump tests, the strain rate was switched back and forth between 10^{-4} and 10^{-2} s^{-1} at selected times using the controller. A well-defined elastic stress increase can be observed in Figure 6.8b, following the strain-rate jump. Both figures show the rate dependence of the flow stress at various values of strain (up to 12%) for the two strain rates 10^{-4} and 10^{-2} s^{-1}. The stress-strain relationship depicted in Figure 6.8b also appears to indicate a small strain-rate history effects, since in the jump test the curve is always between the top constant strain-rate curves (dashed curves).

Temperature has a very important effect on metal properties. At very low temperatures, metals that behave in a ductile fashion can become very brittle. The temperature at which the ductility changes so rapidly is known as the transition temperature. Such strong temperature and strain-rate effects occur more generally in

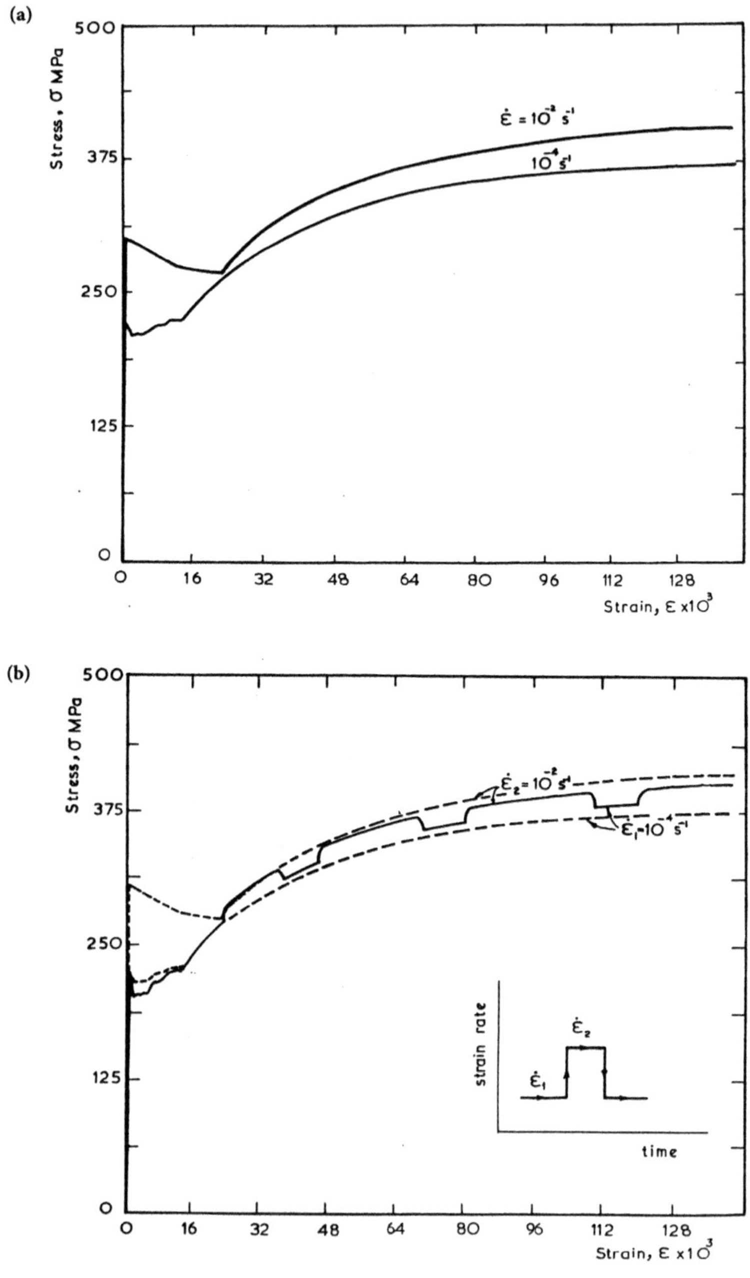

Figure 6.8 (**a**) Effect of strain rate upon the stress-strain relationship of annealed mild steel. (**b**) Steel-rate jumps between nominal strain rates of 10^{-4} and 10^{-2} S^{-1}

metals with bcc arrangement. At the top end of the time-temperature scales is the phenomenon of creep. This aspect will be discussed under a separate heading.

6.3 Plastic Instability

Plastic instability is a mode of failure which is concerned with the thinning and subsequent rupture of relatively thin shells under internal pressure or under the combined effect of pressure and temperature. This is commonly observed in pressure vessels, chemical plants, boilers, diaphragms and storage tanks which operate at high internal pressure and temperature. Figure 6.9a shows an example of a pressure tube before use, while Figure 6.9b shows the same tube having suffered localized severe plastic instability due to the application of high internal pressure [6.3].

The approach commonly adopted in treating plastic instability problems relies upon the application of the incremental theory of plasticity. In this case, elastic strains will be neglected and the Lévy-Mises stress-incremental strain relations are assumed to apply. Furthermore, in most of the cases considered, the principal axes do not rotate relative to the element being strained and the principal stress ratios remain constant throughout the straining process. It is also useful to employ an empirical constitutive relation for the behavior of the material. The one to be used here is of the form $\sigma = A(B + \varepsilon)^n$, where σ and ε are the equivalent stress and equivalent strain, respectively, A is a measure of the basic strength of the material, B gives an indication of the initial state of the material and n is a measure of the rate of strain-hardening.

In order to demonstrate the technique, let us consider the case of a closed thin-walled pressure vessel ($t < r$) which is made of a material whose strain hardening characteristics can be described by the above power law. Let us assume that the cylinder is subjected to a gradually increasing internal pressure P. and it is desired to

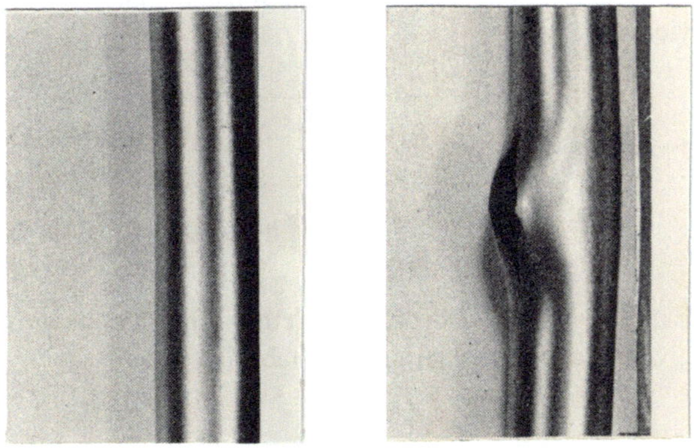

(a) (b)

Figure 6.9 Plastic instability in a straight tube (**a**) before and (**b**) after application of excessive internal pressure

determine the pressure P^* necessary to cause plastic instability. In this case, the state of stress is defined as being,

$$\left.\begin{aligned}\sigma_1 &= \sigma_{\theta\theta} = pr/t\\ \sigma_2 &= \sigma_{zz} = pr/2t\\ \sigma_3 &= \sigma_{rr} = 0\end{aligned}\right\} \tag{6.3}$$

where $\sigma_{\theta\theta}$ is the hoop stress, σ_{zz} is the longitudinal stress and σ_{rr} is the radial stress; r and t are the current values of pressure vessel radius and wall thickness, respectively. Therefore, the corresponding deviatoric stress components are

$$\left.\begin{aligned}S_1 &= \sigma_1/2\\ S_2 &= 0\\ S_3 &= -\sigma_1/2\end{aligned}\right\} \tag{6.4}$$

but since $d\varepsilon_{ij} = S_{ij} d\lambda$, then

$$d\varepsilon_1 = -d\varepsilon_3 \text{ and } d\varepsilon_2 = 0 \tag{6.5}$$

The internal pressure p in the vessel is related to the hoop stress (expression (6.3)) by the relation

$$p = \frac{t\sigma_1}{r} \tag{6.6}$$

At instability, $dp = 0$, giving

$$\frac{d\sigma_1}{\sigma_1} = \frac{dr}{r} - \frac{dt}{t} = d\varepsilon_1 - d\varepsilon_3 = 2d\varepsilon_1 \tag{6.7}$$

Substitution in the equivalent stress $\bar{\sigma}$ and equivalent strain increment $d\bar{\varepsilon}$ expressions yields

$$\left.\begin{aligned}\bar{\sigma} &= \sqrt{\frac{1}{2}\left[(\sigma_1-\sigma_2)^2+(\sigma_2-\sigma_3)^2+(\sigma_3-\sigma_1)^2\right]} = \frac{\sqrt{3}}{2}\sigma_1\\ d\bar{\varepsilon} &= \sqrt{\frac{2}{9}\left[(d\varepsilon_1-d\varepsilon_2)^2+(d\varepsilon_2-d\varepsilon_3)^2+(d\varepsilon_3-d\varepsilon_1)^2\right]} = \frac{2}{\sqrt{3}}d\varepsilon_1\end{aligned}\right\} \tag{6.8}$$

Expressions (6.7) and (6.8) together with the constitutive behavior of the material give the equivalent strain at instability $\bar{\varepsilon}_{ins}$ as

$$\overline{\varepsilon}_{ins} = \frac{n}{\sqrt{3}} - B \tag{6.9}$$

and the equivalent stress at instability $\overline{\sigma}$ as

$$\overline{\sigma}_{ins} = \frac{1}{\sqrt{3}} A \left(\frac{n}{\sqrt{3}} \right)^n \tag{6.10}$$

Now, in order to obtain the instability pressure P_{ins}, it is necessary to determine the instability thickness t_{ins}, instability radius r_{ins} and σ_{ins} at instability. These are determined from the strain definition in the tangential and radial directions. The results after integration give the instability pressure p_{ins} as being,

$$p_{ins} = \frac{2A}{\sqrt{3}} \left(\frac{t_0}{r_0} \right) \left(\frac{n}{\sqrt{3}} \right)^n \exp \sqrt{3} \left(\frac{n}{\sqrt{3}} - B \right) \tag{6.11}$$

Clearly, the same approach can be adopted for thin-walled spherical pressure vessels or circular metal diaphragms.

6.4 Failure by Buckling

Structural members, which carry compressive loads, may be divided into two broad categories depending upon their relative lengths and cross-sectional dimensions. Short, thick members are generally termed columns and these usually fail by crushing when the yield stress of the material in compression is exceeded. Long, slender columns or struts, however, fail by buckling sometime before their compressive yield stress is reached [6.3]. The buckling which is shown in Figure 6.10 occurs due to one or more of the following reasons:

(i) the strut may not be perfectly straight,
(ii) the load by be eccentric, and
(iii) the inhomogeneous behavior of polycrystalline materials.

At load values below the buckling load, the strut will be in a stable equilibrium. In this case, the displacement caused by any lateral disturbance will be totally recovered upon unloading. Details of this mode of failure can be found in many undergraduate and graduate textbooks. They will also include other more complex geometries involving beams, plates and trusses.

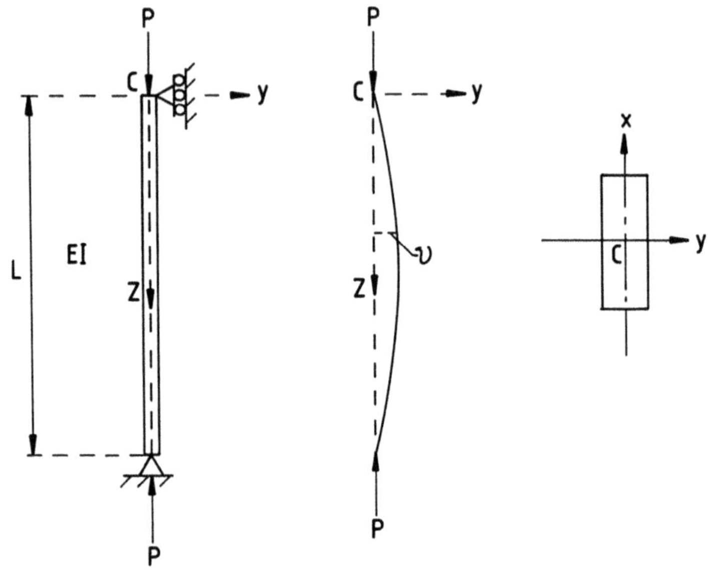

Figure 6.10 Strut subjected to compressive load at the pinned ends

6.5 Fatigue Failure

6.5.1 The Comet Crash: A Case History

Fatigue is undoubtedly the most important failure mode in engineering; several authors have estimated that between 70% and 80% of all failures are due to fatigue. One of the best examples of fatigue failure is the, now historic, Comet investigation. After being grounded for about 3 months, during which some 50 modifications were introduced to anticipate every conceivable danger, the Comet fleet went back into service. A second Comet then disappeared at Elba in 1954. When the tail-plane of the Elba Comet was recovered from the bottom of the sea, it had clearly been hit so violently by a newspaper that, although the paper had long since disappeared, the newsprint could easily be read in a mirror after months of submersion [6.4, 6.5]. There was also a clear impression of an Indian coin. According to de Havilland & Walker [6.4], these were in fact only two of many clues which indicated that the pressure cabin had burst before the tail-plane broke away.

The accident report contained the following comments on the design and testing procedures of the Comet aircraft. For the design of the basic structure of the cabin a multiple of the working pressure difference, P, in excess of current requirements in any country was adopted. The British Civil Airworthiness Requirements called for a "proof pressure of 2P" at which the material may reach its ultimate strength. These requirements were the same as those of the International Civil Aviation Organisations and also for military transport aircraft. de Havilland used a design pressure of

Figure 6.11 Fatigue failure at the forward escape hatch on the port side of a Comet aircraft [6.5]

2·5 P and tested the cabin to 2P. Two test sections of the cabin were built. The front part, 26 ft in length, extended from the nose nearly to the front spar of the wing, and included typical windows, hatches and door. The center part, 24 ft in length, extended from a few feet in front of the front spar to a few feet aft of the rear spar, covering the large cut out containing the wing structure.

There were two reasons for adopting these substantially higher figures. Which they incorrectly perceived. It was believed, and this belief was shared by the Air Registration Board and other expert opinions, that a cabin surviving a test that uses double its working pressure, 2P, would not fail in service under the action of fatigue. Secondly, it was considered that this would ensure a larger margin of safety against the possible failure originating at windows and doors.

Unfortunately, the proof testing of the cabin does not generally shed any light on the fatigue strength of that cabin. This has led to the fatigue testing of a complete full-size pressure cabin and fuselage. For safety, water was used as the testing medium, and the whole fuselage was immersed in a 250,000 gallon tank. Clearly, it was necessary to examine the effect of every conceivable adverse factor in addition to pressure. In particular, the wings were included in the test and were subjected to the appropriate load fluctuations. The test continued for several days, until a large piece of the pressurized cabin was pushed out almost simultaneously. The failure was initiated by a fatigue crack at a corner of one of the forward windows. Figure 6.11 gives a clear view from inside of fatigue failure at the forward escape hatch on the port side of a Comet aircraft [6.5].

6.5.2 Fatigue Crack Initiation and Propagation

Fatigue damage is a localized surface phenomenon due to repeated stress application. Fatigue strength of a component or a structure is affected principally by the properties of the material, the geometry in the form of stress concentration features,

Figure 6.12 Diagrammatic sketch of the different stages adopted in explaining fatigue failure in a single crystal

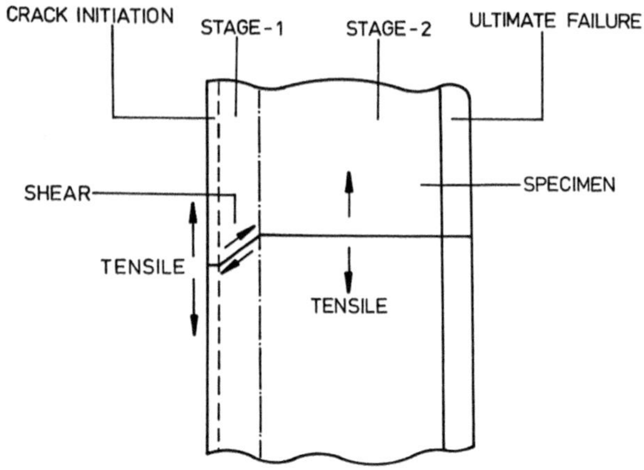

surface integrity in the form of roughness, corrosion pits, decarburization, and state of residual stresses and cyclic loading conditions. The fatigue strength of a component is often surprisingly low and knowledge of the tensile strength does not usually indicate how low [6.6].

It is now generally accepted that, in the ideal case, fatigue fracture can be divided into.

(i) crack initiation,
(ii) stage-I crack growth on planes of high shear stress,
(iii) stage-II crack growth on planes subjected to high tensile stresses, and
(iv) ultimate fracture.

These stages are represented diagrammatically in Figure 6.12.

6.5.3 *Crack Initiation*

Work early this century demonstrated that cyclically loading smooth specimen of ductile metal led to the initiation of cracks in slip bands. Later experiments have shown that crack initiation may involve slip band extrusion or the converse process, that of slip band intrusion [6.6], as demonstrated in Figure 6.13.

Fatigue cracks generally initiate at the free surface of a component or weakly constrained interface, although there are a few special cases where crack initiation at the interior of a specimen has been observed, typically due to hydrogen embrittlement.

The importance of surface integrity is demonstrated by those experiments in which specimens fatigued for a large proportion of their nominal lives ($N_f/50$) have been subjected to surface removal by careful machining or electro-polishing.

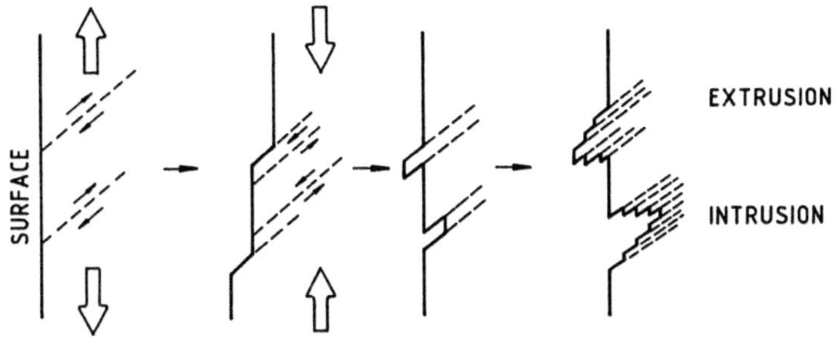

Figure 6.13 Slip band intrusion and extrusion

Removal of about 50 μm restores the life of the specimen to the as-new condition, and the process can be repeated indefinitely.

Such experiments have been carried out on a variety of materials from single-crystal copper to engineering steels. In contrast, attempts to remove fatigue damage by heat treatment have generally been unsuccessful and in some instances have reduced rather than extended fatigue life. It is unfortunate that the bulk of earlier research on crack initiation has been directed toward slip bands; this tended to obscure the fact that many other crack nucleation sites have been reported [6.6]. These include grain boundaries, second phase particles, fragmented deformation twins, among others.

It must be appreciated, however, that in machine components and structures made from commercial metals and alloys there may be present machine tracks, notches, corrosion pits and other stress concentration features such as inclusions, residual stresses and voids in welded joints, which act as pre-existing cracks and remove the necessity for activating any crack initiation mechanisms. Indeed, the linear elastic fracture mechanics (LEFM) approach assumes that all components and structures contain cracks and considers only the conditions necessary for crack propagation.

Considerable confusion arises when attempting to decide when a slip band extrusion/intrusion or other initiation sites become cracks. A useful but quite arbitrary distinction is to separate the microscopic cracks we have so far considered from macroscopic cracks. A figure of 100 μm has been proposed in the technical literature to define the transition from microscopic to macroscopic cracking.

The distinction between microscopic and macroscopic cracks or, also defined as, short and long crack problems is quite important, since the former do not really obey conventional LEFM. In this simple classification, we have relied upon crack size for the distinction. This is not entirely correct; the reasons for the inaccuracy are discussed in detail in a latter chapter.

6.5.4 Stage-I Crack Growth

The term stage-I crack growth is used here to describe the early extension of a slip band crack under cyclic shear stress by mechanisms similar to those involved in crack initiation. Such a crack grows from the free surface into the interior of the component; i.e., the crack is moving into the region where the ratio of the shear stress to normal stress is decreasing. In this region, shear stress controlled slip becomes more difficult and the crack either stops or changes its mode of growth to one governed by tensile stress, i.e. stage-II growth.

The most useful experimental techniques adopted for studying stage-I crack growth have been the sectioning of fatigue cracked specimens and the examination of the fractured surfaces by optical and electron microscopy (fractography). Transgrannular, intergrannular and mixed intergrannular stage-I crack growth have all been reported in the literature.

The extent to which stage-I cracking develops depends on many variables and, since it is generally crystallographic, it is structure sensitive. Planar glide would therefore favor extensive stage-I cracking and wavy glide reduces it. For example, in acutely notched specimens tested at high stress levels, stage-I cracking may be totally absent, while in unnotched specimens, fatigue fracture at low stress levels may involve 90% of the cycles to failure. Even in this latter case, stage-I cracking is not likely to extend more than a few grain diameters leading to the generation of a very small area of fracture. There appears to be special cases in which large areas of stage-I fracture surface may develop (10 mm^2), e.g. torsional fatigue in which crack growth due to the stress gradient lies in the specimen outer fibers.

Since crack initiation occurs in the first few hundreds or perhaps thousands of cycles, it is reasonable to deduce that those factors which influence fatigue limit do so via their effect on stage-I growth. Accordingly, grain size, cold work, precipitate size, distribution stacking fault energy and alloying elements, together with surface treatments such as shot-peening, nitriding and carburizing, should all influence stage-I crack growth.

The transition from stage-I to stage-II cracking is generally attributed to a reduction in the shear stress/normal stress ratio at the crack tip as it grows away from a free surface. The transition in some cases occurs at a grain boundary or, in the case of intergrannular stage-I fracture, at triple boundary point(s). It is also interesting to note that load shedding can cause the reverse problem, a stage-II-stage-I transition, and such a transition seems to be associated with non-propagating fatigue crack.

6.5.5 Stage-II Crack Growth

Stage-II growth may occupy between 10% up to 100% of the fatigue life of a component N_f, depending upon geometry, material and cyclic stress state. Even

Figure 6.14 A typical rotating bend fatigue fracture

when less than 10% of N_f is consumed in stage-II growth, this mechanism will generate by far the largest area of the fatigue fracture surface. Stage-II fatigue fracture surfaces may show the well-known macroscopic progression marks known the beach marks (Figure 6.14) which have been used to diagnose fatigue failure for a very long time. It has been argued that stage-II cracking is largely structure insensitive, and that, for a given stress amplitude, the growth rate will depend upon material mechanical properties, initial crack size and surface treatment.

We may sum up fatigue failure as follows:

(i) Fatigue damage is irreversible.
(ii) Fatigue life N_f involves initiation life N_i and propagation life of cracks N_p.
(iii) Minute cracks or crack-like defects are generated after only a small percentage of the total life to fracture.
(iv) Initial growth of such cracks is slow.
(v) These cracks generally initiate at a free surface; in cases where the crack initiates within the material, a second phase is almost always involved (hydrogen embrittlement).
(vi) Stage-I crack growth may occupy 90% of N_{f}, although the associated fracture area may only be a few grain size diameters.
(vii) Crack initiation and stage-I cracking are structure dependent and surface treatment sensitive.
(viii) Stage-II growth will generate by far the largest area of the fatigue fracture surface area.
(ix) Stage-II cracking is largely structure insensitive.

6.6 Corrosion

In view of its costly and destructive effects, the study of corrosion and its prevention is of vital importance to structural engineers. It involves a large number of variables. However, the factors governing the rate of corrosion may be broadly divided into those relating to (i) the metal or alloy and (ii) the environment. The factors relating to the metal or alloy include its position in the electro-chemical series, contact with dissimilar metals, microstructure, impurities and residual stresses. On the other hand, the factors influenced by the environment include relative humidity, presence of impurities in the atmosphere, rate of supply and distribution of oxygen, rate of flow of liquid, the acidity or alkalinity of that liquid and the applied load. These are only a few of the factors involved, but are sufficient to emphasize the complex nature of the problem. In general two principal forms of corrosion can be distinguished [6.7]:

(i) Direct chemical corrosion, and
(ii) Electro-chemical corrosion.

No attempt is made here to explain the whole complex problem of corrosion, but rather introduce the subject as a mode of failure and as one which is very relevant to the engineer. It is our intention to suggest how, in a comparatively simple cases, possible causes of corrosion might be avoided.

6.6.1 Direct Chemical Corrosion

This type of corrosion usually involves a direct combination between the metal and dry gases such as oxygen, sulphur dioxide and chlorine. At normal temperatures, this kind of corrosion is not troublesome to the structure, since it usually takes the form of metal oxidation, resulting in an oxide film which slows down or prevents further oxidation. At higher temperatures, such as those encountered by boiler tubes when in contact with hot gases, the oxidation continues at a higher rate. As a matter of fact, the rate of oxidation is also influenced by the type of metal, composition of gases and the resistance set up by the formed oxide film [6.7].

6.6.2 Electro-chemical Corrosion

This type of corrosion covers all types of wet corrosion, i.e. where the metal or alloy is in contact with a liquid or even a moist atmosphere. In the electro-chemical corrosion theory, it is assumed that all metals have a tendency to dissolve or corrode when the metal discharges positively charged ions into a solution. This leaves the metal with a characteristic negative charge or potential. The greater the negative potential, the greater is the tendency of the metal to dissolve or corrode.

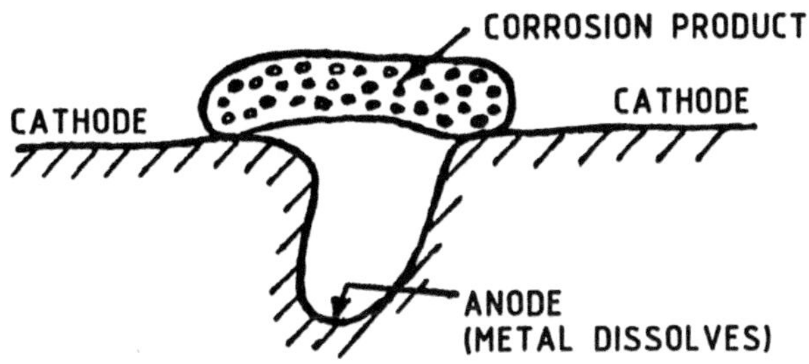

Figure 6.15 Pitting corrosion

The corrosion resistance of metals is governed by their position in the electro-chemical series. As a matter of fact, the tendency of each individual metal to corrode is relatively small, but is greatly increased when it is placed in contact with a dissimilar metal in the presence of a conducting liquid, referred to as an electrolyte. A current will flow between the two, since they are at different potentials. Corrosion of the one higher in the metals table (the anode) will be accelerated, whilst the metal lower in the metals table (the cathode) will be protected. The rate of corrosion is governed by the relative areas of the anode and the cathode.

It must be noted that for electro-chemical corrosion to occur there must be a cathode, an anode and an electrolyte. The formation of an anode and cathode need not necessarily be due to contact between dissimilar materials. Electrolytic corrosion can also occur due to the presence of impurities of second-phase constituents in the structure of an alloy.

Corrosion can also take place in a single metal, if there is a difference in oxygen concentration on the metal surface. This can explain corrosion problems such as pitting and also why metals corrode under scale deposits or along cracks and scratches more readily than elsewhere on the surface. A different aeration cell could be set up when metal sheets are stacked together with water trapped between them.

Pitting is an example of the differential aeration effect. The initial depression or pit in the surface, as shown in Figure 6.15, may be the result of several factors; e.g. a break in the protective film or scale. Once a pit is formed, the corrosion proceeds rapidly since the surface of the metal (cathode) has a greater access to oxygen than the base of the pit (anode), as depicted in Figure 6.15. Corrosion is accelerated by the fact that the surface area of the cathode is considerably greater than that of the anode. The corrosion product accumulates at the mouth of the pit and assists corrosion by making oxygen diffusion more difficult.

6.6.3 *Designing against Corrosion*

When the causes of corrosion are known, the designer can often prevent or reduce their effect by paying attention to certain details of his design. Many components are lubricated and so their surfaces are well protected and the possibility of corrosion

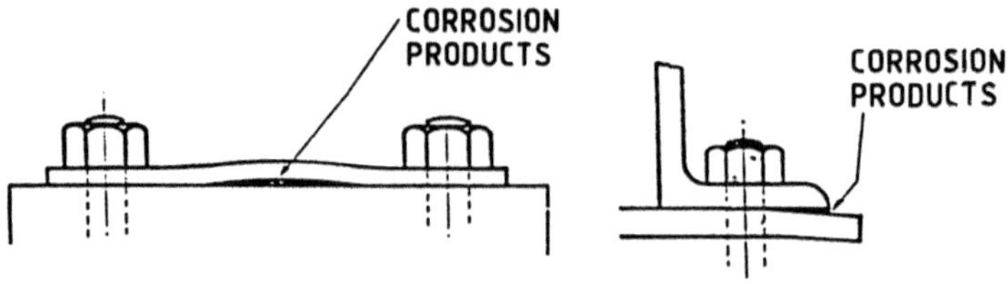

Figure 6.16 Examples of crevice corrosion

does not arise. Other parts however, such as structural steels, bodywork of motor cars, mechanical handling equipment, water pipes and tanks are subjected to either partial or complete immersion in corrosive fluids over varying periods of time, and so require protection of one form or another.

For example, in components which have been immersed completely, corrosion has been found sometimes to be particularly intensive in crevices. This is the result of the slow rate at which oxygen in the electrolyte can be replenished at these sheltered points. Crevice corrosion is particularly dangerous, because it cannot be observed easily; also, bulky corrosion products may exert extra forces unaccounted for between the parts held together to form the crevice, as depicted in Figure 6.16. The remedy to the problem lies in filling the crevices with a suitable jointing compound, preferably impregnated with a non-rotting fabric. The compound used will depend upon the fluids with which it is to come into contact.

Whenever possible, horizontal surfaces exposed to the atmosphere should be avoided, since they tend to hold moisture and dirt. This is particularly important for aluminium and its alloys, which are prone to crevice corrosion, and for magnesium and its alloys. Accumulated dirt will prevent the replenishment of oxygen, with the result that, in the presence of fluids, the dirty areas become anodic and corrosion cells are set up. Figure 6.17 shows desirable and undesirable forms of ledges. In this case, Figure 6.17a gives good drainage and little foothold for dirt particles. Similarly, the dome-shaped tops of storage tanks are preferable to flat tops, as depicted in Figure 6.18.

The interiors of many storage vessels are subjected to corrosive influences, since they are in contact not only with the fluid that they are designed to hold but also with sludge, bacteriological processes and impurities from air.

Suitable drainage is essential for the complete removal of these undesirable substances. For example, water contained in oil or petrol will collect at the bottom unless drainage is efficient and may intensify corrosion because it contains less oxygen than the petrol or oil and sets up a corrosion cell similar to a crevice corrosion cell. Figure 6.19a shows a suitable design of drain, whereas Figure 6.19b indicates the result of allowing the welded piece which receives the screwed pipe to protrude above the tank bottom. A slight slope given to the tank bottom will enable fluid to drain more readily to the outlet. A point worthy of note is that a suitable drain is useless if the tank bottom sags because it is insufficiently supported or stiffened, and the fluid collects in pools. Much of the bodywork of

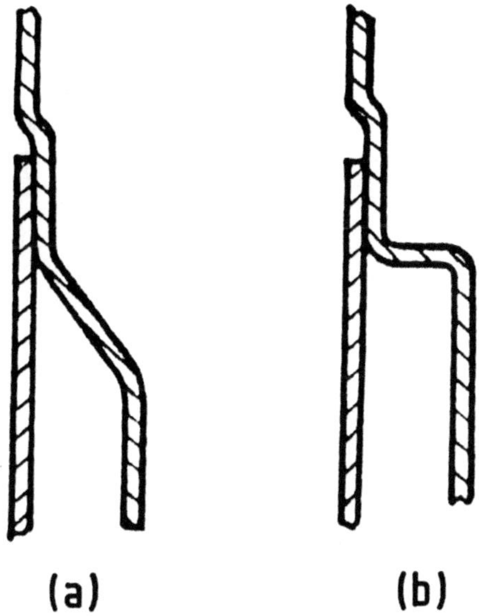

Figure 6.17 (**a**) Desirable form of a ledge. (**b**) Undesirable form of a ledge

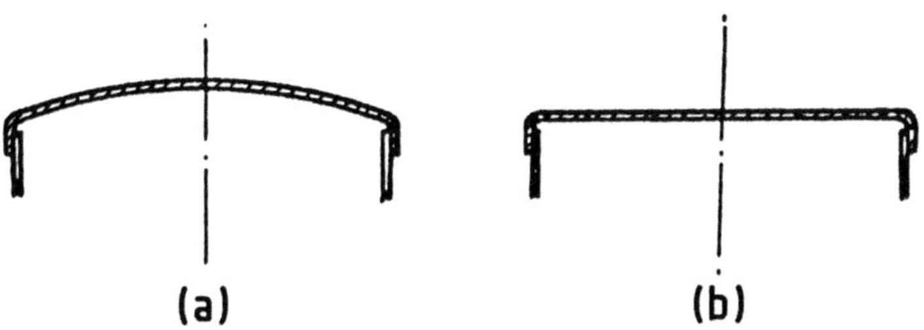

Figure 6.18 (**a**) Desirable dome-shaped top. (**b**) Undesirable flat-shaped top

motor cars can be saved from corrosion by suitably shaping the metal or, if this is not possible, by providing drain holes. In similar ways, the corrosion of car doors and boots can be reduced by providing drain holes and tubes, and by avoiding awkward recesses where fluids might collect. In addition to these design features, the resistance to corrosion can be enhanced by anodizing, plating, painting and surface treatment.

6.7 Stress Corrosion Cracking (SCC)

This mode of failure is due to the joint action of sustained static stress above a threshold value in presence of a corrosive environment. The stresses involved are usually internal stresses produced by some previous treatment such as cold working.

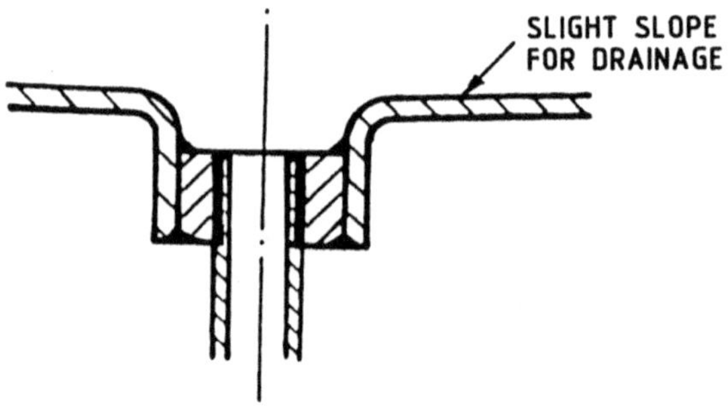

(a)

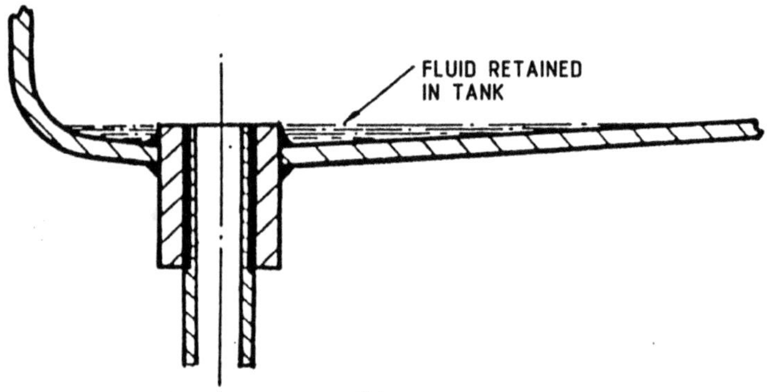

(b)

Figure 6.19 (**a**) Suitable design of drainage. (**b**) Unsuitable design of drainage

It seems that fairly large strains are essential in most cases, so stress corrosion attacks tend to occur in high-strength ferrous and non-ferrous metals carrying large loads. The growth of sharp corrosion pits gives rise to stress concentration effects which may set off fatigue failure or brittle fracture in some ferrous alloys. In this case, failure of the component could occur at stresses well below the normal yield stress of the material and in environments, which would be classified as non-corrosive for unstressed materials. SCC can occur at very low stresses, in some cases as low as 10% of the yield stress of the material [6.8]. A NASA report [6.9] highlighted stress corrosion as being the most serious, dangerous and expensive of the material problems encountered during the Apollo program.

Since the occurrence of SCC requires the exposure of a susceptible alloy to a specific environment at a stress above the threshold value, it follows that control of the problem can be effected by controlling these variables. Clearly, the preventive approach begins by selecting a suitable material, which is not susceptible to SCC. If this is not possible, due to other design considerations, then the designer must consider the geometrical features recommended in the previous subsections. In addition, the designer must avoid geometrical discontinuities. Such geometrical

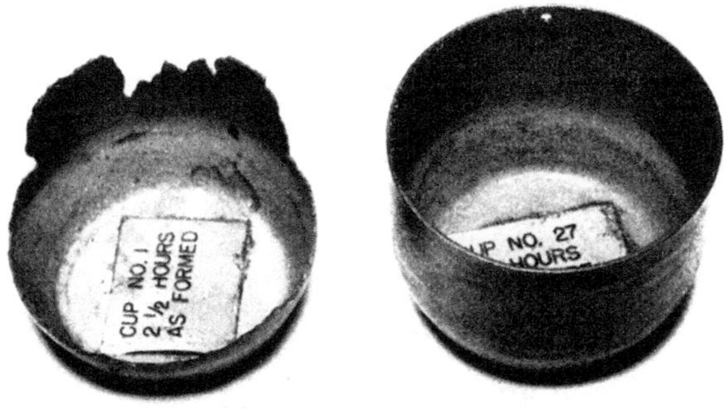

Figure 6.20 Effect of shot-peening upon the resistance to stress corrosion cracking of a cold-drawn brass cup in an ammonia atmosphere. Not peened 21/2 hour in NH3 (left); shot peened 21/2 hour in NH3 (right) [6.10]

discontinues, apart from their influence upon the state of stress, can frequently promote localized changes in the electro-chemistry of the environment by preventing mixing with a bulk environment that may be innocuous so far as SCC is concerned.

Furthermore, since many stress corrosion failures of machine components in service result from sustained tensile residual stress, their removal or reduction to values below the threshold value is generally beneficial. This can be achieved either by stress relieving or via the use of impact surface treatment techniques such as the shot-peening process. Figure 6.20, which is self-explanatory, shows the effect of the shot-peening treatment upon a drawn brass cup in an aggressive ammonia environment [6.10].

Metallic alloys are inherently flawed and are susceptible to slow subcritical crack growth during service. If failure is due to cyclic loading, then the mode of failure is typically caused by fatigue; if it is due to the joint action of sustained static stress and environment then the mode of failure is stress corrosion and the conjoint action of the two is known as corrosion fatigue. Here the term subcritical crack growth implies slow crack growth at loads, which are too small to cause rapid fracture.

Fatigue, the most common cause of failure in the aerospace industry is generally a greater problem in the presence of a corrosive environment. It is expected that a corrosive environment affects not only the initiation, but also the propagation of fatigue cracks. It is worth noting that, in most practical cases, the mechanisms of stress corrosion and corrosion fatigue are distinguishable. For instance, let us consider the effect of corrosion and stress upon the structural components of an aircraft. The wings of a stationery commercial airliner may encounter stress corrosion cracking. The static wind loading together with rain and other corrosive substances constitute a recipe for stress corrosion cracking. As the aircraft begins to move, the wind loadings fluctuate at various frequencies—hence the onset of corrosion fatigue. After take-off the environment may vary; the values and frequencies of load fluctuations will change, but the condition may still be right for stress corrosion and corrosion fatigue. At cruising altitudes, the ambient external temperatures may be as low as −50 °C, a temperature at which corrosion is practically inactive; hence, making the situation of 'pure' fatigue for the main part of the flight.

Finally, in the case of low-strength alloys, corrosion fatigue will occur in a far more extensive range of environments than will promote SCC. This does not mean, however, that all environments, which are corrosive to a particular metal, will promote corrosion fatigue. For instance, aggressive environments are known to retard crack propagation through crack-tip blunting. In this case, the frequency of loading of the crack ligaments in an aggressive environment will play an important role in determining the behavior of cracks.

6.8 Creep and Creep Rupture

Creep is a time-temperature-dependent deformation, which occurs when a material is subjected to load for a prolonged period of time. The familiar load-extension diagram for the response of a tensile specimen signifies that for a given load, a fixed extension will result. If the material creeps, however, the extension will continue to increase due to viscous effects, even though the load remains constant. The variation in the deflection with time is shown in Figure 6.21a. Clearly, creep deformation cannot continue indefinitely, and will lead to creep rupture.

Conventionally, the creep curve of Figure 6.21a has been divided into several parts. There is an initial linear elastic time-independent response. Thereafter, the creep curve is described in terms of three regions: the first of decreasing creep strain rate called primary creep, the second of reasonably constant creep strain rate called secondary or steady state creep and the third of increasing strain rate called tertiary creep preceding rupture. These definitions become more apparent if we plot creep strain rate against time, as depicted in Figure 6.21b.

6.8.1 Creep Mechanisms and Deformation Maps

There are two dominant mechanisms associated with creep, known as dislocation creep and diffusional creep. In the former, dislocations in the crystalline lattice structure of a metal can overcome the natural stiffness of the crystal as well as other obstacles present in an alloy to move through the lattice. At low stresses, this dislocation motion stops or slows down, but creep can continue through the bulk movement of atoms—diffusional flow. The most important mechanism in most engineering structures is dislocation creep. The relative importance of these mechanisms, in particular stress and temperature ranges, can be conveniently summarized on the deformation mechanism map of Figure 6.22. This is an idealized diagram, for a particular material, which also shows regions where instantaneous plastic flow occurs and where deformation is simply elastic. From this map, it can be seen that creep is significant at temperature above about $0{\cdot}3\boldsymbol{T}_{\mathrm{m}}$, where $\boldsymbol{T}_{\mathrm{m}}$ is the melting temperature of the metal [6.11].

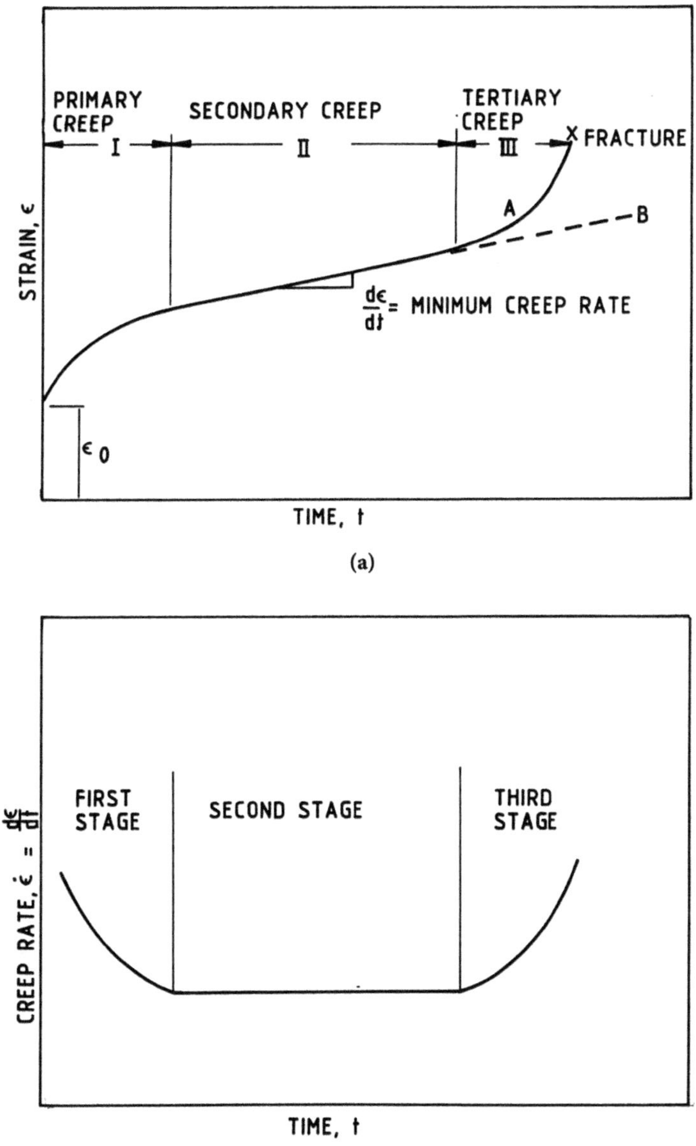

Figure 6.21 (**a**) Creep curve of strain versus time. (**b**) Creep strain rate against time

Creep damage is often manifested in terms of nucleation of voids or holes due to microstructural inhomogeneities, which would grow and coalesce to form cracks. Consequently, the final failure occurs by the interlinking of the cracks through the entire section, as depicted in Figure 6.23. If the applied stresses in the section are

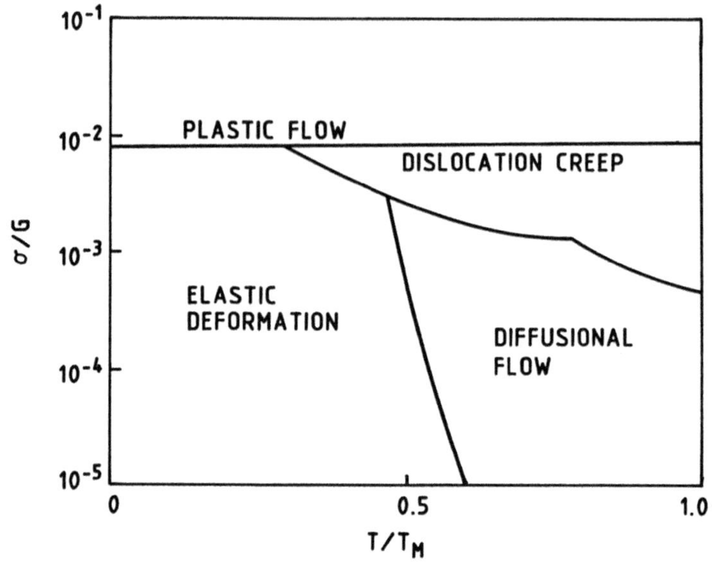

Figure 6.22 Deformation mechanism map

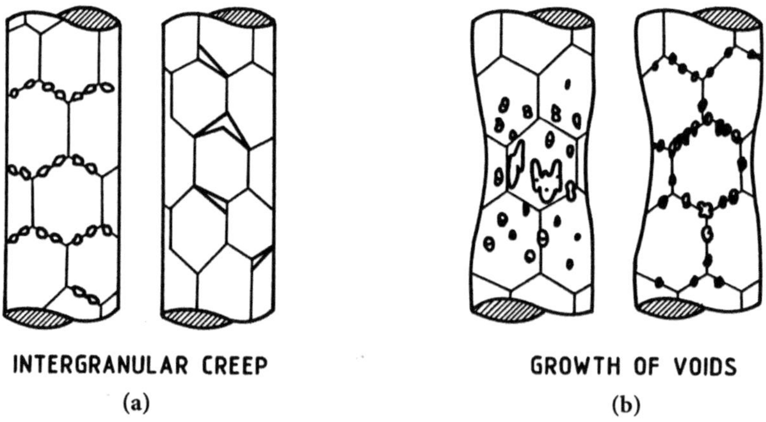

Figure 6.23 Nucleation and coalescence of creep voids to form cracks with (**a**) depicting intergrannular creep and (**b**) growth of voids due to creep

uniform, nucleation of voids and subsequent formation of cracks can also occur relatively uniformly. In this case, a major part of the life is spent in nucleation and growth of voids with a minor fraction on the growth and interlinking of cracks.

Components in service, however, are seldom subjected to uniform stresses. In this case, creep damage becomes non-uniform and localized. As the defect grows into the damaged region, the stress in the remaining ligament gets accentuated, giving rise to accelerated crack growth. In addition, since crack growth generally proceeds from the surface, component life can be further affected by the presence of aggressive environments. Since, as explained earlier, the crack nucleation stage is either circumvented or minimized, the service life of a component in such cases cannot always be predicted from the creep rupture data of uniformly stresses specimens. In fact, several alloys that show very good creep rupture strength show very poor resistance to crack growth.

6.8.2 *Uniaxial Constitutive Formulation of Creep*

Many simplified uniaxial constitutive relations have been proposed to describe the standard creep curves. In this work we will restrict our attention to primary and secondary creep, and describe a few relations, which have found particular application in creep stress analysis. The first step common to nearly all approaches is to separate the elastic and inelastic parts of strain, i.e.

$$\varepsilon^{t} = \varepsilon^{e} + \varepsilon^{c} \tag{6.12}$$

where ε^{t} is the total strain and ε^{c} are the elastic and the inelastic (creep) strain, respectively. The creep strain resulting from a constant-load test may be written as a function of stress σ, time *t* and temperature *T*, as

$$\varepsilon^{c} = \int_{1}(\sigma)\int_{2}(t)\int_{3}(T) \tag{6.13}$$

The function $f_1(\sigma)$ can take any of the following forms:

$f_1(\sigma) = \mathrm{B}\ \sigma^{\mathrm{n}}$	due to Norton
$f_1(\sigma) = \mathrm{C}\ \sinh\ \alpha\sigma$	due to Prandtl
$f_1(\sigma) = \mathrm{D}\ \exp(\beta\sigma)$	due to Dorn
$f_1(\sigma) = \mathrm{A}\ [\sinh(\gamma\sigma)]^{\mathrm{n}}$	due to Garofalo

In the above expressions, symbols other than σ are material constants. It is worth noting that Garafalo-relation contains Norton, Prandtl and Dorn relations as special cases.

For the time dependence, the following expressions have been proposed.

$$\int_2 (t) = t \qquad \text{secondary creep model}$$

$$\int_2 (t) = bt^m \qquad \text{due to Bailey}$$

$$\int_2 (t) = \left(1 + bt^{1/3}\right) \qquad \text{due to Andrade}$$

For the temperature dependence, Arrhenius proposed the following functional relationship,

$$\int_3 (T) = \mathrm{A} \exp(-\,\Delta\mathrm{H}/\mathrm{RT})$$

where ΔH is the activation energy, R is the Boltzmann constant and T is the absolute temperature.

The simplest model, which takes account of the three functions, can be given as

$$\varepsilon^{\mathrm{c}} = \mathrm{B}t^{\mathrm{m}}\,\sigma^{\mathrm{n}} \tag{6.14}$$

which is commonly used in creep stress analysis.

6.8.3 Creep-Fatigue Interaction

We indicated earlier that crack growth of many components operating at high temperatures can occur by fatigue, creep and/or environmental effects. In many instances, mutual interactions among these processes give rise to synergistic effects. The creep contribution to the fracture mechanism, depends upon the operating temperature, hold time, loading frequency, type and magnitude of loading, mechanical properties and microstructure of the component. All these factors are important for developing life prediction criteria for high temperature crack growth.

Recognition of the importance of crack growth behavior in the creep range has led to a significant increase in crack growth studies. Several advances have been made, particularly in the parametric representation of crack growth rates, in the analysis of crack-tip stress fields in the creep range and in the development of theoretical models for predicting crack growth behavior of structural alloys at high temperatures.

References

6.1 J. Campbell, *The Mechanisms of Metallurgical Failure on the Origin of Fracture*, 1st edn. (Elsevier, 2020). https://doi.org/10.1016/C2019-0-02485-5
6.2 S. Turcott, *Decoding Mechanical Failures*, 1st edn. (Steel Empire, Dundas, 2020)
6.3 C. Yoo, S. Lee, *Stability of Structures: Principles and Applications*, 1st edn. (Elsevier, Amsterdam, 2011)
6.4 G. de Havilland, P.B. Walker, in *The Comet Failure, Engineering Progress Through Trouble*, ed. by R.R. Whyte, (Institution of Mechanical Engineers, London, 1975), pp. 51–53
6.5 G.M. Simons, *Comet! The World's First Jet Airliner* (Pen and Sword, Havertown, 2014)
6.6 T. Nicholas, *High Cycle Fatigue – A Mechanics of Materials Perspective*, 1st edn. (Elsevier, 2006). https://doi.org/10.1016/B978-0-08-044691-2.X5000-0
6.7 D. Jones, *Principles and Prevention of Corrosion*, 2nd edn. (Pearson, Harlow, 1995)
6.8 V.S. Raja, T. Shoji (eds.), *Stress Corrosion Cracking – Theory and Practice*, 1st edn. (Elsevier, Amstedam, 2011)
6.9 R.E. Johnson, *Apollo Experience Report: The Problem of Stress Corrosion Cracking*. NASA TN D-7111, Washington, DC, 1973
6.10 S.A. Meguid, *Engineering Fracture Mechanics* (Elsevier Applied Science, London, 1989)
6.11 M.E. Kassner, *Fundamentals of Creep in Metals and Alloys*, 3rd edn. (Elsevier, 2015). https://doi.org/10.1016/C2012-0-06071-1

Chapter 7
Brittle Fracture

Summary This chapter is devoted to the modeling and characterization of brittle fracture of solids. It also covers the Griffith crack and introduces the concept of stress intensity factor as a prelude to linear elastic fracture mechanics.

7.1 Introduction

The term brittle fracture is used to describe fast unstable fractures. This is in contrast to stable fractures of the fatigue type or slow failures such as yielding and plastic deformation. The association of brittle fracture with certain metals arises from a metallurgical feature peculiar to ferrite alloys and certain other metals, such as molybdenum and one form of chromium, which have the same body-centered cubic crystal structure.

These metals are capable of fracture in two ways. The first is by the normal method of extended slip or glide of one part of a crystal over another known as ductile fracture. In such cases, plastic deformation and eventual fracture are largely controlled by the maximum resolved shear stress. The second fracture mode is one of brittle cleavage on totally different planes and is triggered by tensile stresses. The first tends to be slow with much plastic flow, and the second rapid, with speeds of propagation approaching the speed of sound, and without yielding or appreciable plasticity.

Almost all crystalline solids fail by cleavage, if the temperature is sufficiently low. Certain face-centered cubic metals such as aluminium, copper and nickel and their alloys appear to be the only exceptions. It is the objective of this chapter to discuss these important aspects and highlight the parameters which trigger fast and slow fractures.

S. A. Meguid, *Atomistic and Continuum Fracture Mechanics of Solids*,
https://doi.org/10.1007/978-3-031-56085-9_7

7.2 Ductile Fracture

Ductile fracture is normally preceded by extensive plastic deformation. Ductile fracture is slow, and generally results from the formation and coalescence of voids. These voids are often formed at inclusions due to the large tensile stresses set up at the inclusion/metal interface, as seen in Figure 7.1a. Ductile fracture usually follows a transgrannular path but, if the density of inclusions or of pre-existing holes is higher at the grain boundaries than it is within the grains, then the fracture path may follow the boundaries, giving a fibrous or ductile intergrannular fracture. In cases where inclusions are absent, it has been found that voids are formed in severely deformed regions via the presence of localized slip bands and macroscopic instabilities, resulting in either necking or the formation of zones of concentrated shear, as depicted in Figure 7.1b.

The fracture path of a ductile crack is often irregular and the large number of small voids present gives the fracture surface a dull fibrous appearance. The capacity of most metals of engineering interest for plastic deformation and work hardening is extremely valuable as a safeguard against design oversight, accidental overloads or failure by cracking due to fatigue, corrosion or creep.

7.3 Brittle Fracture

This mode of failure is often thought to refer to rapid propagation of cracks without any excessive plastic deformation at a stress level below the yield stress of the material. In practice, however, most brittle fractures show very limited plastic deformation ahead of the crack tip. Brittle fracture may be transgrannular (cleavage) or intergrannular, as depicted in Figure 7.2. It is also worth noting that metals which often show ductile behavior can, under certain circumstances, behave in a brittle fashion leading to fast unstable crack growth. This was clearly demonstrated over the years by some of the unfortunate accidents involving ships, bridges, offshore oil-rigs, gas pipelines, pressure vessels and other major constructions; see e.g. Refs. [7.1–7.4]. Liberty ships, the Tacoma Narrows bridge in the USA, King

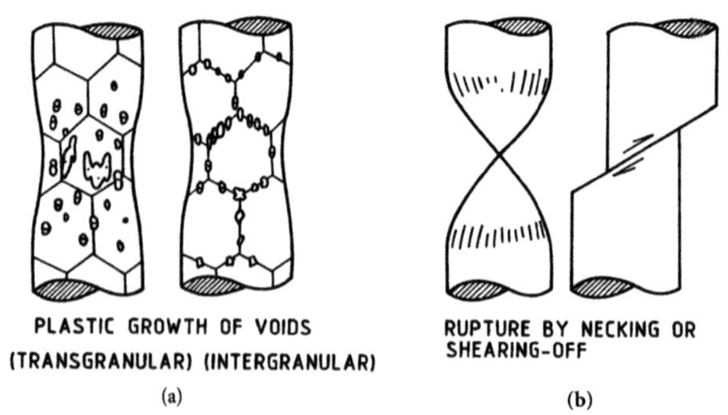

Figure 7.1 Plastic deformation (**a**) by voids growth and coalescence, (**b**) necking or shearing

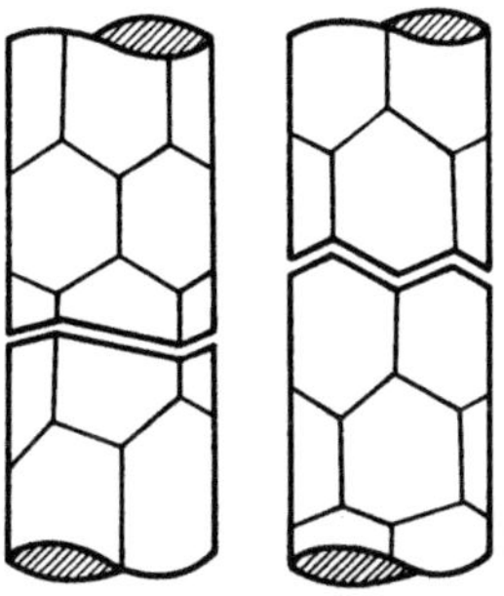

Figure 7.2 Transgrannular and intergrannular brittle fracture

Street bridge in Melbourne, Australia, the Sea Gem drilling rig for North Sea gas and the collapse of the Alexander Kielland oil-rig are only a few examples of the casualties of brittle fracture [7.2–7.4].

Fatigue cracking is not essential for initiating brittle fractures. Other sources include the heat-affected zone cracks in welds that develop in high tensile steels and shrinkage tears in weld metal or in steel castings. Some of these have been responsible for some of the great catastrophic brittle failures of the last 30 years, or so. Nevertheless, the more usual role of these manufacturing and fabrication defects and of corrosion pits or cracking is the initiation of a fatigue crack subsequently followed by brittle fracture. While fatigue cracks are initiated by repeated cyclic loading, brittle fracture, given the right circumstances, is readily set off by a quasistatic stress [7.5].

7.4 Liberty Ships: A Case History

During World War II, Britain turned to the USA for help with basic cargo ships, since traditional shipbuilding methods involving riveting of joints were too slow and the war efforts were hindered by an acute shortage of ships. Henry Kaiser, an American construction engineer, was able to revolutionize the shipbuilding techniques by adopting welded constructions instead of riveted ones. This resulted in speeding the process of building cargo ships that effected a large reduction in construction time. As a matter of fact, the Robert E. Peary was built in only 4 days, which is a remarkable achievement! Liberty ships were soon to be found in almost every convoy and they indeed formed a significant part of the bridge of ships across the Atlantic. They were even used in arctic conditions.

In less than 12 months, many of these ships suffered severe fracture arising mostly from welded square hatches in the upper deck. It later transpired that of the 2700 newly built Liberty ships, 400 suffered fracture of some degree, 90 suffered serious fracture; in 20 cases the fracture was effectively total and 12 ships broke in half. It is interesting to note that, according to the United States Navy, many of the fractures seemed to be related to certain common features. In most cases the fracture was nearly amid-ships and ran approximately in line with a transverse bulkhead.

7.4.1 What Went Wrong?

A ship, at its simplest, can be considered as a girder upon which a system of complex and variable loads act. The unequal distribution of cargo and ballast can cause a ship either to hog or to sag, as shown in Figure 7.3. Liberty ships were mostly subjected to a hogging bending moment resulting in maximum tensile stresses in the top deck. Furthermore, wave bending moments can accentuate this effect, thus producing significant tensile stresses at upper deck levels.

7.4.2 Stress Concentration

Clearly, the presence of stress concentration features, such as notches, holes or sharp edges raises the local stress level. Both the orientation and the geometrical severity of the feature play an important role in the structural integrity of a component/assembly. To demonstrate this argument, let us examine the stress distribution in an infinite plate containing a sharp elliptic hole subjected to the uniaxial tensile loading shown in Figure 7.4. Advancing along AB, the maximum elastic stress σ_{max} according to Inglis is

$$\sigma_{max} = \sigma_0(1 + 2a/b) \tag{7.1}$$

where a and b are the semi-major and semi-minor axes of the elliptic hole, and σ_0 is the far field tensile stress. This maximum tensile stress decreases rapidly and in a

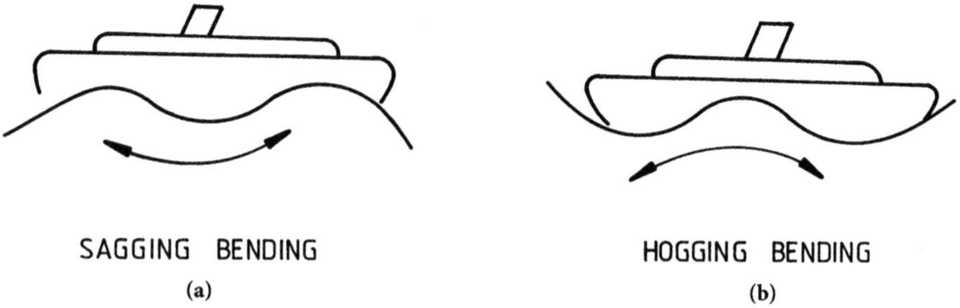

Figure 7.3 Effect of wave bending moment upon longitudinal stresses in cargo ships: (**a**) a sagging configuration and (**b**) a hogging configuration

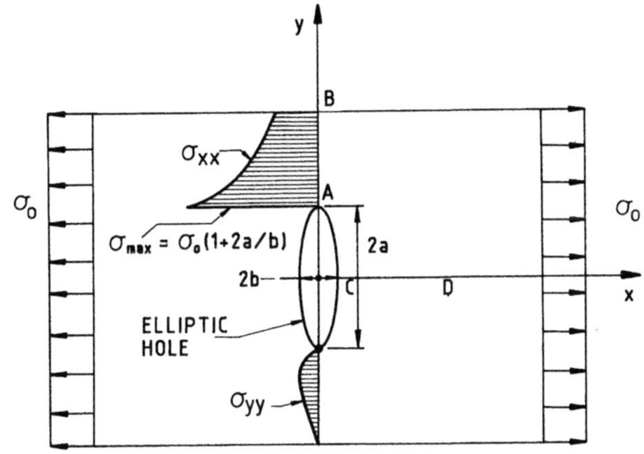

Figure 7.4 Effect of the presence of an elliptic hole upon resulting stress concentration and stress biaxiality

short distance attains approximately its normal value σ_0. If one now advances along CD, Inglis's solution predicts a compressive stress of magnitude σ_0 at point C and that changes to a small tensile stress and then gradually dies out entirely. Obviously, other stresses are brought into existence by the introduction of the geometrical discontinuity.

Let us focus our attention to the maximum tensile stress distribution for varied values of a/b for different orientations. First, when $a/b = 100$, the tension at A is 201 times the far field tensile stress. In the case of a very sharp crack, where a/b exceeds 1000, the presumed maximum elastic stress can be very large; in the case of an ideal crack, when $b \to 0$, $\sigma_{max} \to \infty$. So far we have taken the major axis of the ellipse to be perpendicular to the direction of the tensile stress. If the tensile stress is applied in the direction of the major axis, the concentration of stress is now determined from the expression $\sigma_{max} = \sigma_0(1 + 2b/a)$, with its maximum at point C, accompanied by a compressive stress of magnitude σ_0 at A. Hence, a crack running in the direction of the tensile load does not produce high local stresses, since $b/a = 0$, and

$$\sigma_{max} = \sigma_0 \tag{7.2}$$

In the case of a circular hole, as already explained in Chapter 3, the stress concentration factor at A is:

$$\sigma_{max}/\sigma_0 = 3 \tag{7.3}$$

since $a = b$.

It is clear from the above that both the geometry and the orientation of the stress concentration features played important roles in the brittle fracture of Liberty ships. However, the severity and the speed at which these fractures took place (approaching the speed of sound) cannot be explained solely on the basis of high stress concentration at the square hatches, since they were used in earlier riveted cargo ships.

7.4.3 Crack Arresters

A major advantage of the traditional methods of building a ship is that, if one steel plate fractures, the crack is unlikely to propagate to the next plate. Effectively, as demonstrated in Figure 7.5, the traditional riveted joints acted as crack arresters. Liberty ships, however, were welded, and if a crack were to start in a plate, it would run right through the hull with nothing to stop it, as depicted by Figure 7.6. The ship was effectively a continuous girder of steel. This, however, does not explain the brittle nature of fracture of a supposedly ductile material!

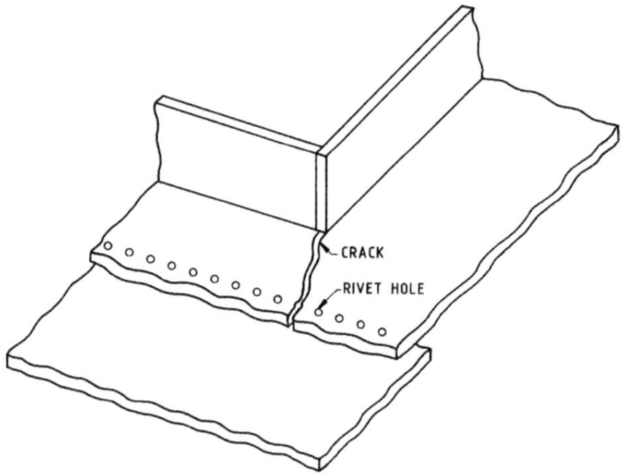

Figure 7.5 Riveted joints act as crack arrester

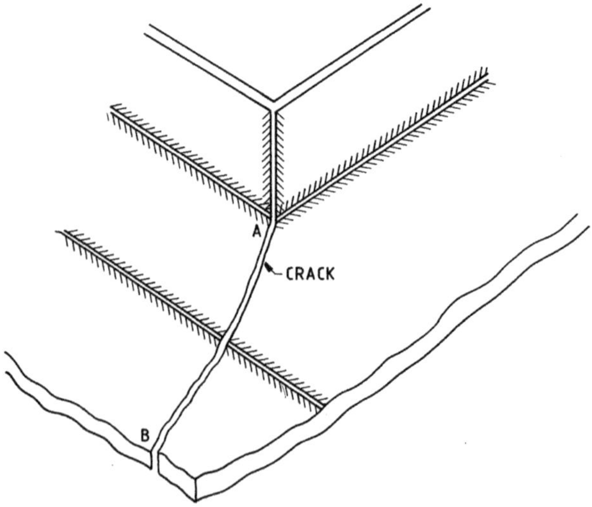

Figure 7.6 Welded joint form a continuous structure with unimpeded crack growth

7.4.4 Impact Transition Curves

Of considerable importance to the above argument is the fact that Liberty ships were made of body-centered ferritic alloys in which the yield stress is far more sensitive to temperature and strain rate changes than it is in face-centered cubic metals such as nickel, aluminium and copper. The presence of cracks and notches will accentuate this effect. Furthermore, the presence of heat-affected zones of high hardness in addition to the presence of stress concentration features which induces triaxiality of the stress state will lead to the suppression of plastic deformation.

It is worth noting that the operating temperature of a structure and the rate of deformation (strain rate) have a marked effect on its failure mechanism. Experience shows that brittle fractures are liable to be more frequent and more extensive with cryogenic temperatures and/or high strain rates. These effects are due mainly to the dependence of the mechanical properties of materials upon strain rate and temperature and to a lesser extent upon the thermal gradients set up by the prevailing temperature distribution. In some steels similar to that used in Liberty ships, conditions may exist where a difference of a few degrees, even within the normal range of atmospheric temperatures, may determine the difference between ductile and brittle behavior.

In order to clarify this dependency, let us examine the effect of temperature and strain rate upon the yield and fracture stresses of a sample. Figure 7.7 indicates that a reduction in temperature and/or an increase in strain rate generally results in a marked increase in the yield stress of the material. The effect on the fracture stress may or may not be pronounced. Generally speaking, however, the fracture stress will increase with a reduction in temperature and/or an increase in the strain rate. Accordingly, even if we test plain samples at low temperatures and/or high strain rates, the fracture stress may be encountered before the yield stress. If a sharp crack

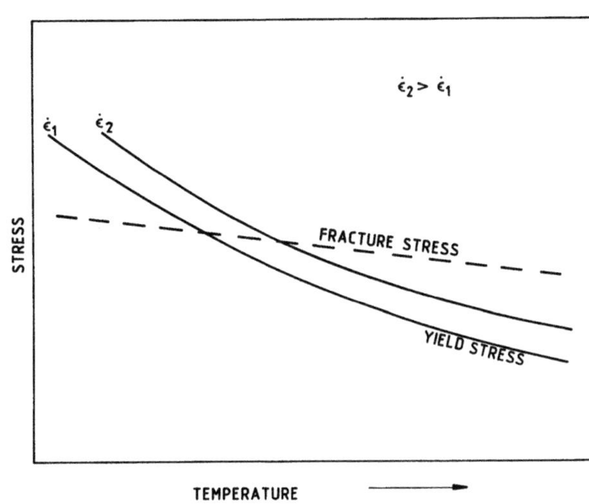

Figure 7.7 Effect of temperature and strain rate upon yield and fracture stresses

Figure 7.8 Schematic of a pendulum impact tester

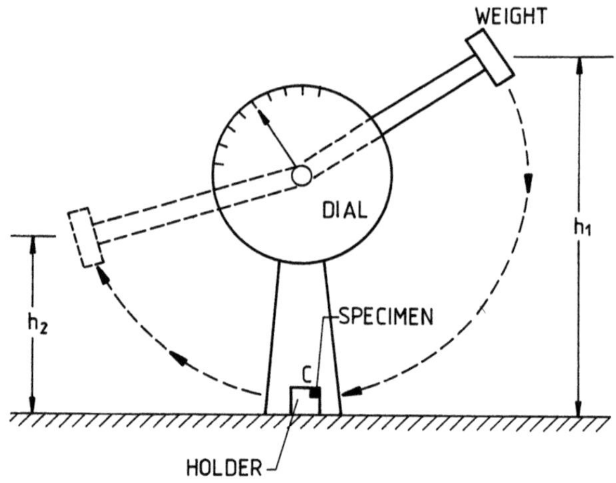

or a notch is present in a relatively thick specimen, then the transverse strain associated with the applied tensile stress may be inhibited by the rigidity of the less stressed material in the vicinity of the stress concentration feature. Thus a transverse tensile stress is set up, so that all three principal stresses are tensile. This triaxiality depresses the maximum shear stress relative to the maximum tensile stress, and this in turn will give rise to cleavage failure.

Though it would indeed be convenient if some single measurable property of metal existed whereby the propensity to brittle fracture in a structure could be confidently predicted, no such single property has yet been discovered. For high strength-low toughness alloys, the fracture toughness of a material can be assessed by a design parameter known as the critical stress intensity factor K_{IC}. It is not possible at present to use this parameter for low strength-high toughness in commonly used thicknesses under slow loading conditions. In this case, elastic-plastic fracture mechanics governs the failure of the component.

For the moment, however, let us content ourselves with pendulum impact tests, which are used to provide an indication of the tendency to brittle fracture. In these tests, the specimen is positioned across the lowest point in the path of a striker mounted on the end of a pendulum, as shown in Figure 7.8. The striker, having been initially lifted to a specific height h_1 and then released, swings against the specimen and fractures it. The striker continues its swing to the other side of the specimen to a height h_2. Clearly, the difference between the two heights multiplied by the weight of the striker corresponds to the amount of energy absorbed in causing the fracture. It is also important to note that the appearance of the fracture surfaces, e.g. the percentage of crystallinity, provides additional clues for assessing the brittleness of the fracture. Such tests are usually carried out on a series of specimens at different controlled temperatures.

Generally, two basic types of tests exist: Izod and Charpy tests. In the Izod test, the specimen is gripped in the form of a vertical cantilever, as shown in Figure 7.9a. Difficulties in maintaining vice jaws as well as the specimen at the required temperature make this type of test undesirable. In the Charpy test, illustrated in Figure 7.9b,

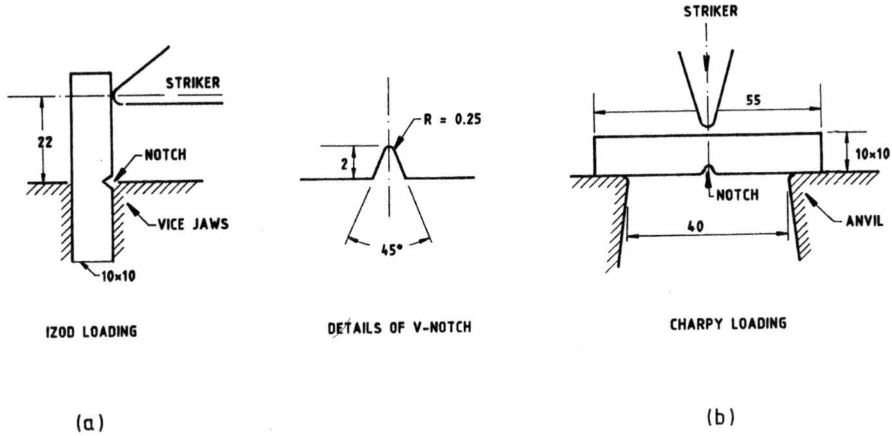

Figure 7.9 Details of the specimen geometry, holder and striker for (**a**) Izod set up and (**b**) Charpy set up

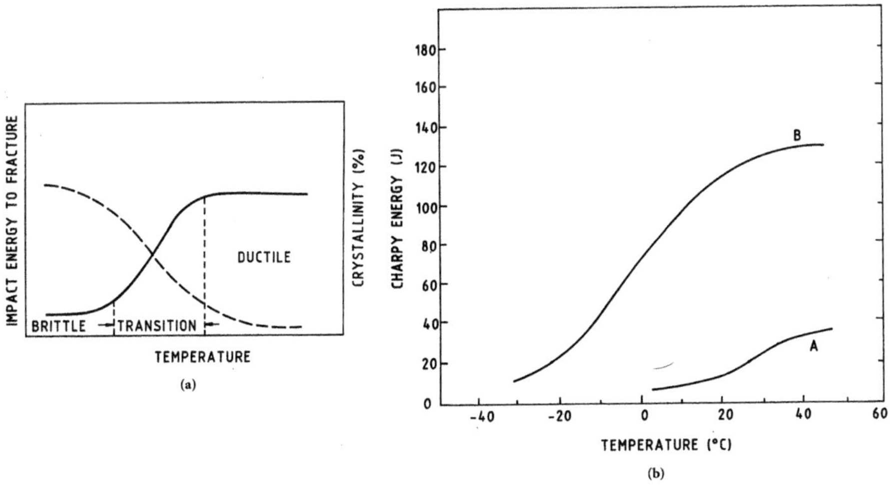

Figure 7.10 (**a**) Impact energy and crystallinity versus temperature. (**b**) Charpy energy versus temperature for (*A*) steel used in Liberty ships, (*B*) high toughness high strength steel

the specimen is a horizontal beam spanning across the path of the striker. In this case, only the specimen itself needs to be brought to the required temperature. Some machines have been developed so as to enable both tests to be performed. It is worth pointing out that differences do exist between the British and ASTM Charpy test requirements. However, the results of the two tests are comparable.

The results of a number of tests are then recorded on a chart similar to that presented in Figure 7.10a, where each point represents the energy value at a given temperature. Several criteria are frequently used to determine the transition of any given steel from tough to brittle behavior, including:

(i) the energy absorbed in fracturing the specimen at a chosen temperature,
(ii) the temperature corresponding to the inflection point, i.e. where the slope of the curve is steepest, and
(iii) the temperature at which the fracture shows some chosen percentage of crystallinity.

Figure 7.10b shows the Charpy energy versus temperature behavior for two contrasting steels: a typical steel, currently used in shipbuilding (B), and the steel used in Liberty ships (A). These curves clearly indicate the poor impact transition qualities of the steel used in Liberty ships. For example, the Charpy transition value for steel used in Liberty ships was approximately 20 J as compared with 70 J for a tougher high strength steel. The effect of temperature on the energy to fracture has been related in low-strength ferritic steels to a change in the microscopic fracture mechanism, from void coalescence at relatively high temperatures to cleavage at relatively low temperatures, as explained earlier in Figure 7.7.

At this stage, it is important to outline some of the limitations of the interpretation of the transition temperature measurements, since the defined transition temperature will depend not only upon the measurement criteria (energy absorbed, ductility and fracture appearance) but also on the thickness and notch severity of the test bar. Therefore, laboratory results must be interpreted with caution if the component thickness is different from that of the test specimen. There are also uncertainties relating to crack length effects. Accordingly, the impact transition temperature of a Charpy specimen will generally be different from that of the structure.

7.4.5 Effect of Welding

In fusion welding, an intense source of heat is applied to the weld preparation zone so as to raise the local surrounding material to its melting point. Normally, a molten filler metal is added to fill the zone and this mixes with any molten parent material. With this localized heating, and efficient conduction of heat from the weld into the neighboring colder areas, thermal residual tensile stresses develop. In this case, the weld may be likened to a small chill casting surrounded by a band of material which is first rapidly heated and then rapidly cooled, thus resulting in the introduction of highly stressed crack-like defects and voids not present in the parent material. Furthermore, a general reduction in notch ductility of the heat-affected zone is noticed because of its rapid cooling. The presence of tensile residual stress in the weld region increases the likelihood of brittle fracture if the material is susceptible to it.

In such situations, stress relieving of welded structures may become essential. For example, the stress relief heat treatment of carbon manganese steels not only relieves the component from the harmful tensile residual stress, but also restores the notch ductility of the regions adjacent to a defect. It must be noted, however, that the stress relief treatment may actually increase embrittlement adjacent to a defect in certain alloy steels, e.g. steels containing vanadium.

7.4.6 *Remedial Measures Taken*

The key factors can then be summarized as being

(i) the presence of certain design features such as square hatches, notches and cut-outs and defects in welds introduced by high stress concentrations and stress triaxiality,
(ii) poor standard of fabrication and in particular of welds carried out by a semi-skilled labour force,
(iii) the use of steels and welds with a fairly low impact toughness (20 J) operating at fairly low temperatures, and
(iv) the corrosive nature and wave loading encountered in marine environment.

It is appropriate at this point to consider the work required to put the ships back into trouble-free service.

In the short term, certain measures were taken to alleviate the situation. For example, the hatch corners were rounded to reduce the concentration of stress. Furthermore, strips of high toughness steel were riveted to the hull which acted as crack arresters. These were introduced in a number of locations, as shown in Figure 7.11, and they proved to be very effective in arresting cracks.

In the long term, however, much research was devoted to producing tougher steels, improved methods of fabrication and more stringent quality control. Research was also directed toward the study of the mechanics of cracked bodies (fracture mechanics) and non-destructive evaluation (NDE) of defects in engineering components.

7.5 Energy Balance for Ideally Brittle Materials

The above case history indicates that structural failure under loading conditions well below the yield stress Y of the component is often attributed to cracks or crack-like defects in the structure. Such failures show that design against yielding is not sufficient to guarantee the mechanical integrity of the component under varied operational loading conditions. Structural integrity studies which concern themselves with the applied mechanics of cracked bodies are known as fracture mechanics. In particular, idealizations based on neglecting crack-tip plasticity are referred to as *linear elastic fracture mechanics* (LEFM).

The basis of LEFM was introduced by Griffith who considered the propagation of cracks in glass. Using fundamental energy balance in classical mechanics and thermodynamics, Griffith proposed that there is a simple energy balance consisting

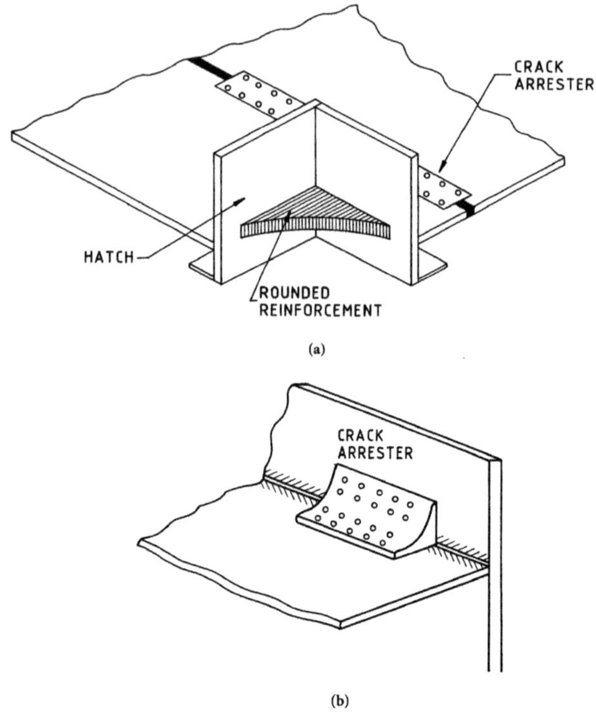

Figure 7.11 Short-term measures taken to improve resistance to cracking: (**a**) rounded hatch corner assisted by riveted crack arrester; (**b**) riveted crack arrester between two welded plates

of the decrease in potential energy within the stressed body due to crack extension U_p and this is counteracted by the increase in surface energy due to the creation of new crack surfaces U_γ. His energy balance approach allows the prediction of the theoretical strength of brittle materials and also provides a relationship between fracture strength and defect size. Griffith's approach, however, does not consider the details of the fracture process at the crack tip. For instance, if one considers a situation where crack extension occurs after a certain amount of plastic deformation at the crack tip, it is then necessary to take into account the energy which is expended in plastic deformation. It is also worth noting that in a dynamic situation, one should consider the kinetic energy term due to fast moving cracks.

7.5.1 Griffith's Energy balance Approach

Consider an infinite plate of unit thickness containing a through thickness central crack of length $2a$. The plate is subjected to a boundary stress a, as depicted in Figure 7.12.

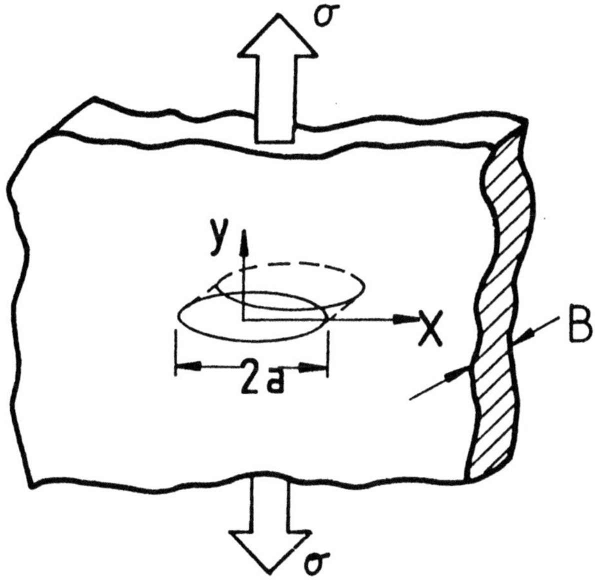

Figure 7.12 An infinite plate containing a through-thickness central crack

In this case one may divide the total energy of the system into four components. The first is the initial elastic strain energy of the loaded but uncracked plate per unit thickness, U_i, where

$$U_i = \int_A \frac{\sigma^2}{2E} dA \tag{7.4}$$

which is constant. The second is the elastic strain energy release caused by the introduction of a crack of length $2a$ and the relaxation of the material above and below the crack U_a. Griffith showed that U_a per unit thickness of the plate can be expressed as,

$$U_a = \pm \frac{\pi \sigma^2 a^2}{E} \tag{7.5}$$

The third energy term considers the work done by the external forces, U_w, and is defined as,

$$U_w = \int_y F dy \tag{7.6}$$

where y is the displacement and F is the applied force per unit thickness.

Generally, two cases of loading conditions may arise: (i) constant displacement or fixed grip and (ii) constant load, as shown in Figure 7.13. In the fixed grip condition,

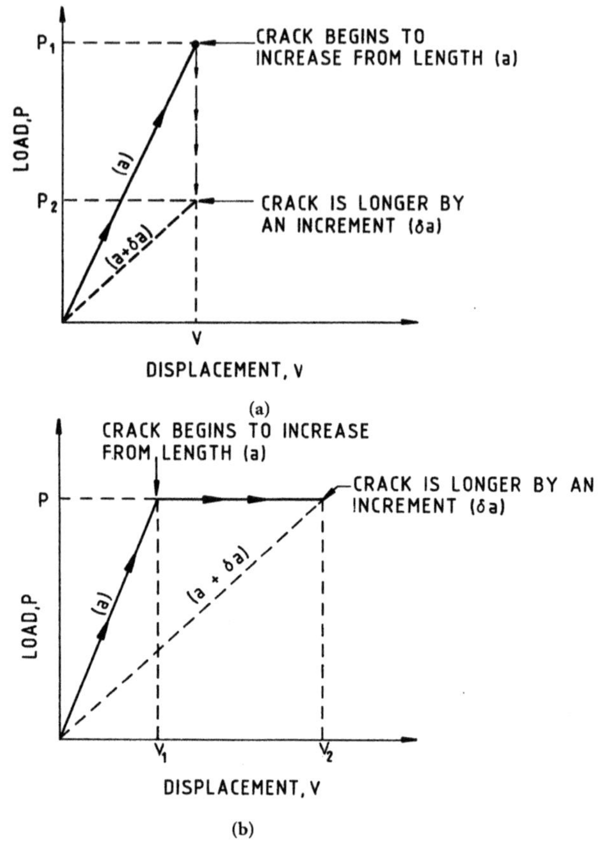

Figure 7.13 Load-extension plots for crack extension: (**a**) constant displacement condition; (**b**) constant load condition

there will be no work done by the external forces, since $dy = 0$. Now, since crack extension is associated with a decrease in component stiffness or increased compliance, the associated load applied by the grips will therefore decrease, thus resulting in a reduction in the elastic strain energy release, U_a, i.e.

$$U_a = -\frac{\pi\sigma^2 a^2}{E} \quad (7.7)$$

In the constant load condition, however, the work done by the load increases the elastic strain energy release, U_a, and accordingly

$$U_a = +\frac{\pi\sigma^2 a^2}{E} \quad (7.8)$$

Therefore, the internal or potential energy of the plate, U_p, can be expressed as,

$$\begin{aligned} U_p &= U_i - U_a \\ &= \int_A \frac{\sigma^2}{2E} dA - \frac{\pi\sigma^2 a^2}{E} \end{aligned} \tag{7.9}$$

for the fixed grip condition, and

$$\begin{aligned} U_p &= U_i + U_a - U_w \\ &= \int_A \frac{\sigma^2}{2E} dA + \frac{\pi\sigma^2 a^2}{E} - \int_\delta F dy \end{aligned} \tag{7.10}$$

for the constant load condition.

It must be observed that U_w must be subtracted from the reversible energy terms, since it does not constitute part of the potential energy of the plate.

Finally, the fourth component corresponds to the increase in the elastic surface energy due to the formation of crack surfaces, and is given by

$$U_\gamma = 4a\gamma_e \tag{7.11}$$

where γ_e is the elastic surface energy of the material per unit thickness. Equation (7.11) implies that U_γ varies linearly with crack length a, since γ_e is assumed to be a material property. Physically, this is equivalent to stating that the energy required to break two atomic bonds ahead of a crack tip is twice that needed to break a single bond. Clearly, this form of energy is irreversible and therefore does not constitute part of the internal or potential energy of the examined system.

We are now in a position to write down an expression for the total potential energy of a cracked plate:

$$U_t = \int_A \frac{\sigma^2}{2E} dA \pm \frac{\pi\sigma^2 a^2}{E} - \int_\delta F dy + 4a\gamma_e \tag{7.12}$$

The respective linear and parabolic variations of U_γ and U_a, r together with the variation of the sum as a function of an introduced crack length $2a$ under the fixed grip condition are shown in Figure 7.14a and of the derivative $\partial U_t/\partial a$ in Figure 7.14b.

Let us now consider the details of a crack in equilibrium under the fixed grip condition. In this case,

$$\frac{\partial U_t}{\partial a} = \frac{\partial}{\partial a}\left[-\frac{\pi\sigma^2 a^2}{E} + 4a\gamma_e\right] = 0 \tag{7.13}$$

since

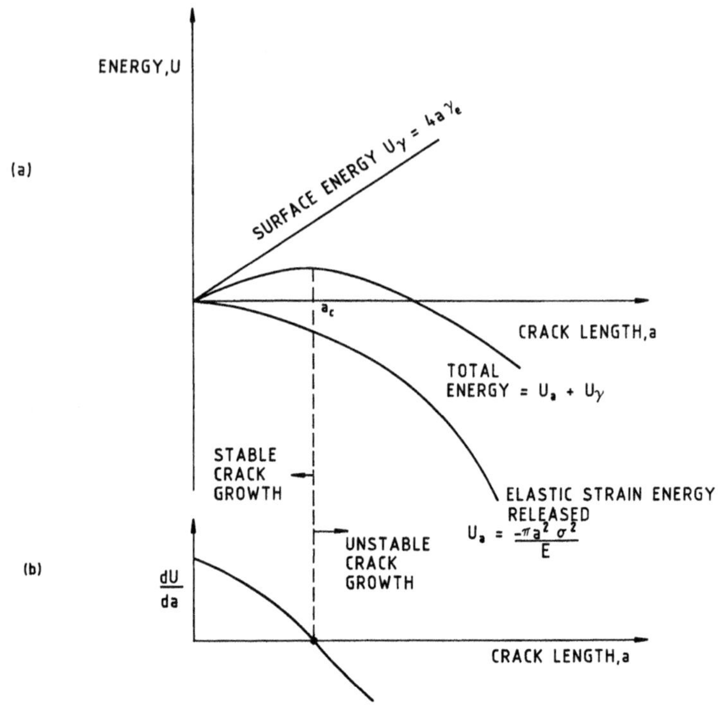

Figure 7.14 Griffith's energy balance approach: (**a**) variation of elastic strain energy release U_a versus crack length; (**b**) change in total energy per unit crack extension against crack length

$$\frac{\partial U_i}{\partial a} = 0$$

It is evident that increasing the crack length in the region $0 \leq a \leq a_c$ requires an energy input to the system, i.e. stable crack growth. However, for $a \geq a_c$ the strain energy released in an incremental crack extension exceeds the energy needed to create new crack surfaces, thus leading to unstable propagation.

Clearly, crack instability is associated with the stationary value of the total potential energy curve depicted i Figure 7.14b. At equilibrium,

$$\left.\begin{aligned} \sigma\sqrt{\pi a} &= \sqrt{2E\gamma_e} && \text{for plane stress} \\ \sigma\sqrt{\pi a} &= \sqrt{2E\gamma_e/(1-\nu^2)} && \text{for plane strain} \end{aligned}\right\} \tag{7.14}$$

The above equations indicate that a stable and unstable crack propagation in ideally brittle materials is governed by the remotely applied stress σ and the square-root of crack length a on the left-hand side and by material properties governed by E and γ_e on the right-hand side. Consequently, the above equation indicates that critical or unstable crack extension in such materials occurs when the product $\sigma\sqrt{\pi a}$ attains a critical value. For the moment, let us call that product the critical stress intensity factor, $K_c = \sigma_c\sqrt{\pi a}$, where σ_c is the fracture stress and is given by

$$\left.\begin{aligned} \sigma_c &= \sqrt{\frac{2E\gamma_e}{\pi a}} && \text{for plane stress} \\ \sigma_c &= \sqrt{\frac{2E\gamma_e}{\pi a(1-\nu^2)}} && \text{for plane strain} \end{aligned}\right\} \tag{7.15}$$

It should be observed from Equation (7.15) that the critical stress (the fracture stress) is inversely proportional to the square root of the effect size and is no longer a material property.

This expression was given incorrectly by Griffith as,

$$\sigma_c = \sqrt{\frac{2E\gamma_e}{\pi a}(1-\nu^2)}$$

Expressions (7.14) can also be written in the following form,

$$\left.\begin{aligned} \frac{\pi\sigma^2 a}{E} &= 2\gamma_e && \text{for plane stress} \\ \frac{\pi\sigma^2 a}{E}(1-\nu^2) &= 2\gamma_e && \text{for plane strain} \end{aligned}\right\} \tag{7.16}$$

where the left-hand side of Equation (7.16) represents the elastic energy per unit crack surface, which is available for crack extension G, while the right-hand side represents the elastic surface energy of crack surfaces for plane stress and plane strain, respectively. This is the energy needed for incremental crack extension and it is a measure of the crack resistance R. Thus, a condition for unstable crack growth exists such that

$$G_c \geq R \tag{7.17}$$

where

$$\left.\begin{aligned} G_c &= \frac{\pi\sigma_c^2 a}{E} && \text{for plane stress conditions} \\ G_c &= \frac{\pi\sigma_c^2 a}{E}(1-\nu^2) && \text{for plane strain conditions} \\ R &= \frac{\partial U_\gamma}{\partial a} && \end{aligned}\right\} \tag{7.18}$$

For the constant load condition, as explained earlier, expression (7.12) will apply with a positive U_a, i.e.

$$U_t = U_i + U_a - U_w + U_\gamma \tag{7.19}$$

Since crack growth instability occurs when

$$\frac{\partial U_t}{\partial a} \leq 0,$$

Accordingly,

$$\frac{\partial}{\partial a}\left(U_a - U_w + U_\gamma\right) \leq 0$$

or alternatively,

$$\frac{\partial}{\partial a}(U_w - U_a) \geq \frac{\partial U_\gamma}{\partial a} \tag{7.20}$$

i.e. $G_c \geq R$.

In the next chapter, we intend to demonstrate the equivalence of fixed grip and constant load conditions. It is also our intention to determine the relationship between G and $\sigma\sqrt{\pi a}$. However, we are now in a position to define G as the energy release rate, and this could be either the strain energy release rate or the potential energy release rate, depending upon whether we are dealing with fixed-grip or constant-load conditions, respectively. Thus G can be defined as the energy released per unit extension of crack front per unit thickness of the body, which is also termed the crack driving force. It must be noted that the work rate here has no reference to time.

7.5.2 Validity of Griffith's Approach

In order to test his theory, Griffith carried out a series of experiments on tubes and spherical vessels made from soda lime glass which were pre-cracked and then annealed to eliminate any residual stresses associated with the pre-cracking process. These vessels were then subjected to internal pressure. Now, by recording the crack size a and stress at fracture a_c, Griffith was able to determine the critical value $\sigma_c\sqrt{a}$ as being within the range 0.25 – 0.28 MPa$\sqrt{\text{m}}$. These values were checked against the expression relating to the surface energy γ_e, i.e. $\sqrt{2\gamma_e E/\pi}$.

Griffith argued that the surface tension of glass is a linear function of temperature. As a result of this argument, surface tension values of γ_e for glass fibers between 1110 ° C and 745 ° C were extrapolated to obtain room temperature γ_e of 0.54 N/m. Taking $E = 62$ GPa, the value of $\sqrt{2\gamma_e E/\pi}$ was found to be 0.15 MPa$\sqrt{\text{m}}$.

Accordingly, with some reservations, the above leads us to conclude that Griffith's theory is valid for materials with little or no preceding plastic deformation near the crack tip. It must be pointed out, however, that most engineering materials sustain limited plasticity at the crack tip. Accordingly, Griffith's expression, if applied as it is, would ultimately give higher values for the thermodynamic surface energy term γ_e. Some modifications must be made to Griffith's expression. This is dealt with under the next heading.

It is also evident from the above that the fracture strength of a given brittle component is dependent upon flaw size. Thus, if we desire to produce a very strong brittle material, we should reduce the inherent flaw size to a minimum. Obviously, one way of achieving this is by reducing the cross-section normal to the applied tensile load. This effectively means that we are reducing the probability of finding a large crack normal to the applied load. This also explains the high fracture strengths attainable in whiskers and fibers.

7.6 Irwin-Orowan Postulate

Thus far, we have treated materials which are ideally brittle and accordingly behave in a purely elastic fashion up to fracture. In practice, however, most engineering materials contain some ductility and accordingly some plastic deformation may be encountered. This will necessitate the introduction of some modifications to the Griffith theory outlined above.

Irwin and Orowan extended Griffith's theory for ideally brittle materials to account for the limited plastic behavior encountered near the crack tip in most engineering materials. Irwin and Orowan postulated that the resistance to crack extension in truly engineering materials is equal to the sum of the elastic surface energy γ_e and the plastic dissipation or plastic work γ_p accompanying crack extension. Consequently, expression (7.16) was modified to

$$\frac{\pi\sigma^2 a}{E} = 2(\gamma_e + \gamma_p) \tag{7.21}$$

For ductile materials γ_p is much greater than γ_e and the resistance to cracking R, in this case, is mainly governed by plastic dissipation. This means that for metals R is mainly plastic energy. The above energy criterion is necessary, but need not be sufficient for crack extension.

Orowan stipulated that strict conditions should be imposed on situations where the above modification is applicable. He argued that the modified approach would only be valid if plastic deformation was confined to a limited zone in the vicinity of the crack tip.

References

7.1 D. Doyle, *Liberty Ships: America's Merchant Marine Transport in World War II*, 1st edn. (Schiffer Military History, Atglen, 2020)
7.2 J. Hodgson, G.M. Boyd, Brittle fracture in welded ships. Trans. Inst. Naval Archit. **100**, 141–180 (1958)
7.3 G.M. Boyd, *Brittle Fracture in Steel Structures* (Butterworths, London, 1970)
7.4 C.F. Tipper, *The Brittle Fracture Story* (Cambridge University Press, Cambridge, 1962)
7.5 A. Necci, S. Tarantola, B. Vamanu, E. Krausmann, L. Ponte, Lessons learned from offshore oil and gas incidents in the arctic and other ice-prone seas. Ocean Eng. **185**, 12–26 (2019). https://doi.org/10.1016/j.oceaneng.2019.05.021

Chapter 8
Linear Elastic Fracture Mechanics

Summary In this chapter, we shall cover the concepts governing the linear elastic fracture mechanics (LEFM) behavior of a continuum. LEFM, as it is known, is concerned with the analysis of cracks in the linear elastic regime. It provides the conceptual, theoretical and experimental tools for characterizing the behavior of cracks in elastic bodies. Unlike perfect elasticity, where the stress state and the stress concentration factor are of concern, in LEFM, both the stress state and the defect size and orientation govern the failure behavior of a continuum. Accordingly, LEFM is a powerful tool that will allow us to determine the safety margins, techniques in crack arrest and life expectancy of a continuum containing defects.

8.1 Introduction

Even when our lives are at risk, we are unable to produce defect-free structures. Fracture mechanics is the applied mechanics framework necessary for describing the behavior of such cracked and severely notched machine components due to the applied thermo-mechanical loads and the environment.

Undoubtedly, great strides have recently been made in the field of fracture mechanics, which have resulted in introducing new design criteria based upon a clearer appreciation of the effect of cracks and crack-like defects on the mechanical integrity of machine components and structures and our ability to monitor their growth accurately.

Fracture mechanics is thus used to formulate quantitatively.

(i) the degree of safety a structure possesses against failure by brittle fracture,
(ii) the conditions necessary for crack initiation, propagation and arrest, and
(iii) the residual life in a machine component subjected to dynamic loading.

Accordingly, fracture mechanics defines the rate at which a crack or a crack-like defect develops before failure in terms of either time or cycles. Such assessments are increasingly used in industry.

S. A. Meguid, *Atomistic and Continuum Fracture Mechanics of Solids*,
https://doi.org/10.1007/978-3-031-56085-9_8

8.2 Modes of Crack-Tip Deformation

Within the framework of LEFM, the stress field in the vicinity of a crack-tip is dependent upon crack length, applied stress σ and a configuration factor Q which accounts for the finite size of the component and its local geometric features. This concept, which utilizes what is known as the stress intensity factor, was developed by Irwin in the 1950s as an extension to Griffith's pioneering work in the 1920s [8.1, 8.2].

Irwin showed that the elastic stress field in the vicinity of a crack can be given by the following series representation of the stress tensor σ_{ij}:

$$\sigma_{ij} = r^{-1/2}\Big\{K_{\rm I}f^{\rm I}_{ij}(\theta) + K_{\rm II}f^{\rm II}_{ij}(\theta) + K_{\rm III}f^{\rm III}_{ij}(\theta) + \text{higher order terms} \qquad (8.1)$$

where r *and* θ are the cylindrical polar co-ordinates defining the position of an element ahead of the crack-tip, as depicted in Figure 8.1; with $K_{\rm I}$, $K_{\rm II}$ and $K_{\rm III}$ are denoted the stress intensity factors corresponding to the three basic modes of crack surface displacement. These modes, which are depicted in Figure 8.2, are known as (i) the opening mode, where the crack surfaces move directly apart, conventionally denoted by the roman numeral I, (ii) the sliding or in-plane shear mode, where the crack surfaces slide over one another in a direction perpendicular to the leading edge of the crack, and (iii) the tearing, or anti-plane shear mode, where the crack surfaces move relative to one another parallel to the leading edge of the crack. Careful superposition of these three modes is sufficient to describe the most general case of crack surface displacement.

It is worth noting that in spite of the fact that many practical problems are of the mixed mode type, the bulk of fracture mechanics work has been devoted to the single mode analysis—primarily mode 1. This is due partly to the simplicity of its application and partly to its severity on the component's resistance to cracking. It is also

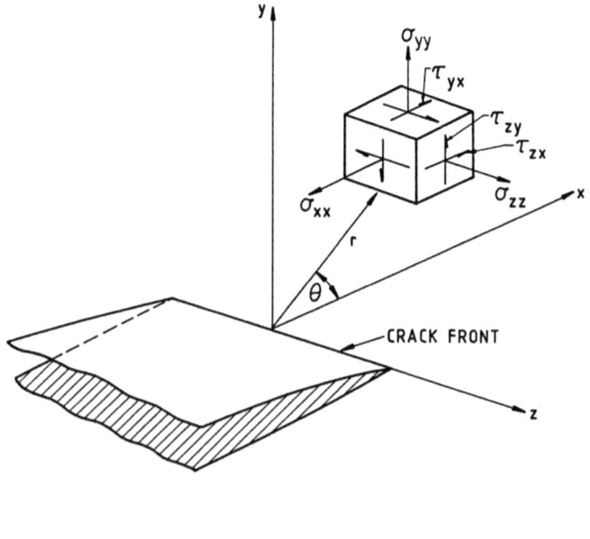

Figure 8.1 Crack-tip co-ordinates and stress state

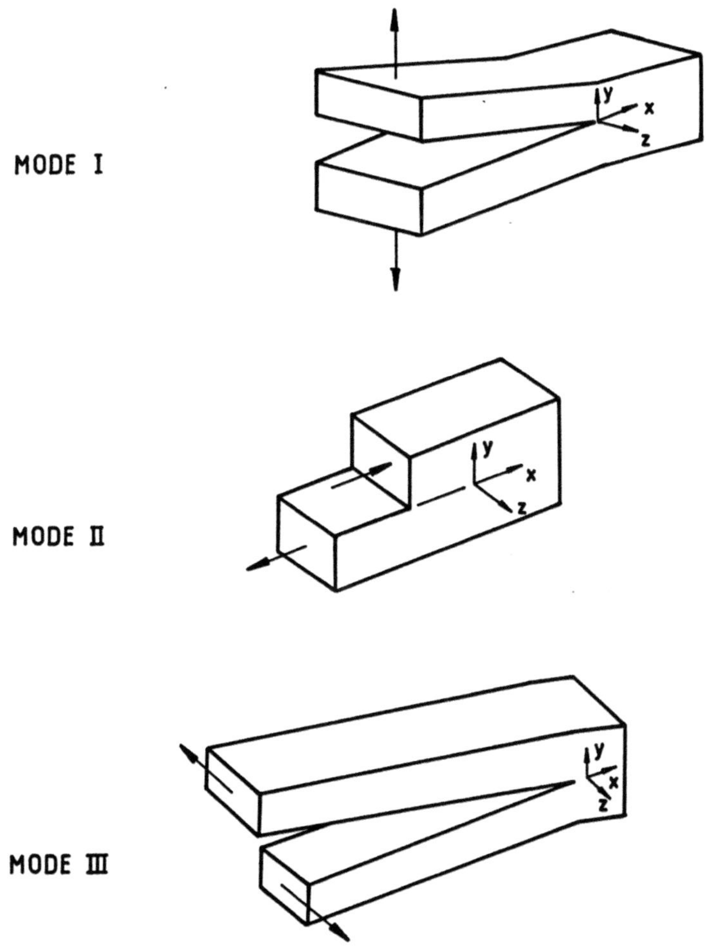

Figure 8.2 Basic modes of loading and the associated crack surface displacements

worth mentioning that cracks that start in a single mode, may in fact become mixed mode later on during their useful life. Furthermore, actual cracks are generally irregular in shapes, and thus a combination of more than one mode may actually exist at any one time.

8.3 Elastic Stress Field Equations: Westergaard Stress Function

A special case of the complex potential formulation of Muskhelishvili for problems possessing x-axis symmetry for both loading and geometry ($\tau_{xy} = 0$ along $y = 0$) is the so-called Westergaard stress function Z, where

$$\phi_{\mathrm{I}} = \mathrm{Re}\ \overline{\overline{Z}}_{\mathrm{I}} + y\,\mathrm{Im}\,\overline{Z}_{\mathrm{I}} \tag{8.2}$$

where

$$\overline{Z}_{\mathrm{I}} = \mathrm{d}\overline{\overline{Z}}_{\mathrm{I}}/\mathrm{d}z, Z_{\mathrm{I}} = \mathrm{d}\overline{Z}_{\mathrm{I}}/\mathrm{d}z \text{ and } z = x + iy$$

In this case, the subscript I indicates that the stress function represents opening mode or mode-I crack-tip deformation.

Now, since

$$\sigma_{xx} = \frac{\partial^2\phi}{\partial y^2}, \sigma_{yy} = \frac{\partial^2\phi}{\partial x^2}, \tau_{xy} = -\frac{\partial^2\phi}{\partial x\partial y}$$

then, in terms of the Westergaard stress function, these stresses become

$$\sigma_{xx} = \mathrm{Re}\ Z_{\mathrm{I}} - y\,\mathrm{Im}\,Z'_{\mathrm{I}} \tag{8.3}$$

$$\sigma_{yy} = \mathrm{Re}\ Z_{\mathrm{I}} + y\,\mathrm{Im}\,Z'_{\mathrm{I}} \tag{8.4}$$

$$\tau_{xy} = -y\ \mathrm{Re}\ Z'_{\mathrm{I}} \tag{8.5}$$

where

$$Z'_{\mathrm{I}} = \mathrm{d}Z_{\mathrm{I}}/\mathrm{d}z$$

The above stress function and the associated stresses are not sufficiently general, and require adjustment because of omission of a real constant in the stress function. This omission can be corrected by the inclusion of a real constant in the expressions for the stress function. However, for the purpose of the present analysis, the omission of that constant does not affect the results [8.2].

8.3.1 Opening Mode Analysis

The above complex function $Z_{\mathrm{I}}(z)$ is frequently used to solve two-dimensional problems in a cracked plate. Let us now consider the case of an infinite plate containing a crack 2a, which is subjected to a biaxial tensile stress σ_0, as shown in Figure 8.3.

It can be shown that a complex stress function ϕ of the form

$$\phi = \frac{\sigma_0}{\sqrt{z^2 - a^2}} \tag{8.6}$$

will satisfy the boundary conditions.

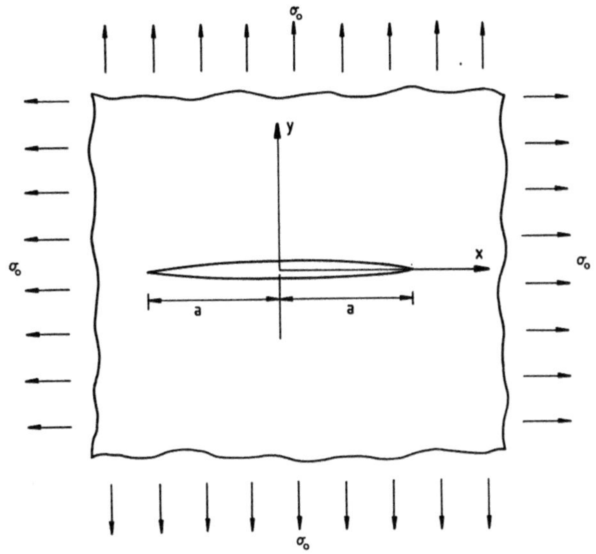

Figure 8.3 Mode I crack in an infinite plate under biaxial tension

$$\sigma_{yy} = 0 \text{ for } -a < x < +a$$
$$\sigma_{yy} = \sigma_0 \text{ for } x \to \infty$$
$$\sigma_{yy} \to \infty \text{for } x = \pm a$$

with

$$Z_{\mathrm{I}} = \frac{\sigma_0 z}{\sqrt{z^2 - a^2}} \tag{8.7}$$

being the Westergaard stress function for mode I.

For convenience, let us translate the origin of the co-ordinate axes of Figure 8.3 to the crack-tip by taking $\xi = z - a = re^{i\theta}$. Accordingly, expression (8.7) becomes

$$Z_{\mathrm{I}}(\xi) = \frac{\sigma_0(a + \xi)}{\sqrt{(a + \xi)^2 - a^2}} \tag{8.8}$$

A first-order approximation of the above expression for ($\xi \ll a$) yields

$$Z_{\mathrm{I}}(\xi) = \frac{\sigma_0 a}{\sqrt{2a\xi}} = \frac{\sigma_0 \sqrt{a}}{\sqrt{2\xi}} \tag{8.9}$$

or alternatively

$$Z_I(\xi) = \frac{\sigma_0\sqrt{\pi a}}{\sqrt{2\pi r}} e^{-i\theta/2} \tag{8.10}$$

Using the stress Equations (8.3), (8.4) and (8.5) we are now able to develop expressions for the stresses in the vicinity of the crack-tip since $\xi \ll a$, and for $r \ll a$. With the knowledge that $Z_I'(\xi) = -\sigma_0\sqrt{a}/2\sqrt{2}\xi^{3/2}$, these stresses can be expressed as

$$\begin{bmatrix} \sigma_{xx} \\ \sigma_{yy} \\ \tau_{xy} \end{bmatrix} = \frac{K_I}{\sqrt{2\pi r}} cos\frac{\theta}{2} \begin{bmatrix} 1 - sin\frac{\theta}{2}\, sin\frac{3\theta}{2} \\ 1 + sin\frac{\theta}{2}\, sin\frac{3\theta}{2} \\ sin\frac{\theta}{2}\, cos\frac{3\theta}{2} \end{bmatrix} \tag{8.11}$$

with $\sigma_z = \nu(\sigma_{xx} + \sigma_{yy})$ for plane strain and $\sigma_z = 0$ for plane stress.

In the above formulation, K_I is the opening mode of the stress intensity factor and is defined as

$$K_I = \lim_{r\to 0} \sigma_{yy}(r, \theta = 0)\sqrt{\pi r} \tag{8.12}$$

or, in terms of the applied stress σ_0 and crack length a, K_I is given by

$$K_I = \sigma_0\sqrt{\pi a}\ \left(\mathrm{MPa}\sqrt{\mathrm{m}}\right) \tag{8.13}$$

Examination of Equations (8.12) and (8.13) for a crack in an infinite plate reveals that the stress intensity factor has the dimensions (stress) × (length)$^{1/2}$ and is generally a function of the gross stress σ_0 and crack length a. In this case, it is considered a square root singularity. Knowledge of K enables the determination of the stress and displacement fields near the crack-tip. Expressions (8.11) show that the stresses are always inversely proportional to the square root of the distance from the crack-tip ($1/\sqrt{r}$) and become singular at the crack-tip. Indeed, developments in fracture mechanics have centered on the realization by Irwin that all elastic stress fields in the vicinity of a crack-tip are similar in geometric distribution and that the term $\sigma_{ij}\sqrt{\pi a}$ controls the magnitude of the local stress.

The physical significance of the stress intensity factor approach is that it can be used as a design parameter to determine e.g. the nature of the fracture process (ductile or brittle) or it can be used to determine the residual life in a component subjected to cyclic loading. It further characterizes subcritical crack growth due to fatigue. The term should not be confused with the stress concentration factor, which is a term used to describe the stress ratio of the maximum over the average or nominal stress at a geometric discontinuity. Furthermore, since the actual fracture process is principally affected by the stress and displacement conditions at the crack-tip, the concept of the stress intensity factor provides the mathematical framework needed for describing the elastic behavior of cracked bodies.

Table 8.1 Error estimate resulting from use of exact and approximate expressions for σ_{yy}

r/a (%)	0.5	1	8	16	30	60
E (%)	0.37	0.74	5.6	10.6	17.5	28.7

In view of the fact that higher order terms were neglected in expressions (8.9) and (8.10), the stress Equation (8.11) are only valid for small values of r at or near the vicinity of a crack-tip. In order to demonstrate the difference between the exact and the approximate solutions, let us compare the exact expression of σ_{yy}^{e} ahead of the crack-tip, as given by expression (8.7); viz.,

$$\sigma_{yy}^{e} = \frac{\sigma_0}{\sqrt{1 - (a/x)^2}} \quad \text{for } x \geq a$$

with the approximate expression of σ_{yy}^{a} as given by expression (8.11).

$$\sigma_{yy}^{a} = \frac{\sigma_0 \sqrt{\pi a}}{\sqrt{2\pi r}} = \frac{\sigma_0 \sqrt{\pi a}}{\sqrt{2\pi (x - a)}} \quad \text{for } x \geq a$$

along the y-axis, i.e. $\theta = 0$. In this case, the percentage error E given by

$$E = \frac{\sigma_{yy}^{e} - \sigma_{yy}^{a}}{\sigma_{yy}^{e}} \times 100 \tag{8.14}$$

provides some of the results calculated for different values of r/a. Clearly, the error increases with increased r/a; approaching 29% when $r = 0.6a$ (Table 8.1).

The displacements u and υ can be similarly obtained from the complex stress potentials, viz.,

$$u = \frac{K_{\mathrm{I}}}{2G} \sqrt{\frac{r}{2\pi}} cos \frac{\theta}{2} \left[\kappa - 1 + 2 \ sin^2 \frac{\theta}{2} \right] \tag{8.15}$$

and the elastic displacement υ along the y-axis is given by

$$\upsilon = \frac{K_{\mathrm{I}}}{2G} \sqrt{\frac{r}{2\pi}} sin \frac{\theta}{2} \left[\kappa - 1 + 2 \ cos^2 \frac{\theta}{2} \right] \tag{8.16}$$

where

$$\begin{aligned} \kappa &= 3 - 4\nu \quad \text{for plane stress} \\ \kappa &= \frac{3 - \nu}{1 + \nu} \quad \text{for plane strain} \end{aligned} \tag{8.17}$$

with G being the shear modulus and ν is Poisson's ratio.

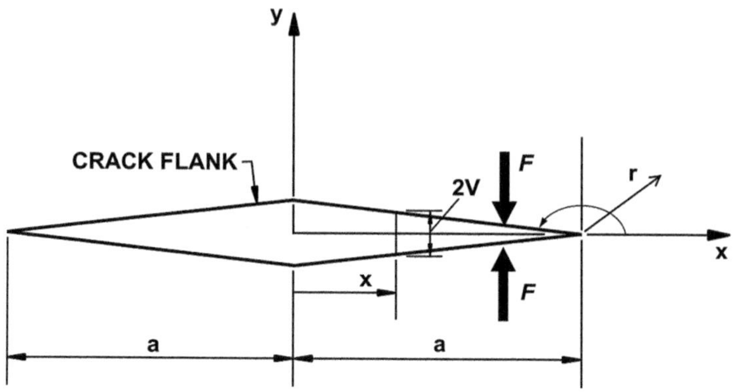

Figure 8.4 Crack flank displacement

A more general expression for the crack flank displacement at any position along the x-axis, Figure 8.4, can be obtained from expression (8.17), as follows:

$$\left.\begin{aligned} v &= \frac{2\sigma}{E}\left(1-\nu^2\right)\sqrt{a^2-x^2} && \text{for plane strain} \\ v &= \frac{2\sigma}{E}\sqrt{a^2-x^2} && \text{for plane stress} \end{aligned}\right\} \tag{8.18}$$

Having established the states of stress and displacement in the vicinity of a crack-tip under mode I loading, the strain-energy release U_a due to the introduction of a crack can now be determined. In this case, the elastic strain energy needed to open up the crack flanks is simply the difference between the strain energy of the uncracked and the cracked plates, i.e.

$$U_a = \frac{1}{2}F(2v) \tag{8.19}$$

Where F is the load required to cause a total crack flank displacement ($2v$). Or alternatively,

$$\begin{aligned} U_a &= \int_{-a}^{a} \sigma \frac{2\sigma}{E}\left(1-\nu^2\right)\sqrt{a^2-x^2}\,\mathrm{d}x \\ &= \frac{\pi\sigma^2}{E}\,a^2\left(1-\nu^2\right) \qquad \text{for plane strain} \end{aligned} \tag{8.20}$$

$$\begin{aligned} U_a &= \int_{-a}^{a} \sigma \frac{2\sigma}{E}\sqrt{a^2-x^2}\,\mathrm{d}x \\ &= \frac{\pi\sigma^2 a^2}{E} \qquad \text{for plane stress} \end{aligned} \tag{8.21}$$

Leading to

$$G_{\mathrm{I}} = \frac{K_{\mathrm{I}}^2}{E}\left(1 - \nu^2\right) \qquad \text{for plane strain} \tag{8.22}$$

and

$$G_{\mathrm{I}} = \frac{K_{\mathrm{I}}^2}{E} \qquad \text{for plane stress} \tag{8.23}$$

Expressions (8.20), (8.21), (8.22) and (8.23) demonstrate clearly the equivalence between the strain energy release rate or the crack driving force G_{I} and the stress intensity factor K_{I}.

8.3.2 Sliding Mode Analysis

A similar approach to mode-I Westergaard can be adopted for the determination of the stress and displacement components for mode-II. In this case, Airy's stress function for the sliding mode takes the form

$$\phi_{\mathrm{II}} = -y \ \mathrm{Re}\ \bar{Z}_{\mathrm{II}} \tag{8.24}$$

Using Airy's definition of the stress components (σ_{xx}, σ_{yy}, τ_{xy}) we get

$$\left.\begin{aligned} \sigma_{xx} &= \frac{\partial^2 \phi}{\partial y^2} = 2\ \mathrm{Im}\ Z_{\mathrm{II}} + y\ \mathrm{Re}\ Z'_{\mathrm{II}} \\ \sigma_{yy} &= \frac{\partial^2 \phi}{\partial x^2} = -y\ \mathrm{Re}\ Z'_{\mathrm{II}} \\ \tau_{xy} &= -\frac{\partial^2 \phi}{\partial x \partial y} = \mathrm{Re}\ Z_{\mathrm{II}} - y\ \mathrm{Im}\ Z'_{\mathrm{II}} \end{aligned}\right\} \tag{8.25}$$

In this case, Westergaard stress function of the form

$$Z_{\mathrm{II}} = \frac{\tau_0 z}{\sqrt{z^2 - a^2}} \tag{8.26}$$

will satisfy the boundary conditions

$$\left.\begin{aligned} \tau_{yx} &= 0 && \text{for} -a < x < +a \\ \tau_{yx} &= \tau_0 && \text{for } y \rightarrow \infty \\ \tau_{yx} &\rightarrow \infty && \text{for } x = \pm a \end{aligned}\right\}$$

In the above expressions τ_0 is the applied boundary shear stress.

For a limiting value of r, the stress and displacement components in terms of the local polar co-ordinate systems are

$$\begin{bmatrix} \sigma_{xx} \\ \sigma_{yy} \\ \tau_{xy} \end{bmatrix} = \frac{K_{\mathrm{II}}}{\sqrt{2\pi r}} \cos\frac{\theta}{2} \begin{bmatrix} \cos\frac{\theta}{2}\left(2+\cos\frac{\theta}{2}\cos\frac{3\theta}{2}\right) \\ \sin\frac{\theta}{2}\cos\frac{3\theta}{2} \\ 1-\sin\frac{\theta}{2}\sin\frac{3\theta}{2} \end{bmatrix} \tag{8.27}$$

For the state of plane strain, the displacement components are

$$\begin{bmatrix} u \\ v \end{bmatrix} = \frac{K_{\mathrm{II}}}{G}\sqrt{\frac{r}{2\pi}} \begin{bmatrix} \sin\frac{\theta}{2}\left(2-2\nu+\cos^2\frac{\theta}{2}\right) \\ \cos\frac{\theta}{2}\left(1-2\nu+\sin^2\frac{\theta}{2}\right) \end{bmatrix} \tag{8.28}$$

For the state of the plane stress, they become

$$\begin{bmatrix} u \\ v \end{bmatrix} = \frac{K_{\mathrm{II}}}{G}\sqrt{\frac{r}{2\pi}} \begin{bmatrix} \sin\frac{\theta}{2}\left(\frac{2}{1+\nu}+\cos^2\frac{\theta}{2}\right) \\ \cos\frac{\theta}{2}\left(\frac{1-\nu}{1+\nu}+\sin^2\frac{\theta}{2}\right) \end{bmatrix} \tag{8.29}$$

where the stress intensity factor for mode-II crack-tip displacement is given by

$$K_{\mathrm{II}} = \lim_{r\to 0} \tau_{xy}(r,\theta=0)\sqrt{2\pi r} \tag{8.30}$$

The strain energy release rate for mode-II, G_{II}, takes on an identical form to that given by expression (8.21) and (8.22), i.e.

$$\left.\begin{aligned} G_{\mathrm{II}} &= \frac{K_{\mathrm{II}}^2}{E}\left(1-\nu^2\right) && \text{for plane strain} \\ G_{\mathrm{II}} &= \frac{K_{\mathrm{II}}^2}{E} && \text{for plane stress} \end{aligned}\right\} \tag{8.31}$$

8.3.3 Tearing Mode Analysis

In this case, as depicted in Figure 8.2, the crack is displaced along the z-axis only. This simplifies matters considerably, since the other displacement components are set to zero; accordingly,

$$\left.\begin{aligned}\varepsilon_{xx} = \varepsilon_{yy} = 0 \text{ and } \gamma_{xy} = \gamma_{yx} = 0 \\ \gamma_{xz} = \gamma_{zx} = \frac{\partial w}{\partial x} \text{ and } \gamma_{yz} = \gamma_{zy} = \frac{\partial w}{\partial y}\end{aligned}\right\} \tag{8.32}$$

Considering an element near the crack-tip, the equilibrium equation gives

$$\frac{\partial \tau_{xz}}{\partial x} + \frac{\partial \tau_{yz}}{\partial y} = 0 \tag{8.33}$$

Evoking the stress-strain relations and replacing τ_{xz} and τ_{yz} by using the strain displacement relations, we get

$$\frac{\partial^2 w}{\partial x^2} + \frac{\partial^2 w}{\partial y^2} = \nabla^2 w = 0 \tag{8.34}$$

If w is taken as

$$w = \frac{1}{G} \text{ Im } Z_{\text{III}} \tag{8.35}$$

then

$$\left.\begin{aligned}\tau_{xz} &= \text{Im } Z'_{\text{III}} \\ \tau_{yz} &= \text{Re } Z'_{\text{III}}\end{aligned}\right\} \tag{8.36}$$

Our task now is to select a stress function Z_{III} capable of satisfying the following boundary conditions:

$$\left.\begin{aligned}&\tau_{yz} = 0 && \text{for } -a < x < a \\ &\tau_{yz} = \tau && \text{for } y \to \infty \\ &\tau_{yz} \to \infty && \text{for } x = \pm a\end{aligned}\right\} \tag{8.37}$$

Again, this function is taken as

$$Z_{\text{III}} = \frac{\tau z}{\sqrt{z^2 - a^2}} \tag{8.38}$$

where τ is the applied boundary shear stress.

In this case, the stresses and displacements are given by the expressions

$$\left.\begin{aligned}\begin{bmatrix}\tau_{xz} \\ \tau_{yz}\end{bmatrix} = \frac{K_{\text{III}}}{\sqrt{2\pi r}} \left[sin\ \theta/2\, cos\, \theta/2 \right] \\ \sigma_{xx} = \sigma_{yy} = \tau_{xy} = 0\end{aligned}\right\} \tag{8.39}$$

and

$$\left.\begin{aligned} w &= \frac{K_{\mathrm{III}}}{G}\sqrt{\frac{2r}{\pi}}\, sin\,\frac{\theta}{2} \\ u &= v = 0 \end{aligned}\right\} \tag{8.40}$$

where the stress intensity factor, K_{III}, can be determined from the now familiar form of

$$K_{\mathrm{III}} = \lim_{r\to 0}\tau_{yz}(r,\theta=0)\sqrt{2\pi r} \tag{8.41}$$

8.4 Superposition of Stress Intensity Factors

The approach adopted so far is based upon the theory of elasticity and thus the stress functions for other boundary value problems with identical geometries and crack mode deformation can be superimposed. Accordingly, the corresponding stress intensity factors for the same mode of crack deformation can be superimposed, as follows:

$$\left.\begin{aligned} K_{\mathrm{I}} &= K_{\mathrm{I}}^{(1)} + K_{\mathrm{I}}^{(2)} + K_{\mathrm{I}}^{(3)} + \cdots + K_{\mathrm{I}}^{(l)} \\ K_{\mathrm{II}} &= K_{\mathrm{II}}^{(1)} + K_{\mathrm{II}}^{(2)} + K_{\mathrm{II}}^{(3)} + \cdots + K_{\mathrm{II}}^{(m)} \\ K_{\mathrm{III}} &= K_{\mathrm{III}}^{(1)} + K_{\mathrm{III}}^{(2)} + K_{\mathrm{III}}^{(3)} + \cdots + K_{\mathrm{III}}^{(n)} \end{aligned}\right\} \tag{8.42}$$

where l, m and n are integers.

To illustrate this, let us consider a crack in an infinite plate subjected to a remote stress σ_0 and a point wedging load F per unit thickness, as shown in Figure 8.5. For the remotely applied stress σ_0, the stress intensity factor $K_{\mathrm{I}}^{(\mathrm{a})} = \sigma_0\sqrt{\pi a}$. For the point-wedging load F, we proceed by utilizing the appropriate Westergaard function.

$$Z_{\mathrm{I}} = \frac{Fa}{\pi z\sqrt{z^2 - a^2}} \tag{a}$$

In this case

$$\sigma_{yy} = \mathrm{Re}\ Z_{\mathrm{I}} = \frac{Fa}{\pi x\sqrt{x^2 - a^2}} \tag{b}$$

Figure 8.5 Combined loading of a crack in a single opening mode deformation

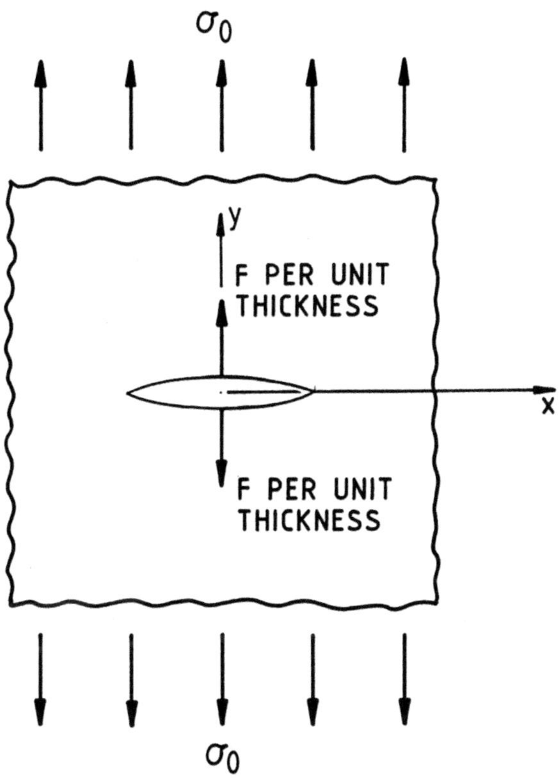

Adopting an approach similar to the one described earlier in the development of the stress intensity factors expressions for the different modes, we let $x = r + a$, where $r \ll a$. Therefore, we may write

$$\sigma_{yy} = \frac{F}{\pi\sqrt{a}}\sqrt{\frac{1}{2r}} = \frac{K_{\mathrm{I}}^{(\mathrm{b})}}{\sqrt{2\pi r}} \tag{c}$$

where

$$K_{\mathrm{I}}^{(\mathrm{b})} = \frac{F}{\sqrt{\pi a}} \tag{d}$$

Consequently, the effective stress intensity factor is the sum of both $K_{\mathrm{I}}^{(\mathrm{a})}$ and $K_{\mathrm{I}}^{(\mathrm{b})}$; viz.,

$$\begin{aligned} K_{\mathrm{I}} &= K_{\mathrm{I}}^{(\mathrm{a})} + K_{\mathrm{I}}^{(\mathrm{b})} \\ K_{\mathrm{I}} &= \sigma\sqrt{\pi a} + \frac{F}{\sqrt{\pi a}} \end{aligned} \tag{e}$$

However, if a crack extends under combined-mode loading, then we may adopt, the strain energy release rate for our computations. Accordingly, the resultant strain energy release rate due to different modes of crack deformation can be superimposed as

$$\left.\begin{aligned} G &= G_{\mathrm{I}} + G_{\mathrm{II}} + G_{\mathrm{III}} \\ &= \frac{1-\nu^2}{E}\left(K_{\mathrm{I}}^2 + K_{\mathrm{II}}^2 + K_{\mathrm{III}}^2\right) \quad \text{for plane strain} \\ &= \frac{1}{E}\left(K_{\mathrm{I}}^2 + K_{\mathrm{II}}^2 + K_{\mathrm{III}}^2\right) \quad \text{for plane stress} \end{aligned}\right\} \tag{8.43}$$

It is worth mentioning that by using the superposition principle, the stress intensity factor for a number of seemingly complicated problems can be readily obtained. To illustrate this, let us consider the case of the crack under internal pressure. The superposition technique adopted here is described in Figure 8.6. The solution in this case is obtained by realizing that

$$K_{\mathrm{I}}^{\mathrm{b}} + K_{\mathrm{I}}^{\mathrm{c}} = 0$$

Accordingly

$$K_{\mathrm{I}}^{\mathrm{c}} = -K_{\mathrm{I}}^{\mathrm{b}} = -\sigma\sqrt{\pi a}$$

Clearly, reversal of the stress sign will provide the required result, i.e. the stress intensity factor for a crack under internal pressure is $\sigma\sqrt{\pi a}.$

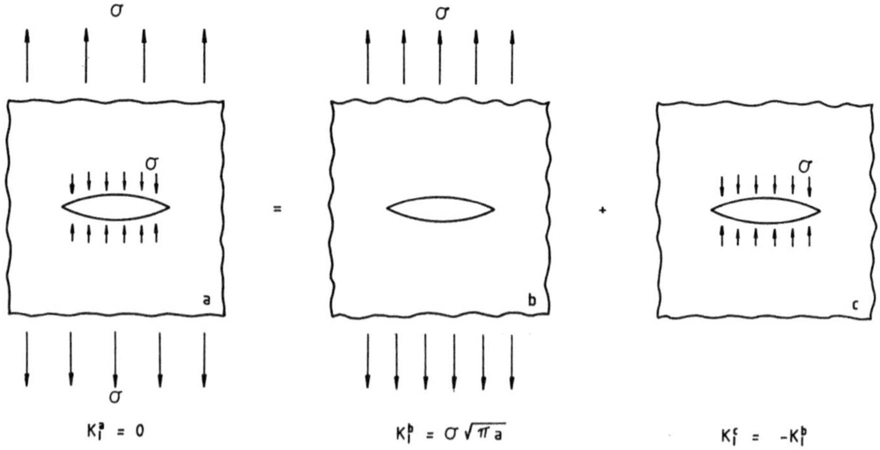

Figure 8.6 Superposition principle for a through-thickness crack in an infinite plate under internal pressure

The above example further demonstrates two important points. The first is that the stress intensity factor for a crack in an infinite plate under remote uniaxial stress σ is the same as the stress intensity factor for a crack subjected to internal pressure of the same magnitude σ. The second is that the stress intensity factor can be derived from either the remote boundary stress or a loaded crack, where the crack loading is equal and opposite to the stresses which exist in an equivalently loaded uncracked plate. Westergaard stress functions have also been used to evaluate the stress intensity factors for several periodic cracks.

8.5 Effect of Finite Size of Component

The stress intensity factor expressions provided above were for an infinite body. Clearly, the finite size of the cracked body is expected to have an influence upon crack-tip stress field. Accordingly, the expressions for the stress intensity factor have to be modified to account for this effect. A more general expression for the stress intensity factor may take the form,

$$K_{\mathrm{M}} = \sigma_0 \sqrt{\pi a}\, f(a/W), \mathrm{M} = \mathrm{I, II \text{ and } III} \tag{8.44}$$

where σ_0 is the applied remote stress, a is a characterizing crack length, w is the width and $f(a/W)$ is a configuration factor which accounts for proximity effects of the boundary surfaces. It is with the determination of this normalized stress intensity factor that most crack analysis methods are concerned. It is worth mentioning, however, that there exists very few closed form solutions to Equation (8.44); most expressions are obtained by empirical and numerical approximations. These are detailed in Section 8.8.

For example, let us consider the center-crack problem. A number of solutions for the finite width plate of Figure 8.7a exist. These are summarized in Table 8.2. Clearly, for large values of W, i.e. $a/W \ll 1$, the configuration correction factors detailed in Table 8.2 reduce to unity and we attain, the infinite plate solution for K_{I}.

Consideration of single and double edge-cracked plates, shown in Figure 8.7b, c, reveals that similar correction factors must be adopted in their analysis. If we consider uniaxial loading, then a non-singular term σ_0 must be subtracted from σ_{xx} expression in Equation (8.11). However, for $-a \le x \le a$, expression (8.11) for σ_{xx} reduces to zero, and σ_0 is no longer negligible. In this case, $\sigma_{xx} = -\sigma_0$. The effects of these stresses have been estimated to increase the stress intensity factor for short edge cracks by about 12%, i.e.

$$K_{\mathrm{I}} = 1.12\sigma_0\sqrt{\pi a} \tag{8.45}$$

The correction factor can also be obtained analytically using the collocation method.

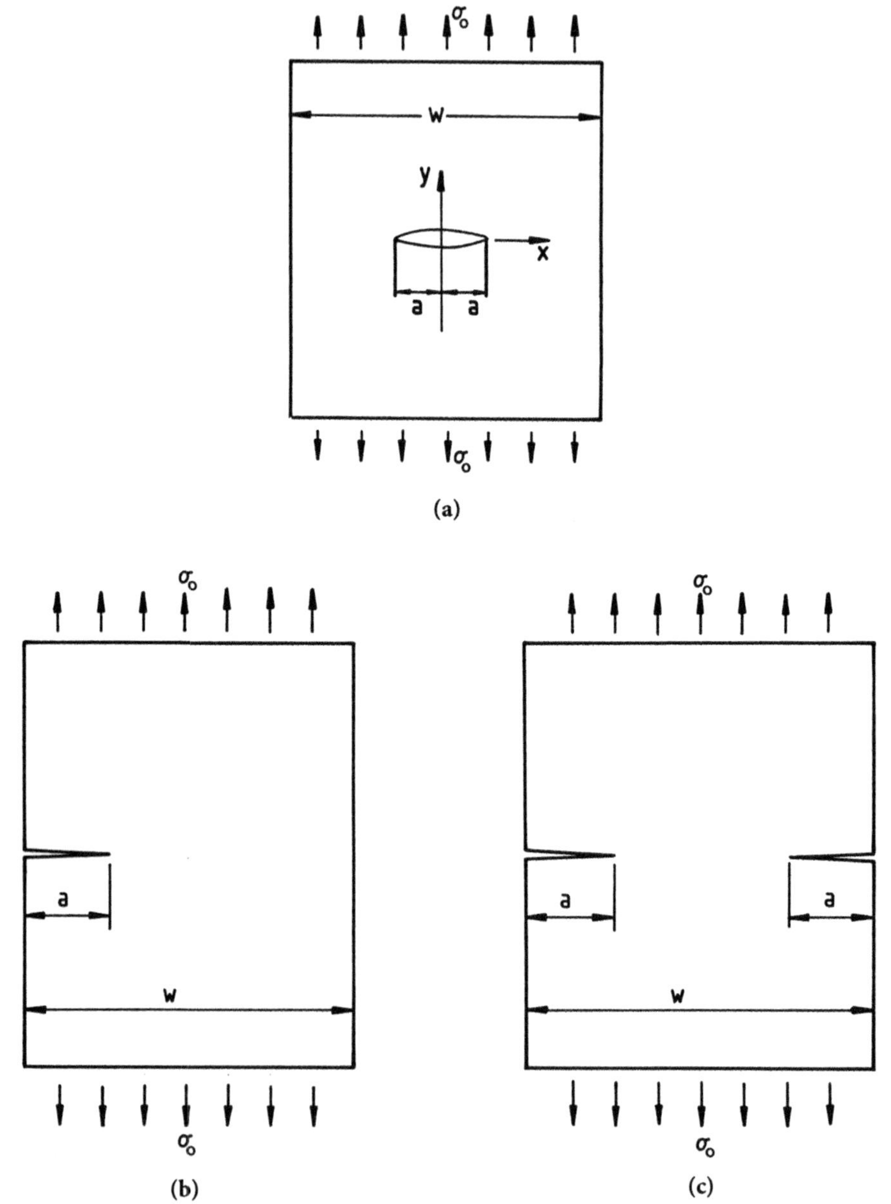

Figure 8.7 (**a**) Finite width center cracked specimen. (**b**) Single edge-cracked specimen of a finite width. (**c**) Double edge-cracked specimen of a finite width

For longer single-edge cracks, a more accurate correction factor is given by:

$$f(a/\mathrm{W}) = 1.12 - 0.231(a/\mathrm{W}) + 10.55(a/\mathrm{W})^2 - 21.72(a/W)^3 + 30.39(a/\mathrm{W})^4 \tag{8.46}$$

which is 0.5% accurate for $a/W < 0.6$.

Table 8.2 Configuration correction factor

Configuration correction factor	Nature of correction
$\sqrt{\frac{W}{\pi a} tan\left(\frac{\pi a}{W}\right)}$	Numerical
$\sqrt{\sec\left(\frac{\pi a}{W}\right)}$	Numerical
$1 + 0.256(a/W) - 1.152(a/W)^2 + 12.200(a/W)^3$	Empirical
$1 + \sum_{k=1}^{7} C_{2k}\left(\frac{2a}{W}\right)^{2k}$	Numerical

For double-edge cracks, the correction factor takes the form

$$f(a/\mathrm{W}) = \frac{1.12 - 0.561(a/\mathrm{W}) - 0.205(a/\mathrm{W})^2 + 0.471(a/\mathrm{W})^3 - 0.190(a/\mathrm{W})^4}{(1 - a/\mathrm{W})^{1/2}} \tag{8.47}$$

which is 0.5% accurate for $a/W < 0.5$. Further details regarding other solutions to cracked finite bodies can be found in the literature.

8.6 Stress Intensity Factors for Semi-elliptic Cracks

The literature contains several crack-tip stress intensity factor solutions for semi-elliptic cracks in plates subjected to tension or bending, as depicted in Figure 8.8. It is unfortunate that these solutions do not agree with each other particularly well. According to existing work in the literature, the basis for selecting the appropriate solution has not been established.

In general, semi-elliptic cracks often initiate at geometrical discontinuities such as notches, cut-outs, corners, among others. There exists several publications that show that semi-elliptic cracks subjected to tensile loadings adopt specific preferred shapes. In finite thickness plates under tensile loads, the cracks grow preferentially in an almost semi-circular shape, while the depth is less than half of the plate thickness and then elongates a little further as they grow toward the back wall. In bending, a similar, though slightly more complex and less well documented picture emerges. Again, it is clear that a fixed relationship exists between a preferred crack shape and its fractional crack depth.

The first contribution in this area was made by Irwin in 1962. His mode-I stress intensity factor variation around an elliptic crack embedded in a plate of infinite depth and surface dimensions shown in Figure 8.9 was given as

$$\mathrm{K_I} = \frac{\sigma_0\sqrt{\pi a}}{I(\theta)}\left[sin^2\theta + \left(\frac{a}{c}\right)^2 cos^2\theta\right]^{1/4} \tag{8.48}$$

where σ_0 is the applied tensile stress and $I(\theta)$ is an elliptic integral of the second kind and is given by

Figure 8.8 Semi-elliptic crack in a plate subjected to tension

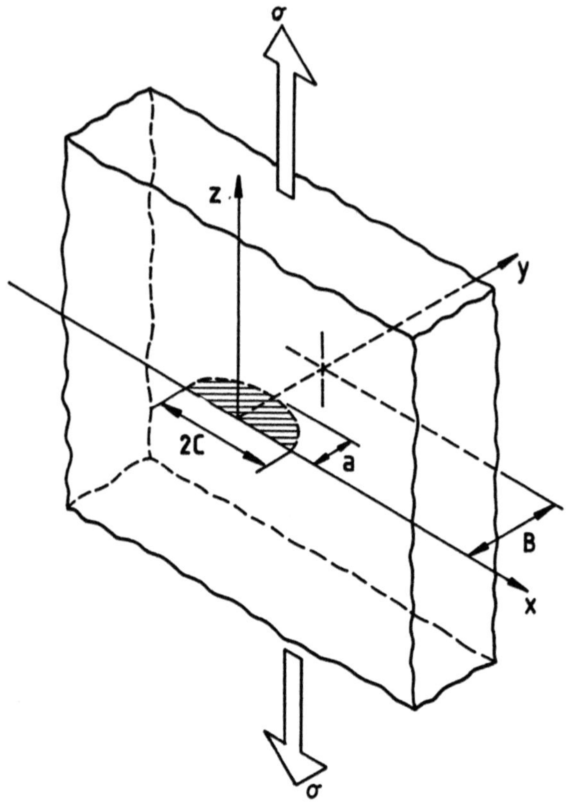

Figure 8.9 An embedded elliptic crack in an infinite solid under mode-I loading

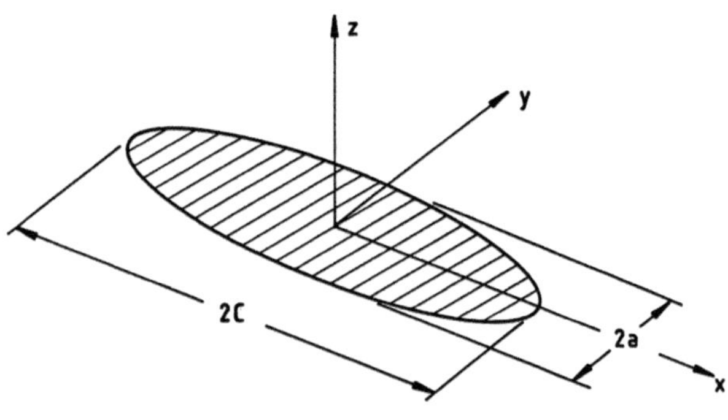

$$I(\theta) = \int_{0}^{\pi/2} \left(1 - sin^2\alpha \, sin^2\theta\right)^{1/2} d\theta \tag{8.49}$$

where $sin^2\alpha = (c^2 - a^2)/c^2$. Values of $I(\theta)$ can be obtained from standard tables. Other researchers [8.1] have suggested the following approximation for $I(\theta)$:

$$I(\theta) = \left[1 + 1.47\left(\frac{a}{c}\right)^{1.64}\right]^{1/2} \tag{8.50}$$

The above expressions indicate that K_I varies along the crack front and has a maximum value of $\sigma_0\sqrt{\pi a/I(\theta)}$ at the ends of the minor axis a, and has a minimum value $\sigma_0\sqrt{\pi a}\sqrt{a/c/I(\theta)}$ at the ends of the major axis.

For a semi-elliptical crack in a semi-infinite plate, a front face correction factor of 1.2 is multiplied by the value of K_I given by expression (8.48). More generally, the literature suggests the following expression for the front surface correction factor:

$$\text{FSC} = \left[1.21 - 0.1\left(\frac{a}{c}\right) + 0.1\left(\frac{a}{c}\right)^4\right]\sqrt{a/c} \tag{8.51}$$

Several solutions exist for semi-elliptic cracks in finite thickness plates of infinite surface area. These are usually seen to be self-consistent for shallow cracks of near semi-circular shape (*a/c* = 1), but diverge with increasing depth and ellipticity.

8.7 Mixed Mode Problems

As mentioned earlier, in spite of the fact that many practical problems are of the mixed mode type, where either the applied loads are complex or the crack may be at some angle to the applied load, the bulk of fracture mechanics work has been devoted to single mode analysis (mode-I). In view of its importance to designers, we intend to provide some coverage to mixed mode problems in this section.

Let us confine our attention to the two-dimensional problem of a through-thickness crack which is oriented at an angle β to the direction of the applied boundary stress σ_0, as depicted in Figure 8.10. One question which arises immediately is whether the approach adopted so far in the analysis of the center and edge crack problems under mode-I loading can be extrapolated to account for the behavior of angled cracked. Unlike the center crack problem, where it is usually assumed and observed that a crack grows in a self-similar manner; i.e. merely extends in length, the angled crack does not possess this similarity. Accordingly, one seeks to obtain both the direction θ along which the crack begins to grow and the magnitude of the applied load σ_0 at which growth starts.

Another important difference between the centered and angle crack problems is in their response to the applied loads. In the earlier analysis of the center-cracked plate, we have always represented the applied stress as tensile, since compressive stresses applied normal to a crack will only close that crack and will therefore not play an important role in the fracture process. However, in the mixed mode problem, the application of compressive stresses will introduce a combined stress field and can produce tensile regions near the crack-tip at some angles to the crack plane, so that fracture can occur in a manner similar to the tensile loading case. Clearly, one must

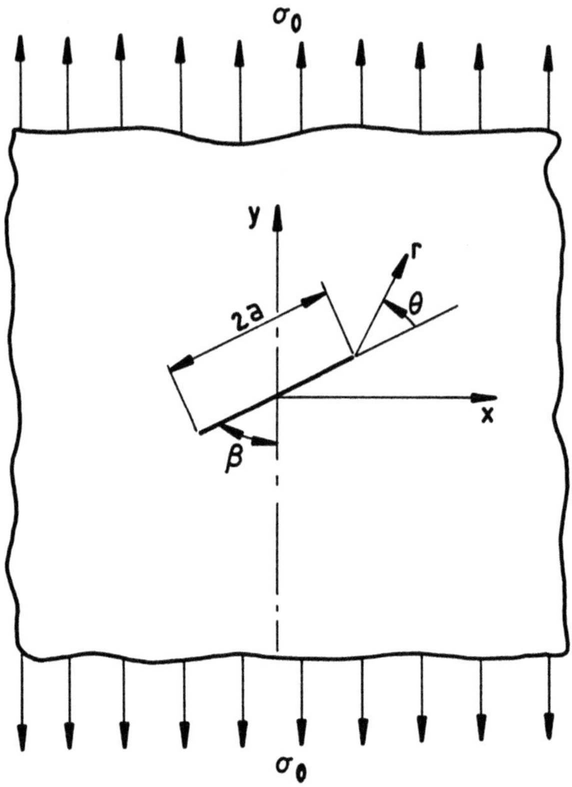

Figure 8.10 Angled crack under tension and associated co-ordinates

observe that, although the applied stress is compressive in nature, the actual cack growth and ultimately the fracture of the component is triggered by the tensile stresses near the crack-tip.

Three approaches are generally considered in predicting the direction θ along which the crack begins to grow for any value of β and the magnitude of the applied load σ_0 at which growth starts.

8.7.1 Criteria for Growth Under Tensile Loading

The simplest approach relies upon the premise that crack growth takes place in a direction perpendicular to the maximum tangential stress at or near the crack-tip. The angled crack extension was treated under monotonic loading with the focus being finding the angle θ which maximizes this stress. The experimental findings and theoretical predictions are given in Figure 8.11. Subsequently, efforts focused on the tangential stress at a small distance from the crack-tip. A considerable amount of experimental data was then generated in support of the predictions, albeit with some scatter.

Figure 8.11 The variation of crack extension angle θ with crack angle β for a tensile boundary stress σ_0

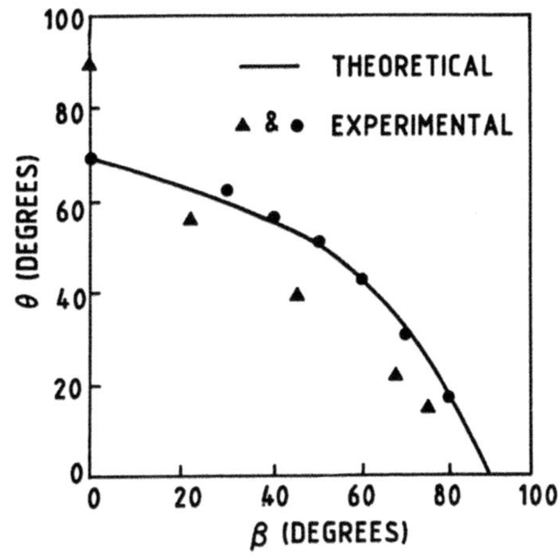

The second approach uses the strain energy density criterion to account for both the onset of crack extension and the direction of this extension. One might argue in favor of a direction for which the strain energy density is minimal on the basis that this corresponds to a maximum in the potential energy of the system. Existing in the literature [8.2] evaluated the strain energy density for three different cases: (i) at the edge of a slender ellipse, (ii) at the tip of a crack, and (iii) at a small distance from the tip of a crack.

The strain energy density in the vicinity of the crack-tip for plane problems is given by

$$\frac{dW}{dr} = \frac{1}{r}\left[a_{11}K_{\mathrm{I}}^2 + 2a_{12}K_{\mathrm{I}}K_{\mathrm{II}} + a_{22}K_{\mathrm{II}}^2\right] \tag{8.52}$$

where

$$a_{11} = \frac{1}{16\pi G}[(3-4\nu-\cos\theta)(1+\cos\theta)]$$
$$a_{12} = \frac{1}{16\pi G}2\sin\theta[\cos\theta-(1-2\nu)]$$
$$a_{22} = \frac{1}{16\pi G}[4(1-\nu)(1-\cos\theta)+(1+\cos\theta)(3\cos\theta-1)]$$

At this stage, we define the right-hand side term within the parentheses in Equation (8.52) as the strain energy density factor S. It should be noted that W tends to infinity as $r \to 0$. As a result, a cut-off point, $r > 0$, has to be introduced to allow for the discontinuity of strain energy density at the crack-tip. Thus, S can be used to describe the variation of the local energy density around the region where the crack is most likely to grow. Existing work in the literature [8.3, 8.4] postulated the following two conditions for angle crack growth:

(i) The direction of crack propagation is given when S is minimum, i.e.

$$\frac{dS}{d\theta} = 0 \text{ and } \frac{d^2S}{d\theta^2} > 0 \text{ at } \theta = \theta_0$$

(ii) The onset of cracking occurs when S reaches a critical value, S_c.

Hypothesis (i) also indicates that crack initiation commences in a radial direction along S.

The stress state in the neighborhood of the crack-tip of the angled crack depicted in Figure 8.12 are given by,

$$\left.\begin{aligned}
\sigma_{rr} &= \left[\frac{K_I}{\sqrt{2\pi r}}\right]\left[\frac{5}{4}cos\frac{\theta}{2} - \frac{1}{4}cos\frac{3\theta}{2}\right] \\
&\quad + \left[\frac{K_{II}}{\sqrt{2\pi r}}\right]\left[-\frac{5}{4}sin\frac{\theta}{2} + \frac{3}{4}sin\frac{3\theta}{2}\right] + \sigma_t cos^2\theta \\
\sigma_{\theta\theta} &= \left[\frac{K_I}{\sqrt{2\pi r}}\right]\left[\frac{3}{4}cos\frac{\theta}{2} + \frac{1}{4}cos\frac{3\theta}{2}\right] \\
&\quad + \left[\frac{K_{II}}{\sqrt{2\pi r}}\right]\left[-\frac{3}{4}sin\frac{\theta}{2} - \frac{3}{4}sin\frac{3\theta}{2}\right] + \sigma_t sin^2\theta \\
\tau_{r\theta} &= \left[\frac{K_I}{\sqrt{2\pi r}}\right]\left[\frac{1}{4}sin\frac{\theta}{2} + \frac{1}{4}sin\frac{3\theta}{2}\right] \\
&\quad + \left[\frac{K_{II}}{\sqrt{2\pi r}}\right]\left[\frac{1}{4}cos\frac{\theta}{2} + \frac{3}{4}cos\frac{3\theta}{2}\right] - \sigma_t sin\,\theta cos\theta
\end{aligned}\right\} \tag{8.53}$$

where

$$K_I = \sigma_0\sqrt{\pi a}\, sin^2\beta \tag{8.54}$$

$$K_{II} = \sigma_0\sqrt{\pi a}\, sin\,\beta cos\beta \tag{8.55}$$

$$\sigma_t = \sigma_0\left(cos^2\beta - sin^2\beta\right) \tag{8.56}$$

and the strain energy density as,

$$W = \left(\frac{1}{2G}\right)\left[\frac{\kappa+1}{8}(\sigma_{rr} + \sigma_{\theta\theta})^2 - \sigma_{rr}\sigma_{\theta\theta} + \tau_{r\theta}^2\right] \tag{8.57}$$

where $\kappa = (3 - \nu)/(1 + \nu)$ in plane stress, ν being Poisson's ratio.

The above analysis is limited to small values of r. These expressions show that mode-I contribution dominates the crack-tip stress field when $\beta > 60°$.

Swedlow [8.3] used the above expressions are used to gain insight into the previously described hypotheses for angled crack growth under tension, i.e. maximum tangential stress and minimum strain energy density. In his work, he used the following numerical values: $E = 70$ GPa, $\nu = 0.3$, $r/a = 0.002$ and $\beta = 70°$.

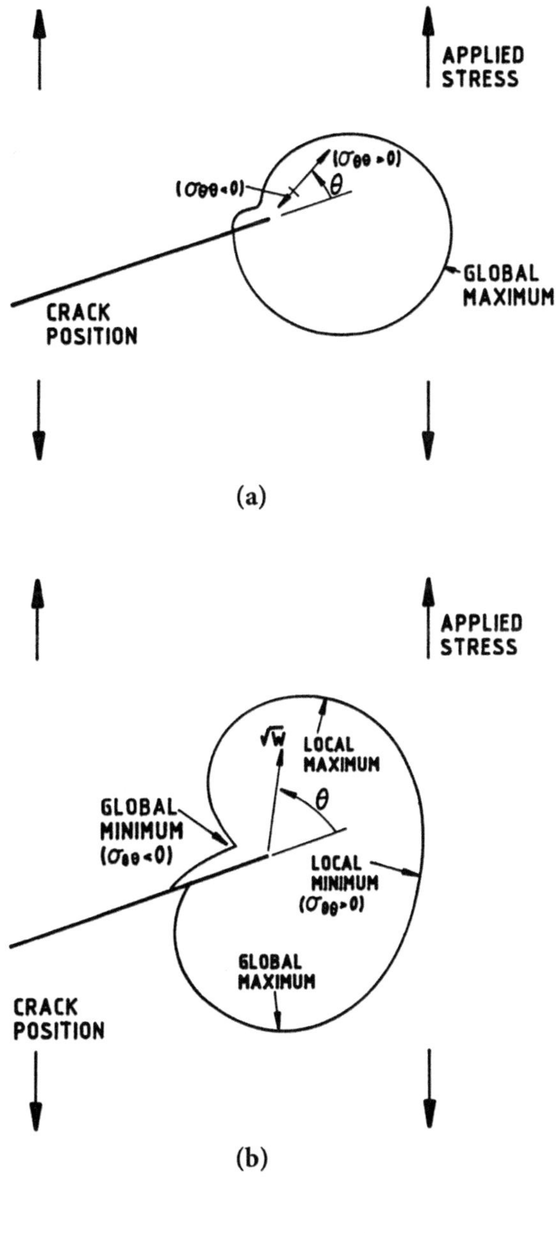

Figure 8.12 (**a**) Locus of tangential stress variations for tensile loading, $\beta = 70°$ and $r/a = 0.002$. (**b**) Locus of strain energy density variations for tensile loading, $\beta = 70°$ and $r/a = 0.002$ [8.2]

He then examined $\sigma_{\theta\theta}$ and W over the range $-180° \leq \theta \leq 180°$. The results were plotted in the form of tangential stress and strain energy loci (Figure 8.12). He made a number of observations. First, there exists only one extreme value for the tangential stress. Second, there are two minima in w, one local and the other global, and there exists no *a priori* basis for selecting one over the other. Third, strain energy density is insensitive to the sign of loading. Fourth, strain energy density is regarded as the integral of field variables over some domain rather than itself being a field quantity.

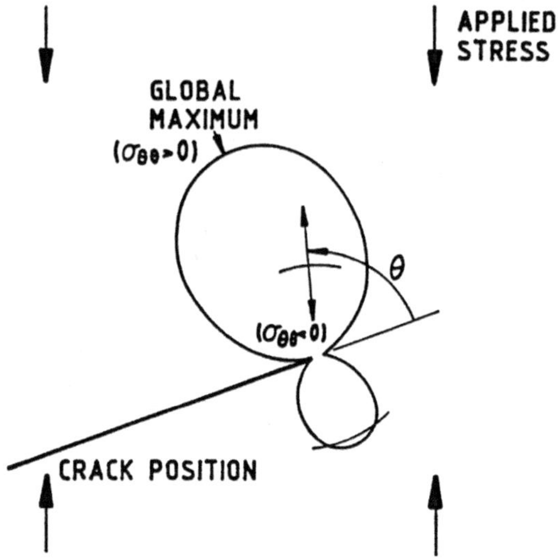

(a)

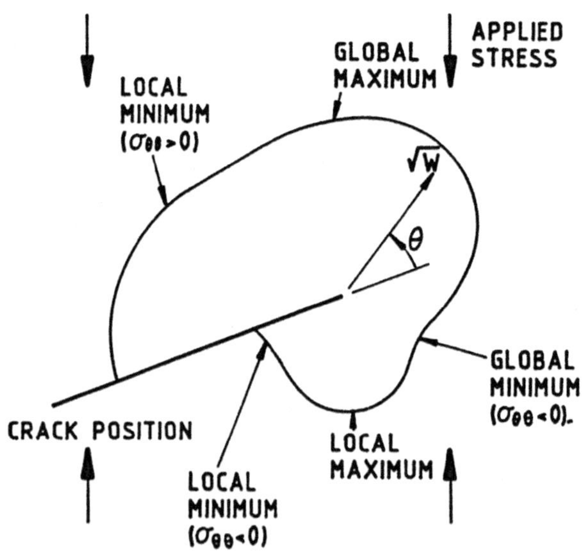

(b)

Figure 8.13 (**a**) Locus of tangential stress variations for compressive loading, $\beta = 70°$, $r/a = 0.002$ and $\mu = 0.3$. (**b**) Locus of strain energy density variations for compressive loading with $\beta = 70°$, $r/a = 0.002$ and $\mu = 0.3$ [8.2]

Finally, Equation (8.53) do not permit the application of $-\sigma_0$ instead of $+\sigma_0$. Expressions (8.53) were later modified to provide permissible expressions for the stress intensity factor under compressive loading and friction along the crack faces (Figure 8.13a, b), as follows:

$$\left.\begin{aligned}
\sigma_{rr} &= \left[\frac{K_{II}}{\sqrt{2\pi r}}\right]\left[-\frac{5}{4}\sin\frac{\theta}{2}+\frac{3}{4}\sin\frac{3\theta}{2}\right] \\
&\quad +\sigma_0\left(\sin^2\theta-\mu\sin2\theta\right)\sin^2\beta+\sigma_t\cos^2\theta \\
\sigma_{\theta\theta} &= \left[\frac{K_{II}}{\sqrt{2\pi r}}\right]\left[-\frac{3}{4}\sin\frac{\theta}{2}-\frac{3}{4}\sin\frac{3\theta}{2}\right] \\
&\quad +\sigma_0(\cos^2\theta+\mu\sin2\theta)\sin^2\beta+\sigma_t\sin^2\theta \\
\tau_{r\theta} &= \left[\frac{K_{II}}{\sqrt{2\pi r}}\right]\left[\frac{1}{4}\cos\frac{\theta}{2}+\frac{3}{4}\cos\frac{3\theta}{2}\right] \\
&\quad +\sigma_0(\sin\theta\cos\theta-\mu\cos2\theta)\sin^2\beta-\sigma_t\sin\theta\cos\theta
\end{aligned}\right\} \tag{8.58}$$

where $\sigma_0 \leq 0$, μ is the coefficient of friction and

$$\left.\begin{aligned}
K_{II} &= \sigma_0\sqrt{\pi a}\left(\sin\beta\cos\beta-\mu\sin^2\beta\right) \\
\sigma_t &= \sigma_0\left(\cos^2\beta-\sin^2\beta\right)
\end{aligned}\right\} \tag{8.59}$$

8.8 Determination of Stress Intensity Factor

In this section, we intend to cover some of the most common methods adopted in the determination of the stress intensity factor in a continuum. At this stage, however, we wish to draw attention to the numerous reviews that exist in the literature concerning this topic. For a detailed account of this section, the reader is advised to consult the literature.

The choice of a method of solution will ultimately depend upon time available, required accuracy in relation to cost and the frequency with which the solutions are to be used.

These methods, which are outlined in Figure 8.14 can be divided broadly into:

(i) theoretical, which is subdivided into analytical and numerical methods, and
(ii) experimental techniques.

8.8.1 Analytical Methods

Muskhelishvili Potential Functions

These methods satisfy all the boundary conditions exactly and have the advantage of leading to explicit expressions for stress intensity factors. Unfortunately, in view of the complexity of the cracked geometries, only a limited number of relatively simple solutions to mostly two-dimensional infinite bodies are possible. These solutions

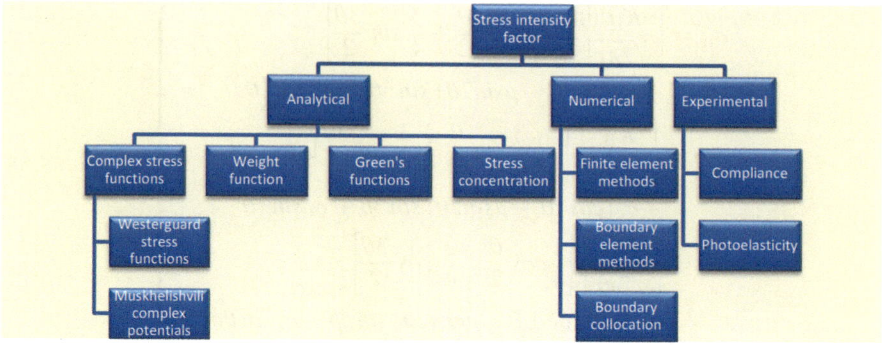

Figure 8.14 Block diagram showing the different methods adopted in the determination of the stress intensity factors

have mostly adopted the complex potentials approach developed by Muskhelishvili, which is detailed in Chapter 3. This complex function approach enables the description of Airy's stress function in terms of two complex potentials, as follows:

$$\phi = \text{Re}\ [\bar{z}\psi(z) + \chi(z)] \tag{3.80}$$

which yields

$$\sigma_{xx} + \sigma_{yy} = 2\psi'(z) + 2\bar{\psi}'(\bar{z}) = 4\ \text{Re}\ \psi'(z) \tag{3.86}$$

$$\sigma_{yy} - \sigma_{xx} + 2i\tau_{xy} = 2[\bar{z}\ \psi''(z) + \chi''(z)] \tag{3.88}$$

From Equation (3.80) and the known properties of $\psi'(z)$, it can be shown that

$$K_I - iK_{II} = 2\sqrt{2\pi}\lim_{z \to z_1}[\sqrt{z - z_1}\ \psi'(z)] \tag{8.60}$$

In this method, conformal mapping can be used to map cracks into holes.

Westergaard Stress Functions

The use of these functions has been clearly demonstrated in Section 8.3. In terms of the Westergaard stress functions, the stress intensity factor can be given by

$$K_M = \sqrt{2\pi}\lim_{z \to a}\left(\sqrt{z - a}\right)Z_M \tag{8.61}$$

where M = I, II or III.

Westergaard stress functions have been developed for a variety of problems. It is worth noting, however, that Westergaard stress functions are only special cases of the complex potentials approach outlined earlier. They have been used in problems possessing x-axis symmetry for both loading and geometry, i.e. $\tau_{xy} = 0$ along $y = 0$.

Bueckner Weight Functions

Another interesting approach is concerned with the use of a weight function which characterizes a cracked geometry. In this case, the stress intensity factor in mode I is given by

$$K_{\mathrm{I}} = \int_a p(x) m(x,a) \mathrm{d}x \tag{8.62}$$

where

$$m(x,a) = \frac{E}{2K_{\mathrm{I}}} \frac{\partial v(x,a)}{\partial a} \quad \text{for plane stress} \tag{8.63}$$

$$m(x,a) = \frac{E}{2(1-\nu^2)K_{\mathrm{I}}} \frac{\partial v(x,a)}{\partial a} \quad \text{for plane strain} \tag{8.64}$$

In the above expressions, $m(x,a)$ is the weight function. This $m(x,a)$ is unique for a given cracked geometry and is independent of the loading condition. To demonstrate the use of the weight function approach, let us consider, for example, the central crack problem shown in Figure 8.15.

$$K_{\mathrm{I}}^{(1)} = \sigma\sqrt{\pi a} \tag{a}$$

$$v(x,a) = \frac{2\sigma}{E}\sqrt{a^2 - x^2} \quad (-a \le x \le a) \tag{b}$$

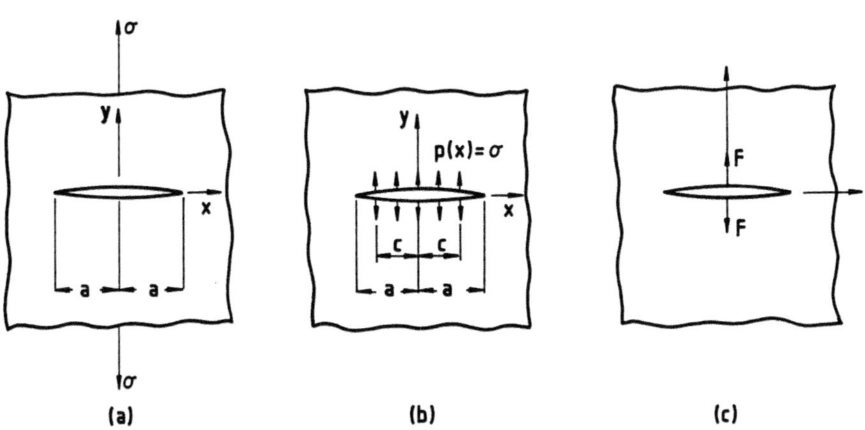

Figure 8.15 Central crack under different loading conditions: (**a**) boundary loading σ, (**b**) centrally located band of pressure applied to crack surface over length 2c and (**c**) point wedging force *F* acting at the center of the crack

Let us now utilize the above results in determining the stress intensity factor for the same geometry but for a centrally located band ($2c$) of pressure applied to the crack surface as depicted in Figure 8.15b. Now, in order to apply the weight function approach, we must determine

$$\frac{\partial v}{\partial a} = \frac{2\sigma}{E}\left(\frac{1}{\sqrt{a^2 - x^2}}\right)a \tag{c}$$

and

$$m(x,a) = \frac{2E\sigma}{2E\sigma\sqrt{\pi a}}\left(\frac{1}{\sqrt{a^2 - x^2}}\right)a \tag{d}$$

i.e.

$$m(x,a) = \left[\frac{a}{\pi(a^2 - x^2)}\right]^{1/2} \tag{e}$$

Accordingly, K_{I} for the second loading is given by

$$K_{\mathrm{I}}^{(2)} = \int_{-c}^{c}\left[\frac{a}{\pi(a^2 - x^2)}\right]^{1/2} p(x)\,dx$$

where

$$p(x) = \sigma \text{ for } -c \le x \le c$$
$$\therefore K_{\mathrm{I}}^{(2)} = 2\sigma\sqrt{\frac{a}{\pi}}\left[\sin^{-1}(c/a)\right] \tag{f}$$

Similarly, for a point wedging force per unit thickness acting at the center of the crack (Figure 8.15c given by the expression)

$$F = 2\sigma c$$

the stress intensity factor $K_{\mathrm{I}}^{(3)}$ is given by

$$K_{\mathrm{I}}^{(3)} = \lim_{c \to 0}\left[2\frac{F}{2c}\sqrt{\frac{a}{\pi}}\sin^{-1}(c/a)\right] \tag{g}$$

giving

$$K_{\mathrm{I}}^{(3)} = \frac{F}{\sqrt{\pi a}} \tag{h}$$

Clearly, the weight function may be adopted in the derivation of additional stress intensity factor solutions, provided that the details of crack-line loading are available. Finally, it is worth mentioning that the weight function approach may be thought of as a form of Green's function for cracked bodies.

Stress Intensity Factors from Stress Concentration Factors

In 1960, Irwin suggested that Neuber's stress concentration results, which pertain to notches of a very small flank angle and root radius, may be used to obtain theoretical expressions for K_{I}. If σ_{max} is the maximum tensile stress at the notch root and ρ is the root radius, then

$$K_{\mathrm{I}} = \frac{\sqrt{\pi}}{2} \lim_{\rho \to 0} \sigma_{\max} \sqrt{\rho} \tag{8.65}$$

with $K_{\mathrm{II}} = K_{\mathrm{III}} = 0$.

Work in the literature adopted the above approach to develop expressions for the stress in the (r, θ) co-ordinates from the focus of an elliptical cut-out in an infinite sheet subjected to uniaxial tension, depicted in Figure 8.16. His expression was developed for small ρ/a. This limitation simplifies the analysis and provides the following expressions for the stresses in the vicinity of the notch-tip.

$$\begin{bmatrix} \sigma_{xx}(r,\theta,\rho) \\ \sigma_{yy}(r,\theta,\rho) \\ \tau_{xy}(r,\theta,\rho) \end{bmatrix} = \frac{K_{\mathrm{I}}}{\sqrt{2\pi r}} \cos\frac{\theta}{2} \begin{bmatrix} 1 - \sin\frac{\theta}{2}\sin\frac{3\theta}{2} - \frac{\rho}{2r}\cos\frac{3\theta}{2}\sec\frac{\theta}{2} \\ 1 + \sin\frac{\theta}{2}\sin\frac{3\theta}{2} + \frac{\rho}{2r}\cos\frac{3\theta}{2}\sec\frac{\theta}{2} \\ \sin\frac{\theta}{2}\cos\frac{3\theta}{2} - \frac{\rho}{2r}\sin\frac{3\theta}{2}\sec\frac{\theta}{2} \end{bmatrix} \tag{8.66}$$

where $K_{\mathrm{I}} = \sigma\sqrt{\pi a}$ and a is the semi-major diameter of the elliptical cut-out.

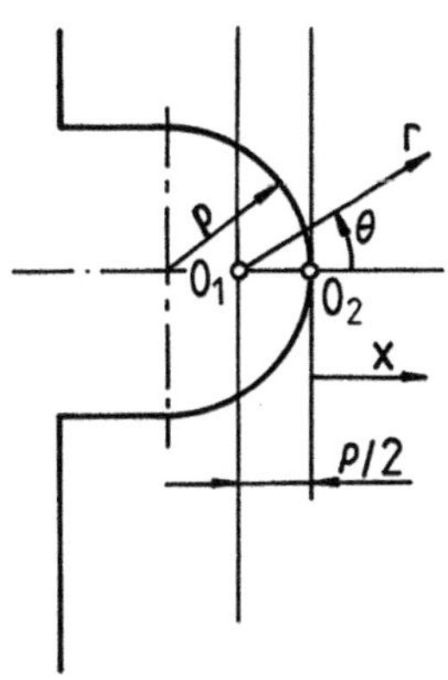

Figure 8.16 Co-ordinate system and stress state for an elliptical cut-out in an infinite plate

8.8.2 Numerical Methods

Since the problems for which exact solutions may be obtained are limited, recourse to numerical methods may be necessary. In view of their elegance, flexibility and applicability to large classes of geometries, materials and loading conditions, we devote our attention to the powerful technique of finite element and boundary element methods.

Finite Element Methods
As explained in Chapter 4, the fundamental concept of the finite element method is that any continuous field variable, such as displacement, stress, pressure or temperature, can be approximated by a discrete model composed of a set of piecewise continuous field variables defined over a finite number of sub-domains, known as elements. These elements are interconnected at specified joints, which are called nodes or nodal points. Since the actual variation of the field variable inside the continuum is not known, some approximating functions are needed to describe its variation. These approximating functions, which are also known as the interpolating functions, are defined in terms of the values of the field variable at the nodal points. When field equations, such as equilibrium or heat balance, for the whole body are written the new unknowns will be the nodal values of the field variable. By solving the field equations, which are generally in the form of banded matrices, the nodal value of the field variable can be obtained throughout the assemblage of elements.

The finite and boundary element methods are conceptually simple understand and apply. They have been used to handle a wide range of two- and three-dimensional cracked bodies. A detailed account for BEM is beyond the scope of this book. All the methods listed below are concerned with the determination of stress intensity factors and contain all the inherent advantages of the finite element method. Various forms of mixed boundary conditions can be analyzed with ease. Furthermore, structures formed by dissimilar materials can be handled. Body forces, thermal loading and shock phenomena can be analyzed. We intend to devote our attention to the following methods:

(i) the displacement approach;
(ii) the stress approach;
(iii) the crack-tip or singular elements approach
(iv) the line integral (energy) approach.

Displacement Approach Using Non-singular Elements
The displacement method involves a correlation of the finite element nodal point displacements with the well-known crack-tip displacement equations. For example, mode-I displacement equations can be given as

$$\upsilon_i = \frac{K_{\text{I}}}{G}\sqrt{\frac{r}{2\pi}}f_i(\theta,\nu) \tag{8.16}$$

where K_I is a function of the geometry of the body containing the crack and of the applied loading conditions. Rearranging Equation (8.16) and substituting nodal point displacements u_i^* at some point, (r,θ) close to the crack-tip, a quantity K_I^* can be computed:

$$K_I^* = \sqrt{\frac{2\pi}{r}} G/[f_i(\theta, \nu)] v_i^* \tag{8.67}$$

From plots of K_I^* as a function of r for fixed values of θ and a particular displacement component (u^* or υ^*), an estimate of K_I can be made by extrapolating back to $r \rightarrow 0$. At this stage, one must observe that nodal displacements are rather inaccurate at a finite distance from the crack. This limitation can be overcome by refining the mesh near the crack-tip; the current multi-point constraint approach in finite elements, shown in Figure 8.17, will allow such as refinement to be made without drastically influencing the processing time and consequently the cost of the analysis. With a suitably refined mesh, it is possible to the tangent extrapolation to estimate the stress intensity factor. In this case, the curves obtained from element displacements rapidly approach a constant slope with increasing distance (r) from the crack-tip, as shown in Figure 8.18. The intercept of the tangent to the constant slope portion of the curve with the K_I^* axis is used as the K_I estimate. Work by the author and his co-workers shows that the most accurate results are obtained from estimates utilizing the displacement υ on the crack surface ($\theta = \pi$; Figure 8.18). For plane strain mode-I conditions, the expression for the nodal stress intensity factor K_I^* reduces to:

$$K_I^* = \frac{E\upsilon^*}{4(1-\nu^2)} \sqrt{\frac{2\pi}{r}} \tag{8.68}$$

Stress Approach Using Non-singular Elements

In a similar fashion to that described above, this approach involves a correlation of the finite element nodal or Gauss point stresses with the crack-tip stress equations, as follows:

$$\sigma_{ij} = \frac{K_M}{\sqrt{2\pi r}} f_{ij}(\theta) \tag{8.69}$$

Rearranging Equation (8.69) gives

$$K_M = \frac{\sqrt{2\pi r}}{f_{ij}(\theta)} \sigma_{ij} \tag{8.70}$$

with M = I, II or III depending upon the examined mode.

Nodal point stresses σ_{ij}^* in the vicinity of the crack-tip can be substituted in Equation (8.70) so that the values of K_M^* can be obtained. Again from plots of K_M^* as

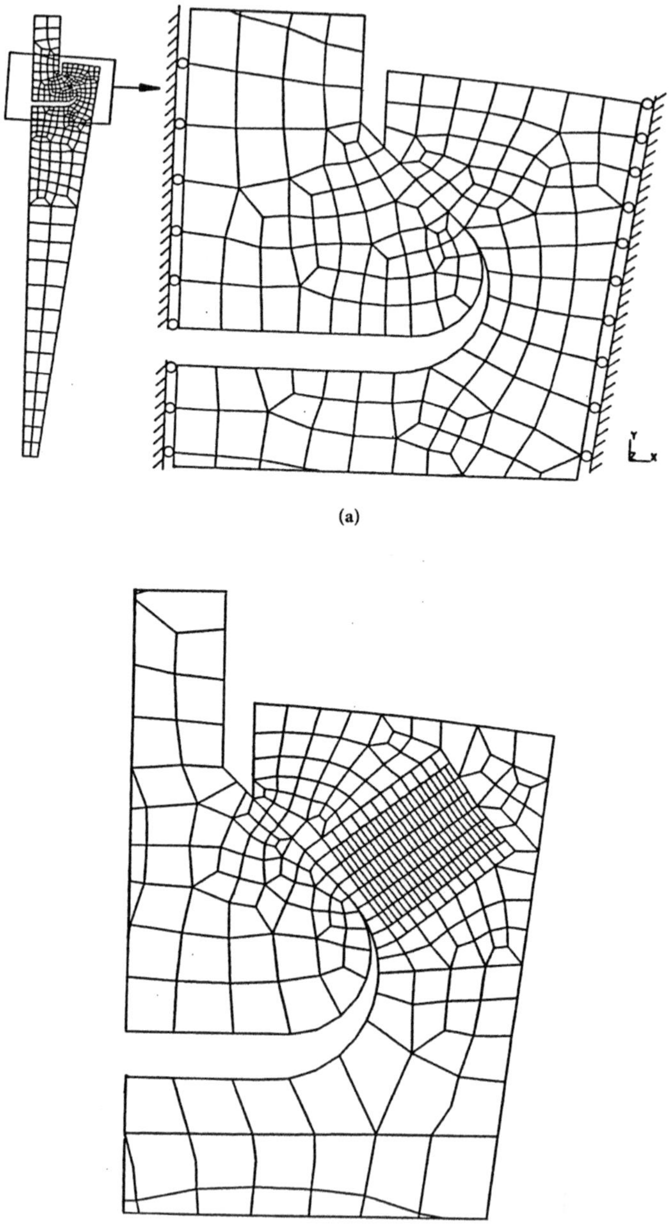

(a)

(b)

Figure 8.17 Techniques adopted in refining mesh: (**a**) conventional approach; (**b**) multi-point constraint (MPC) approach

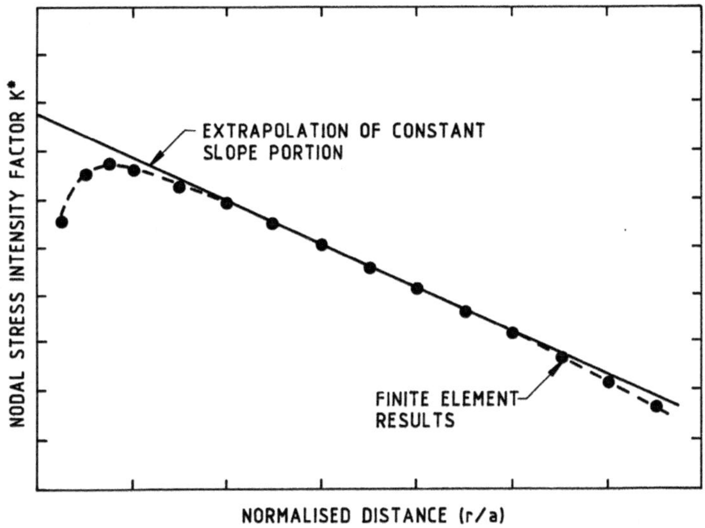

Figure 8.18 Variation of K_I^* with *r/a* for a finite crack in a plate

a function of r near the crack-tip for fixed θ and a particular stress component, an estimate of K_M can be made. In view of the inability of the conventional constant strain elements to represent the stress singularity condition at the crack-tip, the K_M^* curves for $r > 0$ must again be extrapolated back to $r = 0$. Good estimates of K_I have been obtained by this approach for mode-I loading, using σ_{yy} at $\theta = 0$. It is worth noting however that, in most of the cases considered, the displacement approach provided greater accuracy when compared with known closed form solutions.

Singular Elements Approach

A number of special crack-tip elements have been developed from the standard isoparametric element enabling the singularity of the stress field at the crack-tip to be modeled. In this case, modifications to the shape functions are made to for example 8 and 20 node elements by moving the two mid-side nodes to the quarter points at the crack-tip, as shown in Figure 8.19. It is demonstrated that the inverse square-root singularity characteristic of LEFM occurs as a result of this movement of the nodes. Work in this area also reveals that collapsing one side of a quadrilateral element to form a triangular element in two-dimensional problems and one face of a brick element to form a prismatic element in three-dimensional problems provide far better results for the stress intensity factor in elastic fracture than with the conventional rectangular two-dimensional and brick three-dimensional elements, depicted in Figure 8.19. This technique is illustrated in Figure 8.20. The proof of singularity was investigated at the boundary of the elements only. However, it is easy to show that triangular elements possess the same singularity in the interior as well as on the boundary with $\frac{1}{\sqrt{r}}$ singularity representing perfectly elastic behavior. These are achieved by having either one node or multiple independent nodes at the crack-tip.

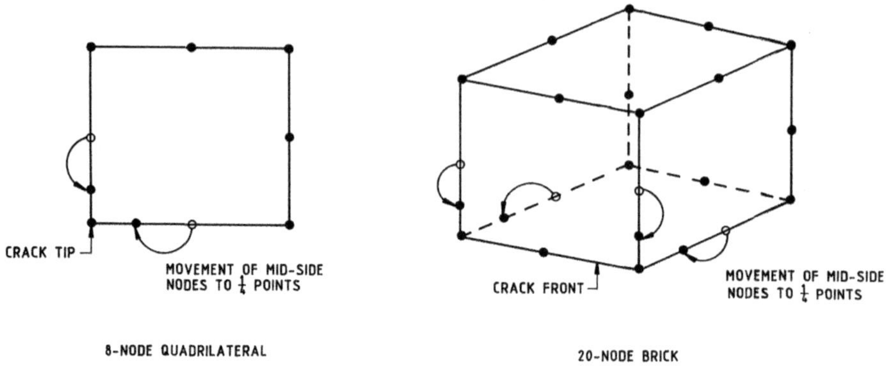

Figure 8.19 Quadrilateral and brick crack-tip elements

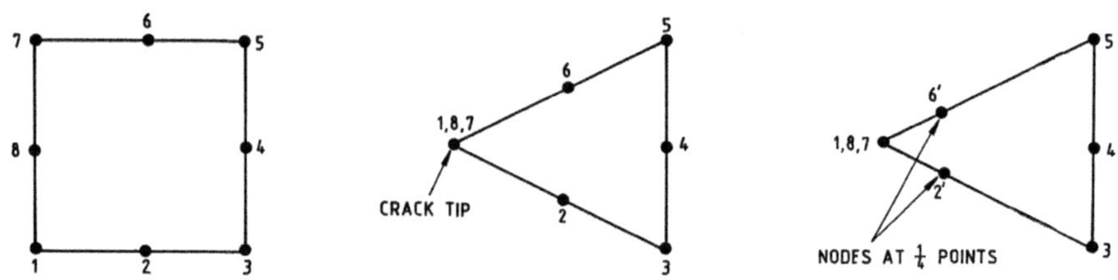

Figure 8.20 Development of crack-tip element

Boundary Element Method

The differential form typically used in describing boundary value problems can be transformed into boundary integral equations over the boundary of the considered domain. The boundary integral equations (BIEs) can be solved numerically by means of piecewise discretization, similar to FEM. whereby the boundary of the domain is divided into sub-boundaries known as boundary elements, as shown in Figure 8.21. The equations of the boundary elements are assembled together to form a system of algebraic equations in terms of the field parameters over the domain boundary. The values of the field function at any point inside the domain can be interpolated in terms of its boundary values.

The boundary element method offers the following advantages over the finite element method:

(i) it reduces the dimensionality (DOF) of the problem by one, resulting in a smaller system of equations and a considerable reduction in the data required for the computation.
(ii) it offers continuous interior modeling within the solution domain, and the values of the solution variables can be calculated at any selected interior point. and
(iii) it is well suited to problems involving large domain.

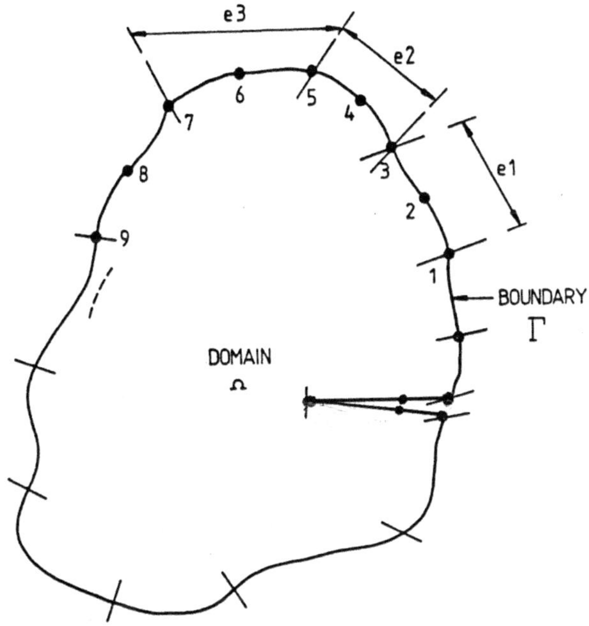

Figure 8.21 Discretization of the boundary

The approach adopted in determining the stress intensity factor is generally similar to that explained using the finite element method. In view of the presence of singularities as well as the complex mathematics involved, the BEM did not overshadow the finite element method.

Boundary Collocation Method

Stress intensity factors in multiply connected two-dimensional bodies subjected to in-plane loading may be determined from a series representation of the complex potentials $\psi(Z)$ and $\chi(Z)$ of Muskhelishvili. The method of solution is to determine a finite number of coefficients by matching the prescribed stresses around the boundary, taking into account the condition for a single-valued displacement. Further details can be found in the appropriate literature.

8.8.3 Experimental Methods

Compliance Measurement Approach

In Chapter 7, we developed expressions for the strain energy release rate per unit thickness G as being,

$$G = \frac{\partial}{\partial a}(W - U_a) \text{ for constant load} \tag{a}$$

$$G = -\frac{\partial U_a}{\partial a} \text{ for constant displacement} \tag{b}$$

Expressions (a) and (b) are applicable for the different modes of crack surface deformation. Let us now apply the above expressions for the cracked body shown in Figure 8.22. In this case expression (a) for constant load becomes

$$G_I = \frac{1}{B}\frac{\partial}{\partial a}\left(p\mathrm{d}v - \frac{1}{2}pv\right) \tag{c}$$

Where pdv is the work done by the external load p and $\frac{1}{2}pv$ is the change in the strain energy of the cracked body. Expression (c) can be written as

$$G_I = \frac{1}{B}\left(P\frac{\partial v}{\partial a} - \frac{1}{2}\frac{\partial}{\partial a}pv\right) \tag{d}$$

If we now introduce the compliance C, which is the amount of displacement u per unit of the applied load p, then expression (d) yields

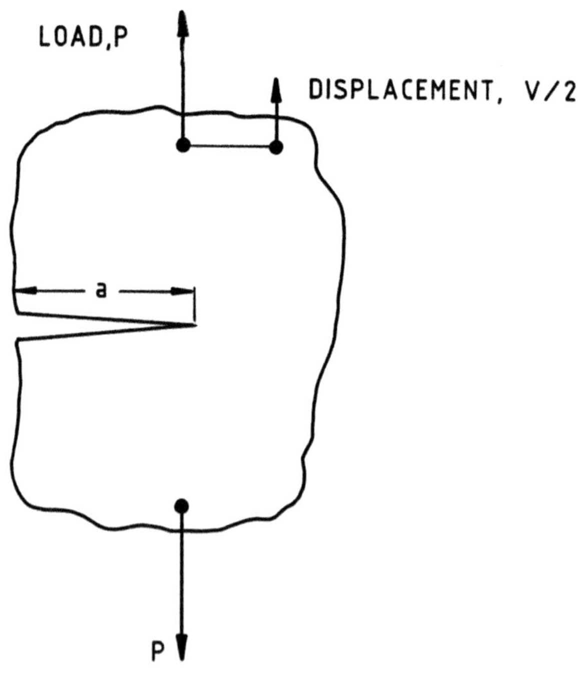

Figure 8.22 Configuration of cracked body, and notation used

$$G_{\mathrm{I}} = \frac{1}{B}\left(p\frac{\partial Cp}{\partial a} - \frac{1}{2}\frac{\partial}{\partial a}Cp^2\right) \tag{e}$$

giving

$$G_{\mathrm{I}} = \frac{p^2}{2B}\frac{\partial C}{\partial a} \tag{8.71}$$

If we now turn to the expression relating to the constant displacement condition, we get

$$G_{\mathrm{I}} = -\frac{1}{B}\left(\frac{1}{2}\frac{\partial}{\partial a}pv\right) \tag{f}$$

i.e.

$$G_{\mathrm{I}} = -\frac{1}{B}\left[\frac{v}{2}\frac{\partial}{\partial a}(v/C)\right] \tag{g}$$

$$= -\frac{v}{2B}\left(\frac{1}{C}\frac{\partial v}{\partial a} + v\frac{\partial}{\partial C}C^{-1}\frac{\partial C}{\partial a}\right) \tag{h}$$

$$= -\frac{v^2}{2B}\left(-C^{-2}\frac{\partial C}{\partial a}\right) \tag{i}$$

$$G_{\mathrm{I}} = \frac{v^2/C^2}{2B}\frac{\partial C}{\partial a} = \frac{p^2}{2B}\frac{\partial C}{\partial a}$$

which is the same as expression (8.71), thus demonstrating the equivalence between constant load and fixed grip conditions. This suggests that G would be the same, whether it was measured using a hard or a soft testing machine.

Expression (8.71) also reveals that a plate with a through-thickness cut or a slit can be used to determine G_{I} and effectively K_{I} according to Equations (8.21) and (8.22). In this case, a set of compliance measurements is taken from a test plate, and the slit is extended by a small increment between each pair of consecutive measurements. The slit length for each increment must be measured accurately. The procedure is repeated until the slit length is greater than the longest crack allowed in the test specimen.

The compliance coefficients are expressed in the dimensionless form "*EBC*," where E is Young's modulus, B is the specimen thickness and C is the compliance at a given slit length a. This set of results is then fitted to a polynomial in a/W, where W is the specimen width, by the well-known statistical methods of best curve fitting. The appropriate choice of the type and the degree of the polynomial function is important, since this function is differentiated with respect to a/W to obtain $\frac{1}{2} \times [\partial(EBC)/\partial(a/W)]$ which is equal to $GEWB^2/p^2$, i.e.

$$\frac{1}{2}\frac{\partial(EBC)}{\partial(a/W)} = \frac{GEWB^2}{p^2} \tag{8.72}$$

The dimensionless quantity $GEWB^2/p^2$ then applies to a specimen of any size that is geometrically similar to the prototype, and of any material, which has a well-defined Young's modulus. Other limitations include that the material is homogeneous, isotropic and should have a wide range of linear elastic behavior. The specimen thickness must be uniform, and at least 10 times the slit width to reduce boundary effects.

From the least-squares polynomial fit we get

$$\frac{G_I EWB^2}{p^2} = \frac{1}{2}\frac{\partial(EBC)}{\partial(a/W)}$$

$$= A_1(a/W) + A_2(a/W)^2 + A_3(a/W)^3 + \cdots + A_n(a/W)^n$$

$$\therefore EG_I = \frac{p^2}{B^2 W}\left[A_1(a/W) + A_2(a/W)^2 + A_3(a/W)^3 + \cdots + A_n(a/W)^n\right]$$

or alternatively

$$K_I = \frac{P}{BW^{1/2}}\left[D_1(a/W)^{1/2} + D_2(a/W)^{3/2} + D_3(a/W)^{5/2} + \cdots + D_n(a/W)^{n/2}\right] \tag{8.73}$$

The above expression is valid for an infinite plate and some correction must be made for the finite width effect. Accordingly,

$$K_I = \frac{P}{BW^{1/2}} Q \tag{8.74}$$

where Q is also a function (a/W). This is the measurement procedure and tables of Q values for the standard specimen configurations can be found in the literature.

It is worth mentioning that judgement must be exercised during the selection of the degree of the polynomial which best represents the data. Too high a degree polynomial produces a better-fit, but at the expense of the smoothness of the derivatives. In general, experience has shown that third to fifth degree polynomials produce sufficiently low residuals and monotonically increasing derivatives.

Figure 8.23 shows a typical single-edge notch (SEN) compliance specimen with a strut gauge and stepped blocks. In this case, the displacement measurements

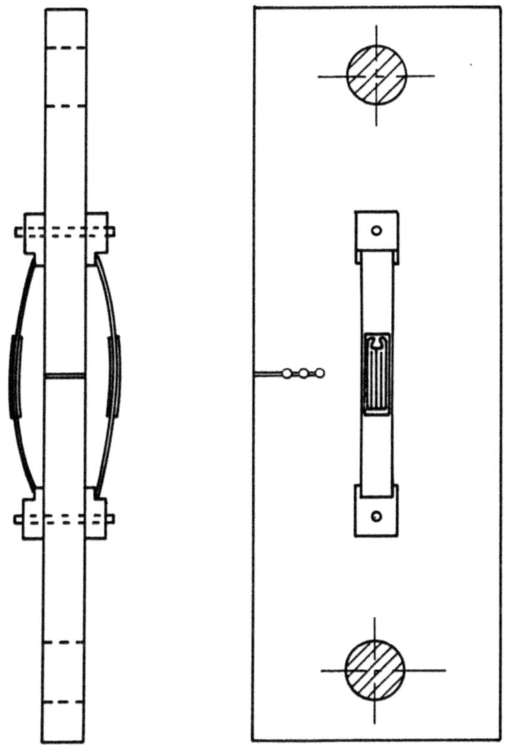

Figure 8.23 Single-edge notch compliance specimen with strut gauges and stepped blocks

are made of the central sections of the specimen to reduce the possible errors associated with large strains occurring near the loading pins. The strut gauges are held on both faces of the specimen in pairs of stepped blocks, which are pinned to the specimen at various gauge lengths. The different devices used to measure the displacement in the compliance calibration vary and include dial indicators, conventional tension-tension transducers, and electrical resistance strain gauges. The choice of these devices is dictated by sensitivity, accuracy, range of measurement and stability.

It is also worth mentioning that the experimental results obtained by the compliance calibration method are compared with predictions obtained by numerical methods for the different configurations of the specimens shown in Figure 8.24. The results show very good agreement between the two.

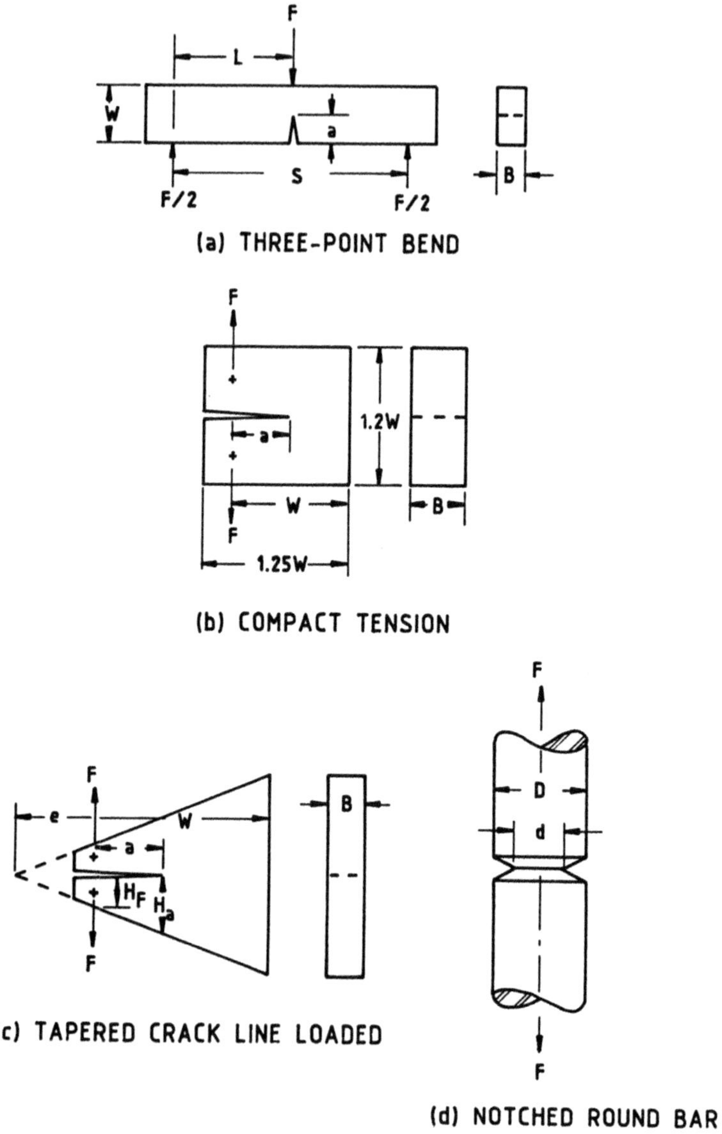

Figure 8.24 Compliance calibration specimens

References

8.1 S.A. Meguid, *Engineering Fracture Mechanics* (Elsevier Applied Science, London, 1989)
8.2 T.L. Anderson, *Fracture Mechanics: Fundamentals and Applications* (CRC Press, Boca Raton, 2017)
8.3 J.L. Swedlow, Criteria for growth of the angled crack, ASTM STP 601, 1976
8.4 Y. Yang, S.J. Chu, W.S. Huang, H. Chen, Crack growth and energy release rate for an angled crack under mixed mode loading. Appl. Sci. **10**, 4227 (2020). https://doi.org/10.3390/app10124227

Chapter 9
Atomistic Fracture Mechanics

Summary In this chapter, we investigate the atomistic fracture behavior a two dimensional material (graphene), accounting for crack tip stress field, using molecular dynamics. We further examine two cases of interacting nanodefects. The first involves an atomistic zigzag crack interacting with atomistic vacancies and in the second we investigate an armchair crack interacting with the same vacancies. We demonstrate that the crack-tip stress field of an existing crack in graphene can be effectively tailored by the strategic positioning of atomic vacancies of varied shapes, locations, and orientations near its tip. The crack-vacancy interactions result in a remarkable improvement in the fracture strength of graphene. The crack healing phenomenon is also examined and discussed in this chapter.

9.1 Two Dimensional Materials

The remarkable isolation of graphene in 2004 led to the development of atomically thin two-dimensional (2D) materials. Graphene is a 2D crystalline form of carbon and is a basic building block for carbon allotropes such as fullerenes, carbon nanotubes and graphite. The carbon atoms in graphene are packed in a planar honeycomb network (Figure 9.1). The unit cell of a single-layer graphene consists of two carbon atoms, separated by 1.42 Å, with a lattice constant of 2.46 Å. Each atom has s, p_x and p_y orbitals and is bonded to three neighboring atoms in the lattice, forming sp^2 atomic network. For details, refer to Refs. [9.1–9.6].

In view of its unique properties, graphene is employed in a variety of engineering applications. Indeed, its extraordinary stiffness, fracture strength, and flexibility, led to its use in flexible electronics, nanoelectromechanical systems, and nanocomposites. See References [9.7–9.12] for a detailed account of the use of graphene in engineering applications. Whilst the macroscopic properties of graphene are readily available in the literature, understanding its atomistic fracture behavior is very challenging. This is because experimental characterization of this material at the atomistic level is very difficult and unreliable. An atomically perfect pristine graphene sample possesses an extraordinary fracture strength of about 130 GPa. However, atomic flaws such as, vacancies, second phase particles, and adatoms are

S. A. Meguid, *Atomistic and Continuum Fracture Mechanics of Solids*,
https://doi.org/10.1007/978-3-031-56085-9_9

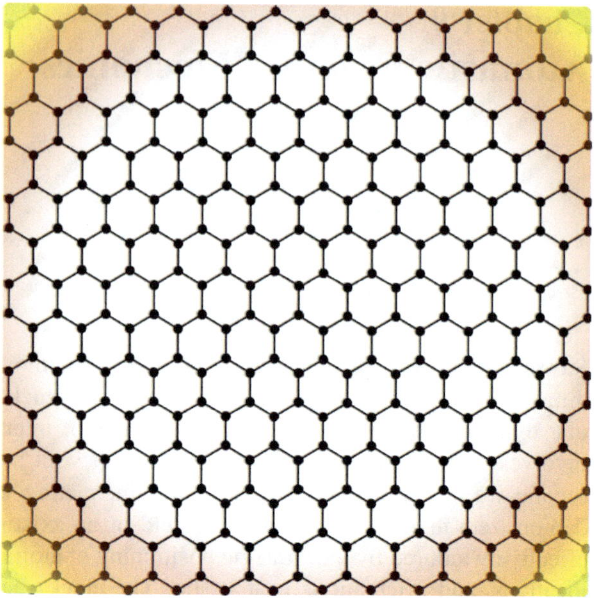

Figure 9.1 A schematic of a single layer with a single atom thickness of carbon forming the molecular structure of graphene

difficult to avoid in the fabrication of graphene. This leads to a major reduction in its fracture strength and ultimately its toughness. Therefore, it is important to study the effect of the presence of and the interactions between atomistic defects upon the fracture behavior of graphene [9.12–9.14].

9.2 The Griffith Crack

Linear elastic fracture mechanics establishes the conditions necessary for crack initiation, growth and arrest in a continuum using the concept of the stress intensity factor (Chapter 7). However, it is unclear how a nanoscale crack behaves at the atomistic level. In this chapter, we employ molecular dynamics simulations to investigate the fracture behavior of a central nanoscale crack in a graphene sheet subject to a tensile deformation. Figure 9.2 shows a typical graphene sheet used for our uniaxial tensile test simulations. The dimensions of the square sheet are assumed to be 400 Å × 400 Å and a central crack of length $2a = 50.8$ Å. In order to avoid the effects of finite dimensions on the MD simulation results, the length and width of the simulated graphene sheets were kept at more than 10 times half the initial crack length. This criterion for selecting the simulation domain ensures that boundaries of the graphene sheet do not interact with the K-dominant region, which is at and near the atomistic crack tip region. In this region, the stress field reaches its maximum at the tip and decays quite rapidly, as we move away from its tip [9.15, 9.16].

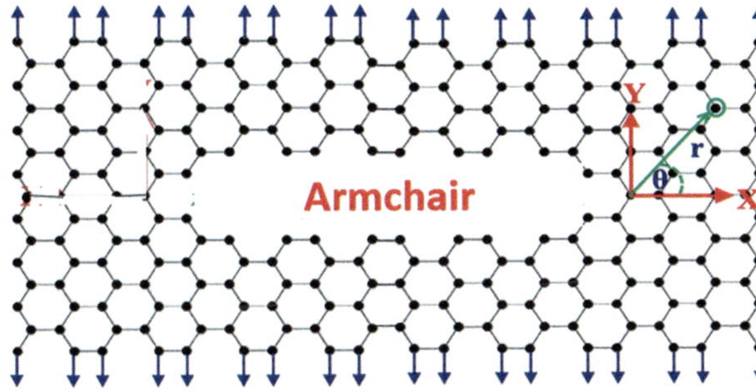

Figure 9.2 A schematic of a square graphene sheet (400 Å × 400 Å) containing an atomistic central crack of length 50.8 Å in atomistic armchair configuration

9.2.1 *Molecular Dynamics Simulations*

Uniaxial tensile test of a graphene sheet was simulated using large-scale atomic/molecular massively parallel simulator (LAMMPS). The adaptive intermolecular reactive bond order (AIREBO) potential was used for the MD simulations. The AIREBO potential consists of three sub-potentials, which are the reactive empirical bond order (REBO) potential, the Lennard-Jones potential, and the torsional potential. The REBO potential evaluates energy stored in atomic bonds. The Lennard-Jones and the torsional potentials include energies due to non-bonded and torsional interactions between atoms, respectively. The REBO potential expresses energy stored in a bond between atom *i* and atom *j* as being,

$$E_{ij}^{REBO} = f\left(r_{ij}\right)\left[V_{ij}^{R} + b_{ij}V_{ij}^{A}\right], \tag{9.1}$$

where V_{ij}^R and V_{ij}^A are the repulsive and attractive potentials, respectively, b_{ij} is the bond order term, which modifies the attractive potential depending on local bonding environment, r_{ij} is the distance between the atoms *i* & *j*, and $f(r_{ij})$ is the cut-off function, which limits the interatomic interactions to the nearest neighbors; expressed as:

$$f\left(r_{ij}\right) = \begin{cases} 1, & r_{ij} < R^{(1)} \\ \frac{1}{2} + \frac{1}{2}\cos\left[\frac{\pi\left(r_{ij} - R^{(1)}\right)}{\left(R^{(2)} - R^{(1)}\right)}\right], & R^{(1)} < r_{ij} < R^{(2)} \\ 0, & R^{(2)} < r_{ij} \end{cases} \tag{9.2}$$

where $R^{(1)}$ and $R^{(2)}$ are the cut-off radii, which are selected to be 1.7 and 2 Å, respectively. The values of cut-off radii were originally defined considering the first and the second nearest neighboring distances of relevant hydrocarbons. As explained in Chapter 2, this cut-off radii introduce non-physical strain hardening in the stress-strain curve of carbon based structures such as diamond and carbon nanotubes. Consequently, modified cut-off radii, ranging from 1.9 to 2.2 Å, have been used in order to eliminate this non-physical strain hardening effects. In this MD work, we used a truncated cut-off function $f_t(r_{ij})$; as given by,

$$f_t(r_{ij}) = \begin{cases} 1, & r_{ij} < R \\ 0, & r_{ij} > R \end{cases} \tag{9.3}$$

where the value of R is 2 Å. Similar truncated cut-off functions have been widely used for the fracture simulations of graphene. For details of how to conduct MD simulations and the parameters that govern the accuracy of the simulations, please refer to Chapter 2 and Ref. [9.17].

In view of the existing symmetry, periodic boundary conditions were implemented in the X and Y directions. The graphene sheet was allowed to relax over 25 ps prior to applying incremental tensile strains. To control the strain rate, the time step was kept at 0.5 fs. Isothermal-isobaric (NPT) ensemble was used for the simulations, where temperature and pressure were controlled using Nośe-Hoover thermostat and barostat, respectively. After the graphene sheet reaches equilibrium, strain was applied to the sheet along the y-direction at a rate of 0.001 ps^{-1}. During the straining stage, the graphene sheet was allowed to relax along the X-direction. The test temperature was kept at 300 K.

9.2.2 Atomistic Stress Intensity Factor

Figure 9.3a, b depict the crack tip geometries of zigzag and armchair cracks. It should also be noted that the armchair crack has a sharp crack tip configuration, whereas the zigzag crack has a blunt crack tip configuration. These two distinct chiralities provide two interesting crack tip configurations, which could result in markedly different crack-tip stress fields [9.18, 9.19]. It is also important to note that if the crack tip is blunt and is formed by consideration of the atomic armchair arrangement of atoms, the atoms forming the crack ligaments are in zigzag formation. Contrary to the blunt zigzag crack tip formation, in the armchair configuration, the crack tip is formed from zigzag atomic arrangement, while the crack ligaments are formed from armchair arrangement, as highlighted by the two perpendicular arrows shown at the bottom of Figure 9.3.

In linear elastic fracture mechanics, the stress state and the crack size are used to derive the stress intensity factor under mode I loading (K_I). Under mode I loading, also known as the opening mode, the crack surfaces are displaced directly apart in tension. Similarly, the stress distribution of the individual carbon atoms at the crack tip can also be used to calculate K_I. In this case, K_I can be calculated as (refer to Figure 9.3 for notations),

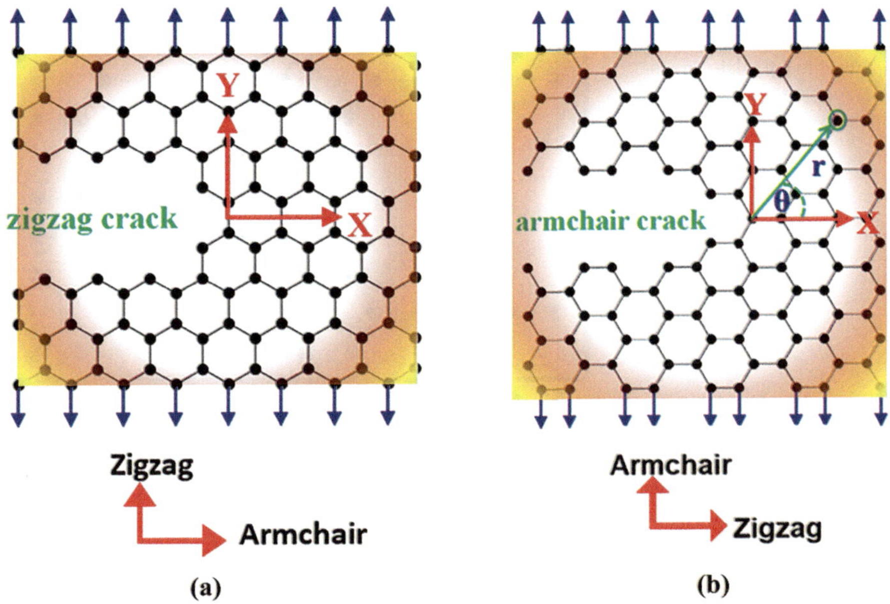

Figure 9.3 Detailed crack geometries depicting (**a**) blunt atomistic zigzag crack tip and (**b**) sharp atomistic armchair crack tip subject to tensile deformation along y-axis. Noteworthy is the fact that the atomistic arrangements of the crack ligaments provide different crack tip configurations

$$K_{\mathrm{I}} = \frac{1}{f(\theta)} \lim_{r \to 0} \sigma_{yy} \sqrt{2\pi r}, \tag{9.4}$$

where $f(\theta)$ is given by,

$$f(\theta) = \cos\left(\frac{\theta}{2}\right)\left[1 + \sin\left(\frac{\theta}{2}\right)\sin\left(\frac{3\theta}{2}\right)\right] \tag{9.5}$$

In contrast to the singular stress state at the tip of a linear elastic crack in a continuum, the stress at the tip of an atomistic crack is finite [9.18]. In addition, the stress state at the atomistic crack tip is perturbed by excessive deformation, excessive stress concentration, and interaction with dangling bonds. It is important to recognize that the behavior of the graphene sheet will be affected by the stress state as well as the defect size. Therefore, K_{I} can be determined using the stress state of the atoms near the crack tip using the following expression,

$$K_{\mathrm{I}} \approx \frac{1}{f(\theta)} \sigma_{yy} \sqrt{2\pi r} \tag{9.6}$$

According to linear elastic fracture mechanics, mode-I stress intensity factor at the onset of crack propagation K_{I} is reduced to,

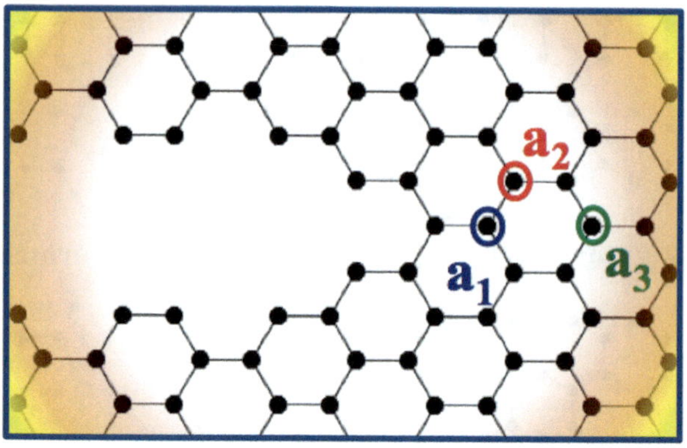

Figure 9.4 Selected three atom locations (a_1, a_2, and a_3) used to calculate an average value for stress intensity factor K_I

Table 9.1 Estimated stress intensity factor K_I using three different locations in close proximity to crack tip

Atom	r (Å)	θ (deg.)	σ_{yy} (GPa)	K_I (MPa $\sqrt{m}$)
a_1	1.40	0.0	85.9	2.54
a_2	2.42	30.0	82.9	2.83
a_3	4.19	0.0	63.4	3.25

$$K_I = \sigma_f \sqrt{\pi a} \tag{9.7}$$

where a is half the crack length and σ_f is the fracture stress, which is obtained by averaging the fracture stress of the far field atoms to ensure a fracture stress that is independent of the defect size. Analogous to the finite element method, we calculated K_I using Equation (9.6) from the knowledge of the stress and the spatial locations of the three discrete atoms a_1, a_2, and a_3, depicted in Figure 9.4. The calculated K_I using the corresponding quantities of the atoms are given in Table 9.1. For the same crack configuration, K_I given by Equation (9.7) is 2.87 MPa $\sqrt{m}$, which is close to the values calculated using Equation (9.6) for the atoms at the crack tip; e.g., for a_1 and a_2.

Let us now examine the length dependence of mode-I stress intensity factor K_I and other interesting properties concerning the fracture behavior of graphene. In this case, five crack lengths, ranging from 31.4 to 70.1 Å, were tested numerically using the same dimensions of the graphene sheet (400 Å × 400 Å). Figure 9.5 shows the variation in K_I with the crack length. It can be deduced from the figure that the variation in K_I is not significant beyond a crack length of approximately 50 Å.

According to Griffith's thermodynamic criterion of ideally brittle materials, σ_f can be expressed in terms of Young's modulus (E), the surface energy (γ), and the crack length (a); such that,

$$\sigma_f = \sqrt{\frac{2\gamma E}{\pi a}} \tag{9.8}$$

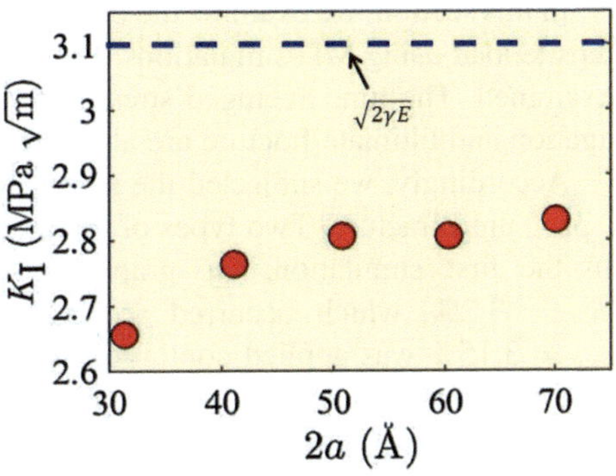

Figure 9.5 Change in K_I with crack length (2a) using Equation (9.6) as compared with Griffith's crack [9.21]

By combining Equations (9.7) and (9.8), an expression for the fracture toughness (K_{IC}) can be deduced in terms of γ and E; as follows,

$$K_{IC} = \sqrt{2\gamma E} \tag{9.9}$$

for plane stress condition. Using MD simulations, we determined Young's modulus E for graphene to be 1 TPa and the equilibrated potential energy of a carbon-carbon bond of graphene (PE_{bond}) to be 4.916 eV, which are in excellent agreement with existing experimental values [9.20]. If breaking of an individual carbon-carbon bond results in a crack advance Δa, then γ can be calculated as,

$$\gamma = \frac{PE_{bond}}{2\Delta at}, \tag{9.10}$$

where t is the thickness of the graphene sheet.

For a zigzag crack (Figure 9.3a), Δa is 2.418 Å, which yields the a value of $2\gamma = 9.58$ J/m^2. This value is in agreement with the ones calculated using density functional theory. Substituting the computed values of E and γ into Equation (9.9), K_{IC} was calculated to be 3.1 MPa$\sqrt{m}$. Figure 9.5 shows that K_{IC} obtained using Equation (9.9) is somewhat higher than the K_I given by Equation (9.7). Recent MD simulations in [9.21] revealed that Griffith's criterion of fracture is applicable to graphene to within some 15% accuracy, when the crack length is around 100 Å, which agrees quite well with the current simulation results.

9.2.3 *Single Crack Tip Stress Field*

In characterizing the fracture behavior of graphene, the stress experienced by the individual carbon atoms at the crack tip is very useful in determining the fracture process zone. Furthermore, the presence of atomistic vacancies in close proximity to the crack tip would influence the crack tip stress field.

In this section, we examine the behavior of a single atomistic crack under uniaxial tensile load using MD simulations. Both the zigzag and armchair configurations are examined. The time averaged stresses acting on the atoms at incipient crack propagation and ultimate fracture are also considered.

Accordingly, we subjected the graphene sheet to a tensile strain at a rate of 10^{-3} ps^{-1} until fracture. Two types of loading were considered in the MD simulations. In the first simulation, the graphene sheet was strained incrementally up to $\varepsilon_{yy} = 3.2\%$, which occurred at 32.25 ps. In the second simulation, a strain $\varepsilon_{yy} = 3.15\%$ was applied continuously over a period of 31.5 ps, and followed by equilibrating the graphene sheet under constant strain. The sheet was allowed to relax under the applied strain of 3.15% for 2.5 ps. It fractured at 34 ps. It is worth noting that from 32 ps to 33.5 ps, the stresses of the individual atoms were calculated at each time step (over 3000 time steps). The far field stress variations of the two simulations for a graphene sheet containing a zigzag crack were captured and compared in Figure 9.6a.

The stress distribution of a graphene sheet containing an armchair crack was also obtained by adopting the same procedure. Figure 9.6b, c show the stress distribution at the tip of armchair and zigzag cracks just prior to their propagation. The stress variation along the x-direction is compared in Figure 9.6d. The respective peak stress at the tip of a zigzag and an armchair cracks at the onset of crack propagation are determined using a strain of 0.1%, which leads to the initiation of crack propagation, corresponding to 112 and 130 GPa tensile stresses, respectively. It should also be noted that crack propagation in a zigzag and an armchair cracks occurs at strains of around 3.15% and 3.5%, respectively. This is due to the different bond arrangements at their crack tips. As shown in Figure 9.3a, there is a bond perpendicular to the crack at the zigzag crack tip. During the tensile test, this isolated bond is subjected to a higher stress compared to other bonds. In contrast, the two inclined bonds at the tip of the armchair crack (Figure 9.3c) accommodate part of the applied load by adjusting the bond angles, which allows the bonds at the crack tip to carry higher stresses.

It is worth noting that the crack tip stress field is finite and does not suffer from the singularity issue experienced by cracks in an elastic continuum.

9.3 Interacting Nanodefects

Let us now examine the case of an atomistic crack interacting with vacancies of varied arrangements. Again, the analysis is conducted using MD. Further information concerned this topic, together with the case of an atomistic crack interacting with an inhomogeneity, can be viewed in Refs. [9.21, 9.22].

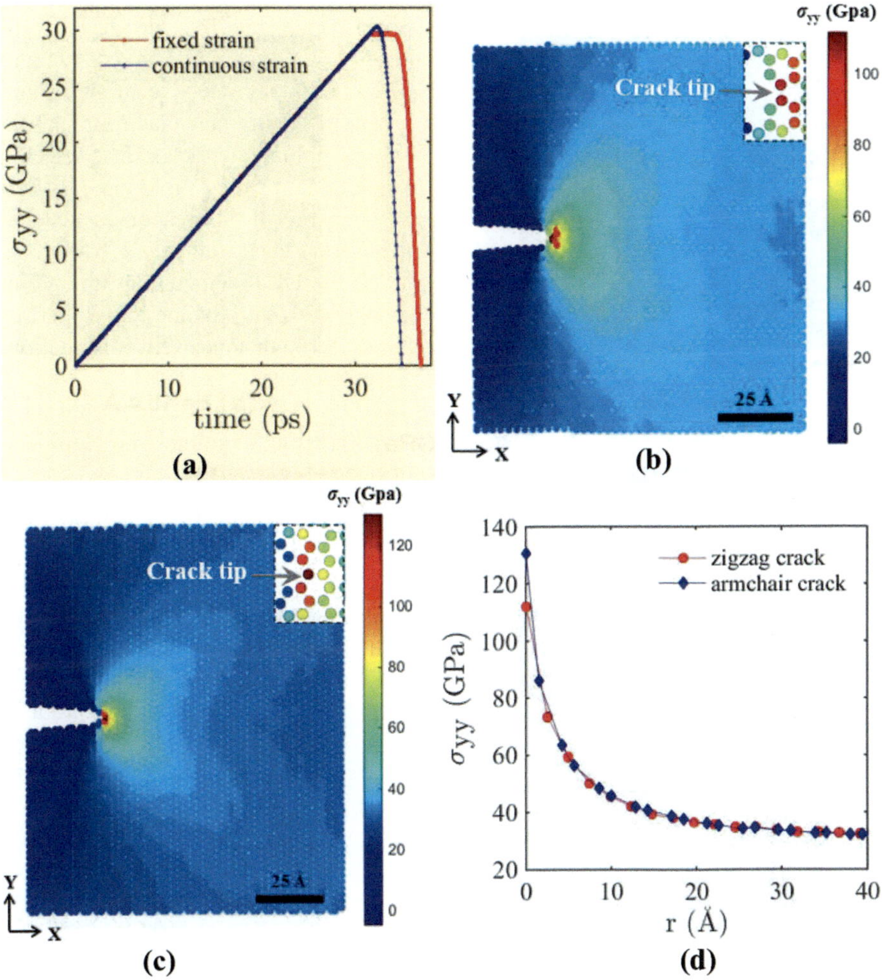

Figure 9.6 Stress distribution at the crack tips of two atomic arrangements of graphene. (**a**) Time variation of the far field stress of a zigzag crack of length 50.8 Å. The blue curve shows the stress variation when the strain $\varepsilon_{yy} = 3.5\%$ is applied incrementally until fracture, while the red shows the stress variation when a fixed strain $\varepsilon_{yy} = 3.15\%$ is applied, and the sheet was allowed to relax. (**b**) and (**c**) show averaged stress distribution at the crack tips of zigzag and armchair configurations at ε_{yy} of 3.15% and 3.5%, respectively. The respective crack lengths of the armchair and zigzag configurations are 53.1 and 50.8 Å, (**d**) comparison of crack tip stress field σ_{yy} for armchair and zigzag configurations [9.21]

9.3.1 Atomistic Crack Interacting with Collinear Vacancies

Let us consider the influence of two symmetrically placed vacancies ahead of the crack tip, ($\theta = 0$). This configuration of vacancies creates very high stress amplifications (~60%), when the vacancy is close to touching the crack tip (see

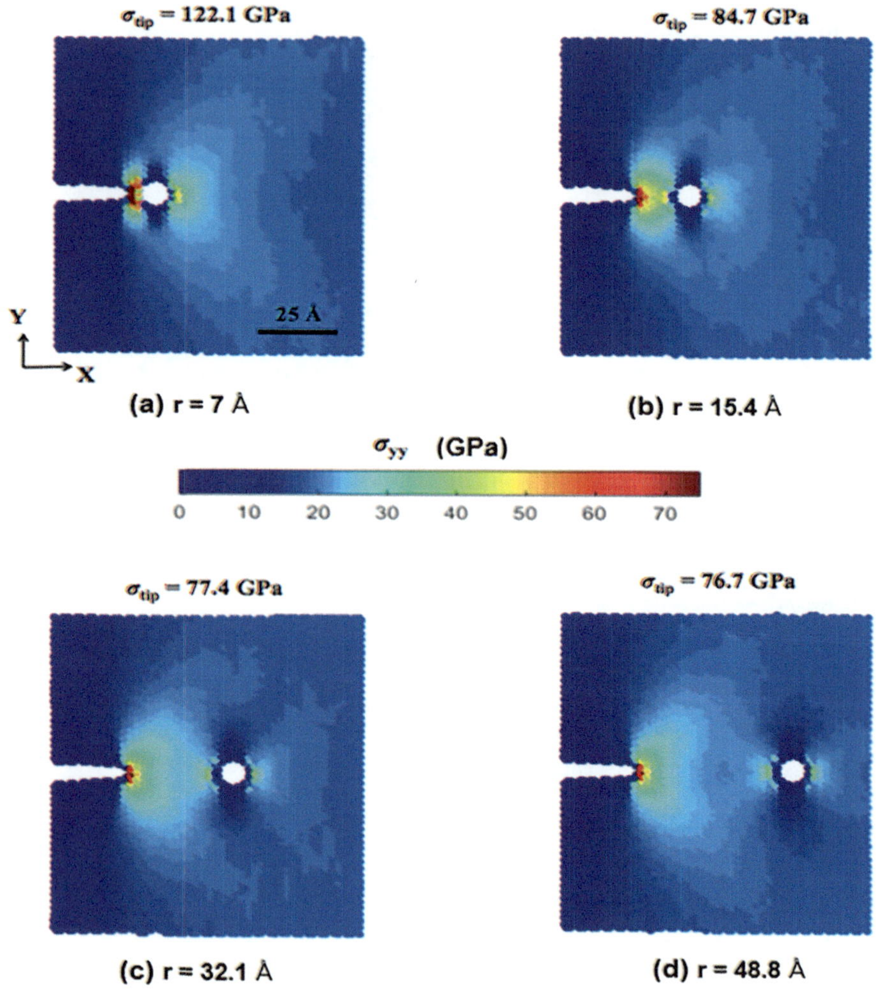

Figure 9.7 Stress trajectories (σ_{yy}) corresponding the armchair crack tip due to an applied tensile strain ε_{yy} of 2%: (**a**, **b**) major atomic interactions because of closeness of the atomic vacancies to the crack tip, while (**c**) shows much reduced interaction with the crack tip and (**d**) shows no interaction with the crack tip

Figure 9.7a). It can be seen in Figure 9.7d that the interaction between the stress fields created by the crack and the vacancy is minimum, and that the stress at the crack tip approaches the state of stress of a single non-interacting crack, which is 76.5 GPa. Similar observations can be made in the case of the zigzag crack. Among the studied cases, the highest stress amplification at the zigzag crack tip was 28%, which was reached when a vacancy was placed 8.4 Å ahead of the crack tip. The atomic configurations of the zigzag crack tip and the interacting vacancy results reveal significantly less amplification effects than observed with the armchair crack.

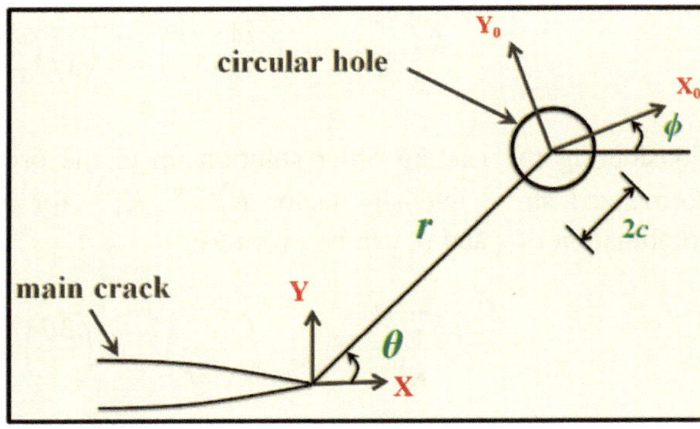

Figure 9.8 An arbitrarily located and oriented a circular hole near the tip of a main crack

The above shows that the toughness of a graphene sheet can be enhanced by introducing defense holes in the regions that encourage shielding (reduce the stress state ahead of the crack tip) effects. Meguid and his collaborators studied the interaction between a main crack and an arbitrarily located circular hole under mode I loading [9.23–9.25]. They considered a plane problem of an infinitely extended solid containing a semi-infinite main crack and a circular hole located near its tip, as depicted in Figure 9.8. The origins of the two rectangular coordinate systems XY and X_0Y_0 are taken at the tip of the main crack and at the center of the circular hole, respectively. The orientation angle of the hole is ϕ and its diameter is $2c$. The distance between the tip of the main crack and the center of the hole is denoted r. The inclination angle between the X-axis and the line connecting the tip of the main crack and the center of the hole is θ.

Meguid and his collaborators used the superposition of a number of subproblems, each containing a single defect, to develop a closed form solution. In the absence of the circular vacancy, the singular stress field near the tip of the main crack under mode I loading can be described using the stress intensity factor $K_{\mathrm{I}}^{(\mathrm{c})}$. However, the presence of the circular hole influences the stress field and the corresponding new stress intensity factor is assumed to be $K_{\mathrm{I}}^{(\mathrm{c-h})}$, which is given by:

$$K_{\mathrm{I}}^{(c-h)} = K_{\mathrm{I}}^{(c)} + \Delta K_{\mathrm{I}} \tag{9.11}$$

where ΔK_{I} takes into account the stress disturbance due to the presence of the hole in close proximity to the crack tip. The problem was analyzed using the complex potentials of Muskhelishvili together with an appropriate superposition procedure [9.23–9.25]. They obtained a closed-form solution for $K_{\mathrm{I}}^{(\mathrm{c-h})}$ using a perturbation technique; as explained in [9.23]. Their work revealed that when a collinear circular hole is located ahead of the main crack, i.e. $\theta = 0$, the solution for the normalized stress intensity factor under mode I loading $K_{\mathrm{I}}^{(\mathrm{c-h})}/K_{\mathrm{I}}^{(\mathrm{c})}$ can be explicitly expressed, up to the order $(c/r)^4$, as follows:

$$\frac{K_{\mathrm{I}}^{(c-h)}}{K_{\mathrm{I}}^{(c)}} = 1 + \left(\frac{1}{2}\right)\left(\frac{c}{r}\right)^2 + \left(\frac{1}{4}\right)\left(\frac{c}{r}\right)^4 + \cdots \tag{9.12}$$

Considering the leading order solution up to the order $(c/r)^2$, a solution for the normalized stress intensity factor $K_{\mathrm{I}}^{(\mathrm{c-h})}/K_{\mathrm{I}}^{(\mathrm{c})}$ for a general case, i.e. for any combination of r and θ, can be given as:

$$\frac{K_{\mathrm{I}}^{(c-h)}}{K_{\mathrm{I}}^{(c)}} = 1 + \left(\frac{c}{\sqrt{2}r}\right)^2 cos\left(\frac{3\theta}{2}\right) cos\left(\frac{\theta}{2}\right) \tag{9.13}$$

It should be noted that Equation (9.13) was derived for the case where a circular hole is interacting with the main crack. However, in the current MD problem, two symmetrically located holes interact with the crack tip. The closed-form solution developed by Meguid and his collaborators can be extended to treat the problem of an arbitrarily selected number of holes interacting with the main crack [9.23]. In this case, the interaction among the circular holes should also be considered. Assuming the interaction between the two holes to be negligible, we modified Equation (9.11) in such a way that it takes into account the presence of two symmetrically located holes at the crack tip as follows:

$$K_{\mathrm{I}}^{(c-h)} = K_{\mathrm{I}}^{(c)} + 2\Delta K_{\mathrm{I}} \tag{9.14}$$

Accordingly, the solution for the normalized stress intensity factor $K_{\mathrm{I}}^{(\mathrm{c-h})}/K_{\mathrm{I}}^{(\mathrm{c})}$ for the current problem can be given by:

$$\frac{K_{\mathrm{I}}^{(c-h)}}{K_{\mathrm{I}}^{(c)}} = 1 + \left(\frac{c}{r}\right)^2 cos\left(\frac{3\theta}{2}\right) cos\left(\frac{\theta}{2}\right) \tag{9.15}$$

In order to validate the closed-form solutions for $K_{\mathrm{I}}^{(\mathrm{c-h})}/K_{\mathrm{I}}^{(\mathrm{c})}$, i.e. Equations (9.12) and (9.15), let us compare $K_{\mathrm{I}}^{(\mathrm{c-h})}/K_{\mathrm{I}}^{(\mathrm{c})}$ obtained from the closed-form solution with $\sigma^{(\mathrm{c-h})}/\sigma^{(\mathrm{c})}$ obtained from MD simulations for the same defect sizes and configurations. It should be noted that it is impractical to obtain the stress intensity factor from MD simulations, when a hole is located near the main crack tip. This is because the presence of the hole influences the stress field at the crack tip. The normalized crack tip stress $\sigma^{(\mathrm{c-h})}/\sigma^{(\mathrm{c})}$ should be comparable to a corresponding normalized stress intensity factor, as evident from Equation (9.6). We calculated the percentage discrepancy of $K_{\mathrm{I}}^{(\mathrm{c-h})}/K_{\mathrm{I}}^{(\mathrm{c})}$ and compared it with $\sigma^{(\mathrm{c-h})}/\sigma^{(\mathrm{c})}$ as follows,

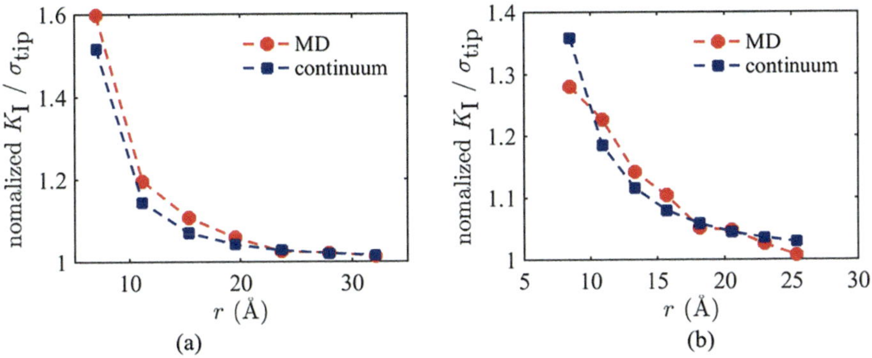

Figure 9.9 Comparison between the results obtained from the MD simulations (normalized by σ_{tip}) and the continuum model (normalized K_{I} from Equation (9.12)) when $\theta = 0$ for: (**a**) armchair crack and (**b**) zigzag crack [9.21]

$$\text{error}\% = \frac{\left[\left(\frac{\sigma^{(c-h)}}{\sigma^{(c)}}\right) - \left(\frac{K_{\text{I}}^{(c-h)}}{K_{\text{I}}^{(c)}}\right)\right]}{\left(\frac{\sigma^{(c-h)}}{\sigma^{(c)}}\right)} \times 100 \tag{9.16}$$

We first obtained a converged value for $K_{\text{I}}^{(\text{c}-\text{h})}/K_{\text{I}}^{(\text{c})}$ using the continuum closed-form solution for the case where a collinear circular hole is located ahead of the crack (Equation (9.12)). As shown in Figure 9.9, the agreement between the closed-form solution and the MD simulations is close. The maximum discrepancy for armchair and zigzag MD simulations and the closed-form solution were found to be in the range 5–6%, respectively.

Figure 9.9 shows the spatial distribution of the normalized stress intensity factor for various (r, θ) combinations where $K_{\text{I}}^{(\text{c}-\text{h})}/K_{\text{I}}^{(\text{c})}$ was obtained using the approximate closed-form solution given by Equation (9.15). It can be seen in Figure 9.9 that the discrepancy is less than 10% for many spatial locations of the holes. In the case of the armchair crack, 27% of the studied cases have less than 5% discrepancy, and it is 21% for the zigzag crack. For smaller values of r, the closed-form solution over predicts the shielding effect at higher θ values and under predicts the amplification effect at lower θ values.

The difference in results between MD predictions and the closed-form solution is attributed to several reasons: (a) the neglect of the higher order terms of the solution provided in Equation (9.15), (b) the closed-form solutions assume a singular stress field near the tip of the main crack, whereas MD does not (refer to Figure 9.5d), (c) the interaction between the holes has been neglected in the solution given by Equation (9.15); (d) the closed-form solution assumes a semi-infinite crack, whereas a finite central crack has been considered in the MD simulations.

9.3.2 Atomistic Crack Interacting with Multiple Vacancies

The situation envisaged is that of an atomistic central crack interacting with two symmetric atomistic vacancies on each side of the central crack tips, as depicted in Figure 9.10. The system is subjected to a uniaxial tensile deformation in the form of extensional displacement (strain) along the Y-axis. The atomistic crack length and vacancy diameter were taken to be 50.8 and 12.1 Å, respectively. The spatial location of the center of the hole with respect to the crack tip is given in terms of the polar coordinates r and θ, as described earlier.

In view of the symmetric nature of the vacancies, the crack propagation path of an atomistic central crack in a graphene sheet is expected to be self-similar. Furthermore, the presence of four symmetrically located vacancies, should also lead to a self-similar growth for the zigzag and armchair configurations. Figures 9.11 and 9.12 show snapshots of the respective time histories of the crack propagation profiles for both zigzag and armchair configurations.

The zigzag configuration of Figure 9.11 displays a symmetric central crack growth guided by the two symmetric vacancies. Three additional observations can be made. First, is the observed excessive crack opening displacement of the crack ligaments (Figure 9.11a). Second, as the deformation continues, some carbon-carbon bonds break, leading to incremental crack growth (Figure 9.11b). Third, with the added extensional deformation and the higher intensity of the stresses at the crack tip process zone, the fracture of the graphene sheet develops (Figure 9.11c).

Unlike the zigzag crack growth, the armchair crack growth deviated from symmetry, in spite of symmetric loading and overall atomic geometry. Figure 9.12 shows the two-stage propagation of the armchair crack. At the tip of the armchair crack, there are two inclined bonds (Figure 9.12c), and the crack propagation initiates by breaking one of these bonds. Figure 9.12a shows the initiation of the crack propagation by breaking a bond at $\varepsilon_{yy} = 3.2\%$. When $\varepsilon_{yy} = 3.3\%$, the crack path deviates upward toward the two upper atomistic vacancies; forming two

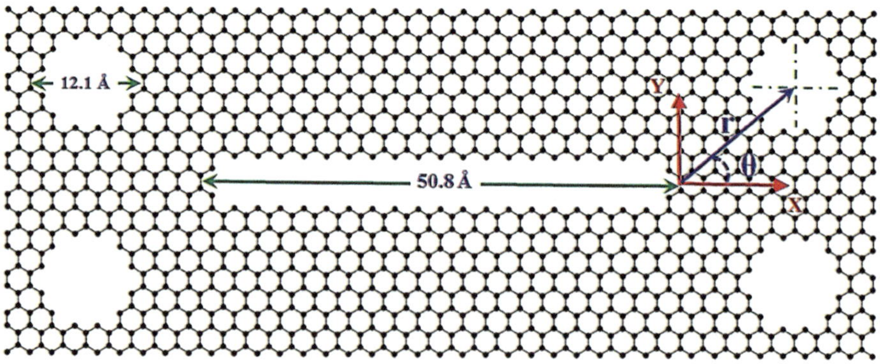

Figure 9.10 An atomistic zigzag crack interacting with four vacancies subject to a uniaxial tensile deformation along Y-direction

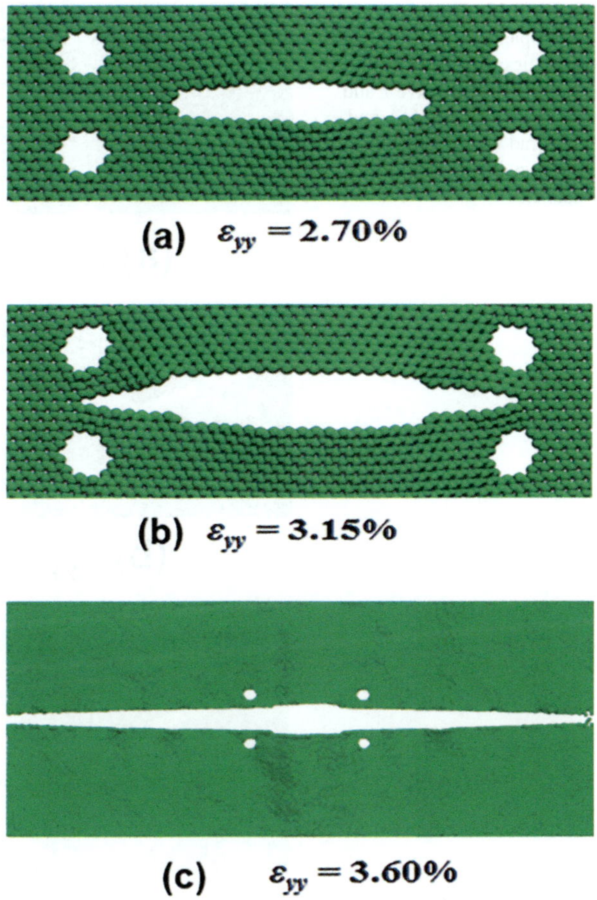

Figure 9.11 Snapshots of the self-similar propagation of a central zigzag crack, length 2a = 50.8 Å, the respective r and θ of the vacancy are 17.9 Å and 0.627 rad (Figure 9.3b) interacting with four symmetric vacancies

carbon-carbon chains at the top vacancies. Formation of the carbon-carbon chains in armchair atomistic configurations had previously been observed in experimental investigations; see, e,g., Refs. [9.26–9.29]. As shown in Figure 9.13c, the partially fractured graphene continues to carry load until the extensional strain $\varepsilon_{yy} = 4\%$, leading to the ultimate fracture. It is worth nothing that such crack branching does not prevail during the propagation of the zigzag crack, which, as stated earlier, has also been reported in the literature [9.26–9.29]. This is due to the different bond arrangements at the crack tips of a zigzag and an armchair cracks.

Figure 9.12c depicts two symmetrically inclined bonds at the tip of the armchair crack providing two potential crack propagation paths. In contrast, there is a bond perpendicular to the zigzag crack at its tip (see Figure 9.12a) resulting in a self-similar crack propagation path, which is a direct extension of the initial crack path along the x-direction. Even though there are two potential crack propagation paths for the armchair crack, thermal fluctuations of the atoms at the crack front dictates

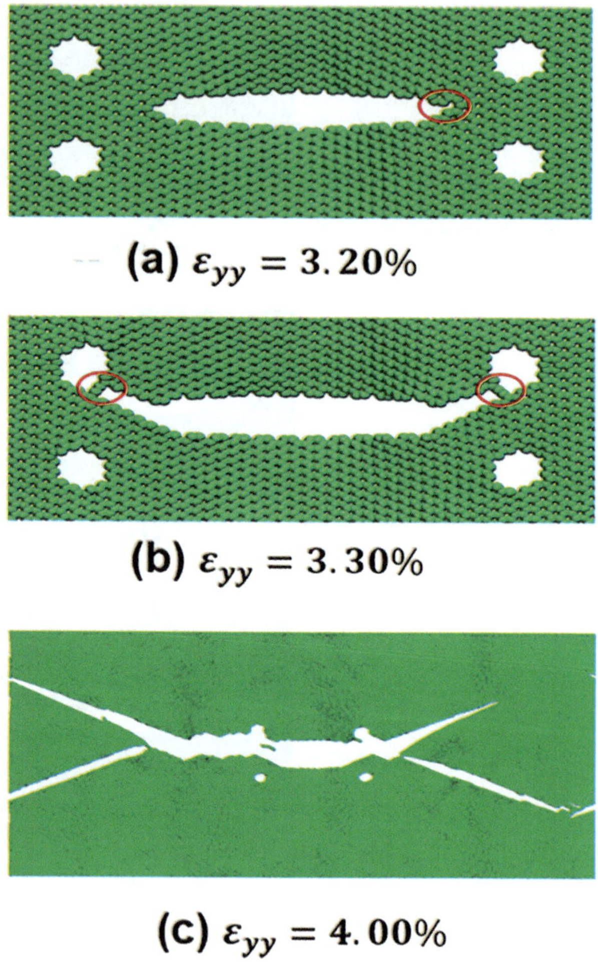

Figure 9.12 Snapshots of the propagation of a central armchair crack of length 53.1 Å; the respective r and θ of the holes are 16 Å and 0.856 rad which interact with two symmetric vacancies

the preferred propagation path, which lead to the crack branching observed in Figure 9.12.

9.3.3 *Crack Healing*

Self-healing of cracks emanating in graphene has drawn much attention from the research community in recent years. This is because such defects are difficult to avoid in many practical applications and they can lead to the deterioration of the electromechanical properties of graphene.

Figure 9.13 shows the healing of an initially propagated zigzag crack interacting with the previously examined vacancy configurations, where the stress-strain curves

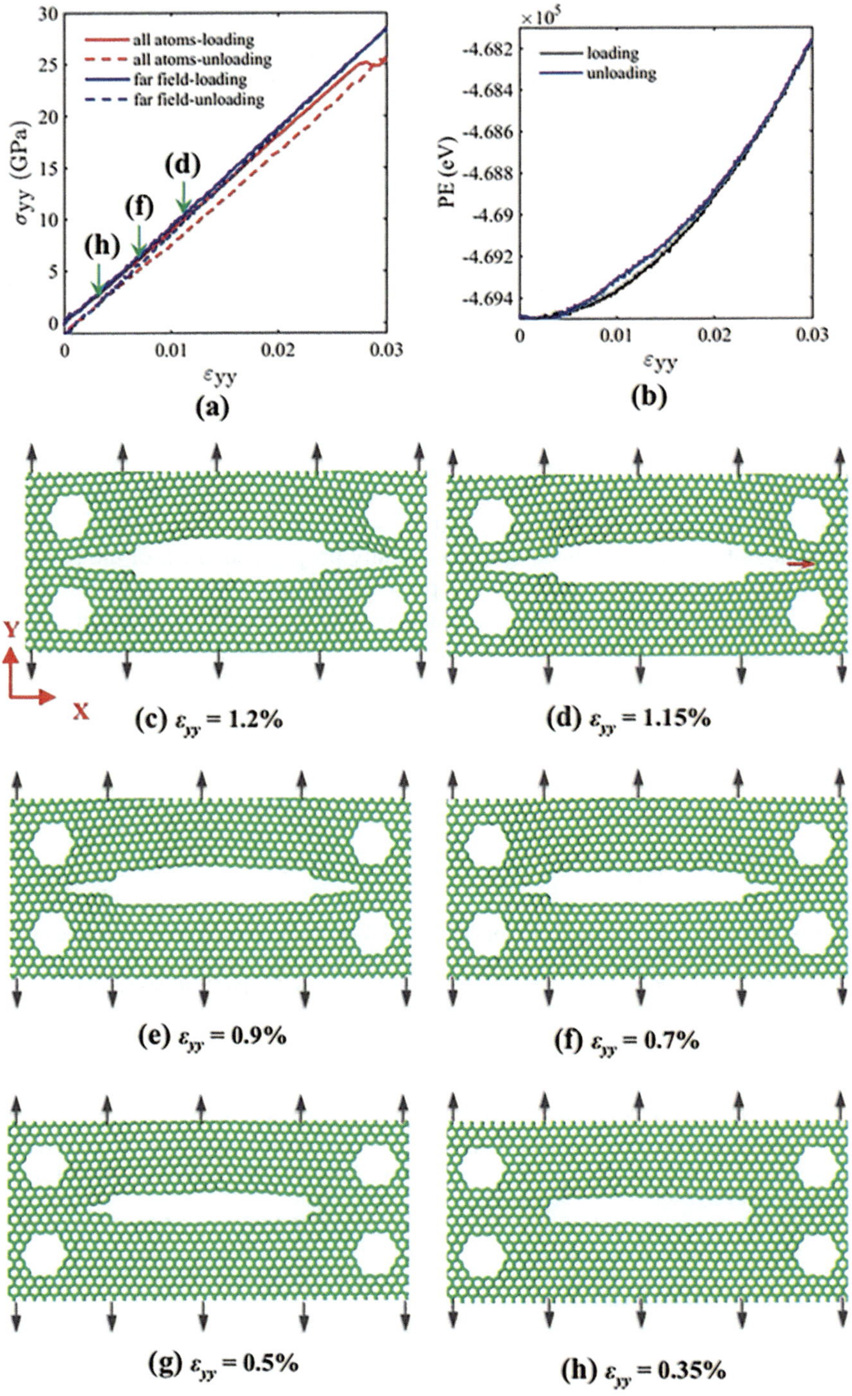

Figure 9.13 Healing of an initially propagated zigzag crack: (**a**) stress-strain curve, (**b**) the potential energy (PE)-strain relationship of the graphene sheet, (**c**) propagated crack just prior to the healing, (**d**) initiation of self-healing; the location of healing is shown by a red arrow. Finally, (**e**–**h**) various stages of crack healing [9.21]

during the loading and the unloading stages are given in Figure 9.13a. The sudden decrease in the average stress (red curve) around $\varepsilon_{yy} = 2.8\%$ is due to the initial crack propagation. It is worth noting that upon complete unloading of the tensile deformation, the graphene sheet experiences a compressive stress of approximately 1 GP, and it is released during the subsequent equilibration of the sheet. Figure 9.13b shows that the potential energy during the unloading is slightly higher than the loading stage between ε_{yy} ~0.5% and 2%. This is due to the fact that the crack length at the beginning of the unloading stage is slightly higher than that at the loading stage, as a result of the initial crack propagation. However, the crack length decreases during the unloading stage due to crack healing.

As depicted in Figure 9.13d, crack healing initiates at $\varepsilon_{yy} = 1.15\%$, which is 36.5% of the final fracture strain of the graphene sheet. The far field stress at the initiation of the crack healing is about 10 GPa. Figure 9.13e–h show the various stages of the crack healing. Interestingly, the initially propagated crack healed completely even when a strain $\varepsilon_{yy} = 0.35\%$ is applied to the graphene sheet. It should be noted that crack healing of an initially propagated crack of the two stage crack propagation occurs only in the presence of interacting vacancies. In the absence of vacancies, a single stage crack propagation will ensue leading to the complete fracture of graphene ensues. In addition to the arrest of crack propagation, the presence of vacancies reduces the crack-tip stress field during the unloading stage, thus facilitating the crack healing phenomenon. Further details can be obtained from the literature; see, e.g., [9.21].

References

9.1 K.S. Novoselov, V.I. Fal'ko, L. Colombo, P.R. Gellert, M.G. Schwab, K. Kim, A roadmap for graphene. Nature **490**, 192–200 (2012). https://doi.org/10.1038/nature11458

9.2 M.A. Rafiee, J. Rafiee, I. Srivastava, Z. Wang, H. Song, Z.-Z. Yu, N. Koratkar, Fracture and fatigue in graphene nanocomposites. Small **6**, 179–183 (2010). https://doi.org/10.1002/smll.200901480

9.3 Y.-S. Jun, S. Sy, W. Ahn, H. Zarrin, L. Rasen, R. Tjandra, B.M. Amoli, B. Zhao, G. Chiu, A. Yu, Highly conductive interconnected graphene foam based polymer composite. Carbon **95**, 653–658 (2015). https://doi.org/10.1016/j.carbon.2015.08.079

9.4 Y. Xu, A.S. Fleischer, G. Feng, Reinforcement and shape stabilization of phase-change material via graphene oxide aerogel. Carbon **114**, 334–346 (2017). https://doi.org/10.1016/j.carbon.2016.11.069

9.5 T. Sainsbury, S. Gnaniah, S.J. Spencer, S. Mignuzzi, N.A. Belsey, K.R. Paton, A. Satti, Extreme mechanical reinforcement in graphene oxide based thin-film nanocomposites via covalently tailored nanofiller matrix compatibilization. Carbon **114**, 367–376 (2017). https://doi.org/10.1016/j.carbon.2016.11.061

9.6 Y. Cui, S.I. Kundalwal, S. Kumar, Gas barrier performance of graphene/polymer nanocomposites. Carbon **98**, 313–333 (2016). https://doi.org/10.1016/j.carbon.2015.11.018

9.7 P. Ruffieux, S. Wang, B. Yang, C. Sánchez-Sánchez, J. Liu, T. Dienel, L. Talirz, P. Shinde, C.A. Pignedoli, D. Passerone, T. Dumslaff, X. Feng, K. Müllen, R. Fasel, On-surface synthesis of graphene nanoribbons with zigzag edge topology. Nature **531**, 489–492 (2016). https://doi.org/10.1038/nature17151

9.8 J. Cai, P. Ruffieux, R. Jaafar, M. Bieri, T. Braun, S. Blankenburg, M. Muoth, A.P. Seitsonen, M. Saleh, X. Feng, K. Müllen, R. Fasel, Atomically precise bottom-up fabrication of graphene nanoribbons. Nature **466**, 470–473 (2010). https://doi.org/10.1038/nature09211

9.9 L. Talirz, H. Söde, T. Dumslaff, S. Wang, J.R. Sanchez-Valencia, J. Liu, P. Shinde, C.A. Pignedoli, L. Liang, V. Meunier, N.C. Plumb, M. Shi, X. Feng, A. Narita, K. Müllen, R. Fasel, P. Ruffieux, On-surface synthesis and characterization of 9-atom wide armchair graphene nanoribbons. ACS Nano **11**, 1380–1388 (2017). https://doi.org/10.1021/acsnano.6b06405

9.10 Y.-C. Chen, T. Cao, C. Chen, Z. Pedramrazi, D. Haberer, D.G. de Oteyza, F.R. Fischer, S.G. Louie, M.F. Crommie, Molecular bandgap engineering of bottom-up synthesized graphene nanoribbon heterojunctions. Nat. Nanotechnol. **10**, 156–160 (2015). https://doi.org/10.1038/nnano.2014.307

9.11 F. Banhart, J. Kotakoski, A.V. Krasheninnikov, Structural defects in graphene. ACS Nano **5**, 26–41 (2011). https://doi.org/10.1021/nn102598m

9.12 A. Eichler, J. Moser, J. Chaste, M. Zdrojek, I. Wilson-Rae, A. Bachtold, Nonlinear damping in mechanical resonators made from carbon nanotubes and graphene. Nat. Nanotechnol. **6**, 339–342 (2011). https://doi.org/10.1038/nnano.2011.71

9.13 C. Chen, S. Rosenblatt, K.I. Bolotin, W. Kalb, P. Kim, I. Kymissis, H.L. Stormer, T.F. Heinz, J. Hone, Performance of monolayer graphene nanomechanical resonators with electrical readout. Nat. Nanotechnol. **4**, 861–867 (2009). https://doi.org/10.1038/nnano.2009.267

9.14 L.G. Cançado, A. Jorio, E.H.M. Ferreira, F. Stavale, C.A. Achete, R.B. Capaz, M.V.O. Moutinho, A. Lombardo, T.S. Kulmala, A.C. Ferrari, Quantifying defects in graphene via Raman spectroscopy at different excitation energies. Nano Lett. **11**, 3190–3196 (2011). https://doi.org/10.1021/nl201432g

9.15 A.A. Griffith, The phenomena of rupture and flow in solids. Philos. Trans. R. Soc. A (Lond.) **221**, 163–198 (1921). https://doi.org/10.1098/rsta.1921.0006

9.16 S.A. Meguid, *Engineering Fracture Mechanics* (Elsevier Applied Science, London, 1989)

9.17 A.R. Alian, S.A. Meguid, A critical study of the parameters governing molecular dynamics simulations of nanostructured materials. Comput. Mater. Sci. **153**, 183–199 (2018). https://doi.org/10.1016/j.commatsci.2018.06.028

9.18 M.A.N. Dewapriya, S.A. Meguid, Atomistic modeling of out-of-plane deformation of a propagating Griffith crack in graphene. Acta Mech. **228**, 3063–3075 (2017). https://doi.org/10.1007/s00707-017-1883-7

9.19 M.T. Yin, M.L. Cohen, Structural theory of graphite and graphitic silicon. Phys. Rev. B **29**, 6996–6998 (1984). https://doi.org/10.1103/PhysRevB.29.6996

9.20 H. Yin, H.J. Qi, F. Fan, T. Zhu, B. Wang, Y. Wei, Griffith criterion for brittle fracture in graphene. Nano Lett. **15**, 1918–1924 (2015). https://doi.org/10.1021/nl5047686

9.21 M.A.N. Dewapriya, S.A. Meguid, Atomistic simulations of nanoscale crack-vacancy interaction in graphene. Carbon **125**, 113–131 (2017). https://doi.org/10.1016/j.carbon.2017.09.015

9.22 M.A.N. Dewapriya, S.A. Meguid, R.K.N.D. Rajapakse, Atomistic modelling of crack-inclusion interaction in graphene. Eng. Fract. Mech. **195**, 92–103 (2018). https://doi.org/10.1016/j.engfracmech.2018.04.003

9.23 S.X. Gong, S.A. Meguid, A general solution to the antiplane problem of an arbitrarily located elliptical hole near the tip of a main crack. Int. J. Solids Struct. **28**, 249–263 (1991). https://doi.org/10.1016/0020-7683(91)90209-X

9.24 S.A. Meguid, S.X. Gong, P.E. Gaultier, Main crack-microcrack interaction under mode I, II and III loadings: Shielding and amplification. Int. J. Mech. Sci. **33**, 351–359 (1991). https://doi.org/10.1016/0020-7403(91)90074-D

9.25 S.X. Gong, S.A. Meguid, Microdefect interacting with a main crack: A general treatment. Int. J. Mech. Sci. **34**, 933–945 (1992). https://doi.org/10.1016/0020-7403(92)90063-M

9.26 C. Lee, X. Wei, J.W. Kysar, J. Hone, Measurement of the elastic properties and intrinsic strength of monolayer graphene. Science **321**, 385–388 (2008). https://doi.org/10.1126/science.1157996

9.27 P. Zhang, L. Ma, F. Fan, Z. Zeng, C. Peng, P.E. Loya, Z. Liu, Y. Gong, J. Zhang, X. Zhang, Others, fracture toughness of graphene, nature. Communications **5**, 3782 (2014). https://doi.org/10.1038/ncomms4782
9.28 T. Zhang, X. Li, H. Gao, Fracture of graphene: A review. Int. J. Fract. **196**, 1–31 (2015). https://doi.org/10.1007/s10704-015-0039-9
9.29 G. López-Polín, J. Gómez-Herrero, C. Gómez-Navarro, Confining crack propagation in defective graphene. Nano Lett. **15**, 2050–2054 (2015). https://doi.org/10.1021/nl504936q

Chapter 10
Multiscale Modeling of Nanocomposites

Summary In view of the fact that the atomistic basic length is the Angstrom and the basic time scale is femtosecond, material characterization necessitates the building up of hierarchal length and time scales. Each scale is treated differently and is governed by the physics of that particular scale. Typically, length scale hierarchy in material characterization would trigger quantum mechanics, density functional theory, molecular dynamics, micromechanics, finite element and continuum mechanics. The objective of multiscale endeavors is to bridge these different scales to facilitate either the prediction of the effective properties of a complex material system(s) or provide the atomic hierarchal structure necessary to achieve specific properties. The main focus of this chapter is to provide a fundamental understanding of the concepts and principles upon which multiscale modeling is based and demonstrate its applicability by examining three case studies.

10.1 Introduction

Currently, a large number of modeling techniques exist in the literature. These techniques aim not only to simulate material behavior at the particular scale of interest, but also in developing new materials with highly desirable properties. These scales can range from angstrom to the much coarser continuum levels of meters. Figure 10.1 shows the hierarchy of the different length and time scales. It consists of quantum mechanics (QM) and density functional theory (DFT), molecular dynamics (MD), micromechanics (MM), finite element and continuum mechanics [10.1–10.3].

The different methods of analyses to be considered should be accurate and best suited for the physics governing the specific length and time scales of the system. Decidedly, modeling inaccuracies would arise if the governing equations for a length scale is improperly applied to a different length scale, where the governing equations are vastly different. It is also impractical and very costly to model an entire continuum using purely atomistic approaches due to the enormous number of degrees of freedom involved and the computing challenges. Many engineering problems are characterized in terms of multiple scales and require multiscale

S. A. Meguid, *Atomistic and Continuum Fracture Mechanics of Solids*,
https://doi.org/10.1007/978-3-031-56085-9_10

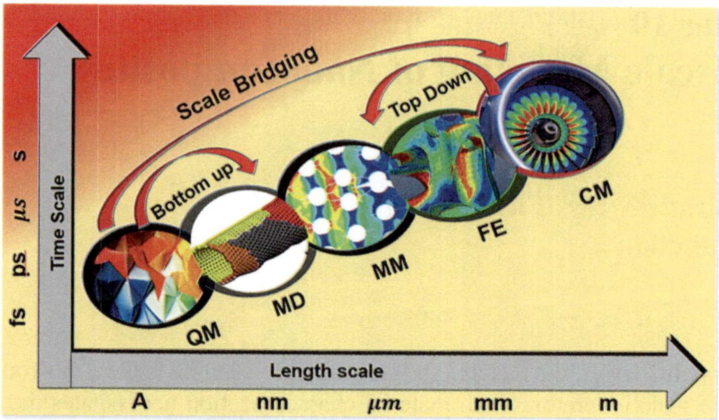

Figure 10.1 Illustrative figure depicting varied length scales in material characterization and common techniques adopted in treating them. Multiscale modeling bridges these techniques. Key: *QM* quantum mechanics, *DFT* density functional theory, *MD* molecular dynamics, *MM* micromechanics, *FE* finite element and *CM* continuum mechanics

techniques to bridge their varied length and time scales. In this chapter, we shall devote attention to varied length and time scales with the focus being multifunctional nanocomposites. The time scale comes in this class of problems when dealing with strain rates that need to be integrated to provide velocities and ultimately deformation.

10.2 Numerical Characterization of Multifunctional Nanocomposite

10.2.1 Atomistic Modeling

Atomistic numerical methods, such as DFT and MD, have been extensively used to model multifunctional nanocomposites. In the case of DFT, a very limited number of atoms is used to characterize the electrical properties of multifunctional nanoreinforced composites. For this reason, it will not be discussed beyond this point. In the case of MD, discussed in details in Chapter 2, the classical equations of motion for each atom are integrated step-wise in the time domain. The time steps can range from 1 to 10 femtoseconds and the corresponding simulation time can range from several picoseconds to nanoseconds, depending on the system being investigated. Due to the large number of degrees of freedom involved, MD is computationally intensive. It is worth noting that even the use of state-of-the-art parallel supercomputers can only handle a limited number of atoms ($\sim 10^9$), corresponding to less than one cubic micron [10.1]. Therefore, the atomistic level approach presently can only model short nanotube segments and a small number of short polymer chains with relatively high strain rates to reduce the computing time. In spite of these limitations, MD has been very helpful in the atomistic characterization of the elastic properties and the fracture behavior of multi-phase material systems.

10.2.2 Micromechanics

To adequately address scale up issues in multiscale analysis, it is also desirable to couple atomistic models with well-established micromechanical techniques to describe the mechanical behavior of polymer nanocomposites. In this case, the problem is often formulated at the atomistic level using a representative volume element (RVE) that contains a fiber surrounded by a polymeric matrix, which is subsequently homogenized into a representative fiber having uniform properties. The representative fiber is then used to describe the nanophase and its immediate surrounding in the micromechanical description.

Several micromechanics schemes exist. Many are based on the eigenstrain and Eshelby tensor. They include: the inhomogeneity problem, effective medium theory, two scale composite method, coherence fields method, multiparticle effective field method, and various extensions of the Mori-Tanaka scheme, among many others [10.2]. They have been implemented in treating all composite morphologies; including, fully aligned, two-dimensional in-plane random, three-dimensional random orientations, interphases and interfaces.

10.2.3 Finite Element Method

It is worth noting that the finite element method has been successful in component/system characterization. There exists considerable amount of work in the literature to support this statement. The method has been given its due attention in Chapter 4.

10.2.4 Continuum Mechanics

The traditional framework in mechanics has always been the continuum. Under this framework, material systems are assumed to be composed of an infinitely divisible continuous medium, with a constitutive relation that remains the same for a wide range of system sizes. The underlying atomic structure of matter is neglected altogether and is replaced with a continuous and homogeneous material representation. Continuum approaches have been applied to study nanoscale materials. However, traditional continuum based models cannot accurately describe the influence of the size and surface effects of nanofillers upon the mechanical properties, bond formation, bond breakage and their interactions in composite systems. This is because they lack the appropriate constitutive relations that govern their behavior at this scale. At the nanoscale, traditional continuum mechanical concepts do not maintain their validity and gross oversimplifications do arise from the use of a purely continuum model. Existing works show that the classic relationship between Young's modulus and the shear modulus in the elastic theory of continuum mechanics is not retained for a SWCNT [10.1–10.3]. Specifically, they showed that if the classic formula is used, the shear modulus would be grossly overestimated for SWCNTs with large chiral angles and underestimated for small ones.

10.2.5 From Atomistic to Continuum

The mechanical deformation and the failure of many engineering materials are in fact multiscale phenomena and their observed macroscale behavior is governed by processes that occur on different length and time scales. These processes are often dependent on each other and affect the overall deformation and fracture behavior of these materials. It is therefore necessary to model these material systems using the physics and constitutive laws governing each individual length and time scales. Since the various scales of any material system depend on each other, it is essential to use multiscale techniques to formulate a system, wherein the scales of interest are bridged into a coupled, unified and informed approach leading to the computation of the effective properties of that system. Clearly, the degree to which these scales are coupled and unified would depend upon the physics of the material system being investigated. Hence, numerous multiscale approaches currently exist in the literature.

Most multiscale modeling techniques adopt either sequential [10.3] or concurrent [10.4] approaches to treat this class of problems, where it is common to employ MD for atomistic and finite element methods (FE) for continuum scales. The sequential approach assumes that the problem considered can be easily separated into processes that are governed by different length and time scales. In doing so, the simulations are running independently of each other and a complete separation of both length and time scales do exist. Typically, two length scales are developed, one fine and the other is coarse. A characteristic output of the finer scale simulation is used as boundary conditions for the coarser scale. A schematic of the sequential approach is given in Figure 10.2. This is demonstrated in the second case study of this chapter.

As an alternative, concurrent methods are adopted. Concurrent methods, such as the quasicontinuum method [10.4], perform the entire multiscale simulation simultaneously and continually feed information from one length scale to the other in a dynamically controlled fashion. Concurrent methods are better suited in representing scales with a strong dependence. The transfer of information ensures consistency amongst the field variables between the different simulation techniques. This two-way transfer of field variables is achieved through the use of a well-thought out transition region depicted in Figure 10.3.

Figure 10.2 Schematic of a sequential multiscale modeling technique

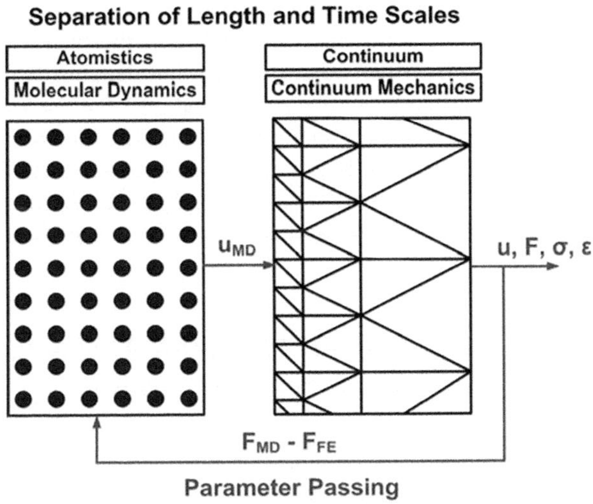

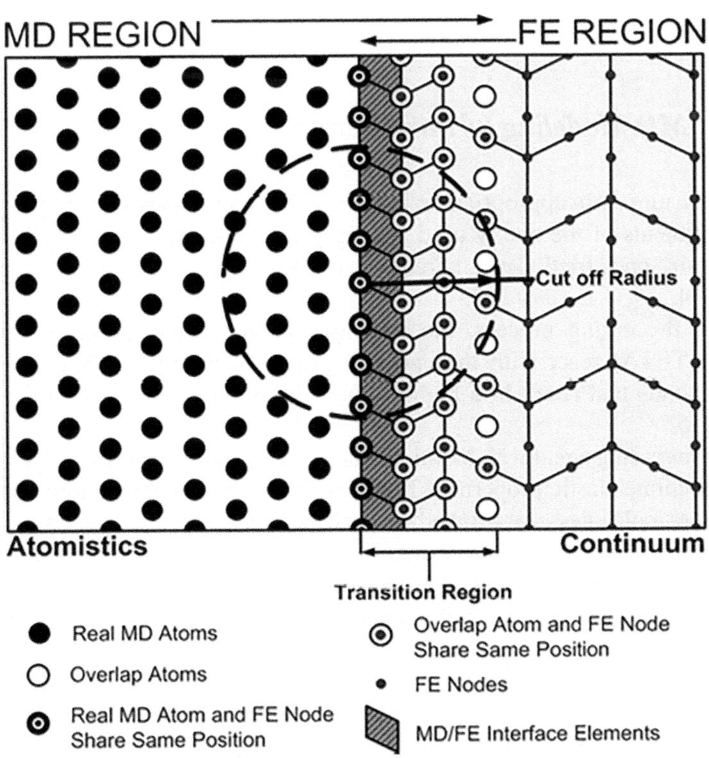

Figure 10.3 Transition region used to couple MD and FE methods in concurrent multiscale techniques

Unfortunately, most concurrent multiscale modeling techniques suffer from one or more of the following difficulties: (i) the necessity of meshing traditional FE regions down to the atomic scale, thus leading to physical inconsistencies and numerical difficulties, (ii) contamination of the solution due to wave reflection resulting from the improper description of the transition zone, and (iii) the FE/MD energy mismatch leading to erroneous non-physical effects in the transition region. A number of excellent reviews detailing the different multiscale modeling techniques, their merits and limitations can be viewed in the scientific literature.

10.3 Case Study #I: Multiscale Modeling of Mechanical Properties of Nanocomposites

In the following, we demonstrate how to apply multiscale modeling techniques to predict the effective properties of CNT-reinforced epoxy. A comprehensive and a detailed account of multiscale modeling of composites can also be found in references [10.1–10.5].

10.3.1 MD Modeling of Pure Epoxy

At this juncture, it is appropriate to conduct MD simulations of the epoxy matrix. The constituents of the epoxy used are diglycidyl ether of bisphenol A (DGEBA) epoxy resin and triethylene tetramine (TETA) curing agent, as depicted in Figure 10.4.

During the curing process, the hydrogen atoms in the amine groups of the hardener (TETA) react with the epoxide groups of the resin (DGEBA) forming covalent bonds that result in a highly cross-linked epoxy structure, as depicted in Figure 10.5.

The resin/curing agent weight ratio in the epoxy polymer was set to 2:1 in order to achieve accurate elastic properties. The cross-linked polymer structure consisted of 80 DGEBA molecules cross-linked with 40 molecules of curing agent TETA (see Figure 10.5a). The cross-linked structure was used to form a 3D structure, as shown in Figure 10.5b. This cross-linked epoxy structure was then used to build the epoxy system in the subsequent MD simulations. The neat epoxy model was generated by randomly placing 5 cross-linked structures in a cubic simulation box of size 150 Å × 150 Å × 150 Å to form a system containing 400 chains of DEGBA and 200 chains of TETA, as shown in Figure 10.5c. The total number of atoms in the simulation box is 25,300. In the current case study, the volumetric bulk modulus (K) and shear modulus (G) of the cross-linked epoxy were determined by applying the volumetric and three-dimensional shear strains, respectively.

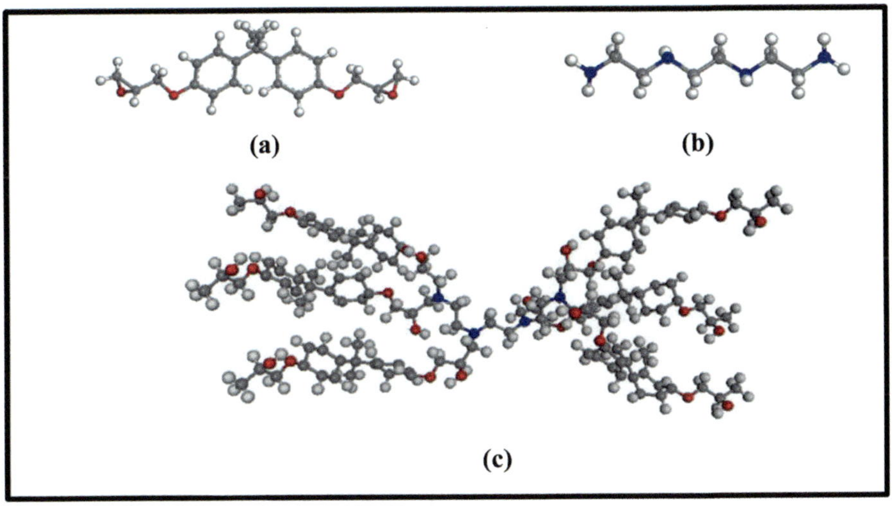

Figure 10.4 Chemical composition of (**a**) epoxy resin (DGEBA), (**b**) curing agent (TETA), and (**c**) cured epoxy oligomer. Hydrogen, carbon, oxygen and nitrogen atoms of the epoxy chain are respectively presented by white, gray, red and blue colors

In all simulations, strain increments of 0.25% were applied along a particular direction by uniformly expanding or shearing the simulation box and updating the atoms coordinates to fit within the new dimensions. After each strain increment, the MD unit cell was equilibrated using the NVT ensemble at 300 K for 10 ps. Then, the stress tensor is averaged over an interval of 10 ps, to reduce the effect of the numerical fluctuations. These steps were repeated in the subsequent deformation increments. The procedure was stopped when the total strain reaches 2.5%. The bulk and shear moduli, Young's modulus (E) and Poisson's ratio (ν) were determined using the following expressions:

$$E = \frac{9KG}{3K + G} \text{ and } \nu = \frac{3K - 2G}{2(3K + G)} \tag{10.1}$$

The predicted elastic properties of the epoxy using MD simulations are summarized in Table 10.1. The predicted values of the moduli are consistent with the experimentally measured values of a similar epoxy [10.6].

10.3.2 Representative Volume Element

Meguid and his collaborators showed in numerous articles [10.1, 10.2, 10.5] that the homogeneous dispersion of a few weight percentages of CNTs in an epoxy matrix can significantly improve the tensile and shear strength as well as the toughness of

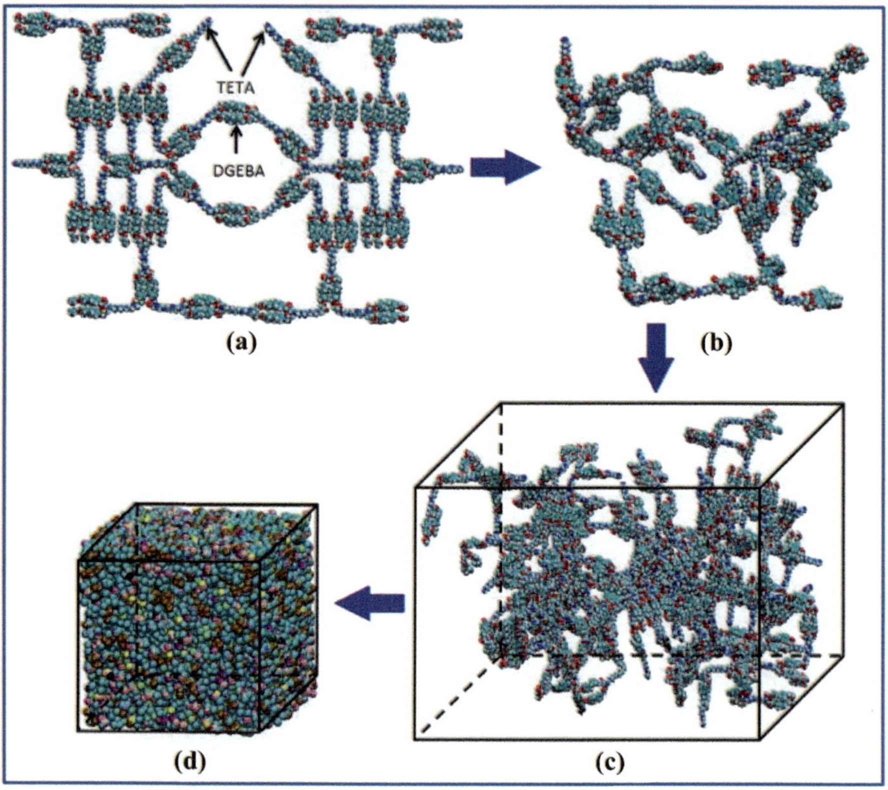

Figure 10.5 Steps used in the preparation of the polymer matrix (**a**) cross-linked structure, (**b**) distorted structure in the 3D unit cell, (**c**) five 80:40 cross-linked structure randomly placed in a simulation box of size 150 Å × 150 Å × 150 Å, and (**d**) the equilibrated system in a simulation box of size 61 Å × 61 Å × 61 Å after 25 volume reduction steps [10.5]

Table 10.1 Elastic moduli of thermoset epoxy

	Young's modulus (GPa)	Shear modulus (GPa)	Bulk's modulus (GPa)	Poisson's ratio
MD predictions	3.2	1.1	4.9	0.39
Experimental results [10.6]	2.9	1.0	4.4	0.35

the resulting synthesized nanocomposites. However, at higher CNT concentrations the mechanical properties of the nanocomposite were found to deteriorate due to the formation of CNT agglomerates that lead to relatively poor interfacial properties and the high stress concentration that results from the aggregated agglomerates.

In order to develop a relationship between the nano-micro-continuum scales, we must select a random sample of a given volume that is representative of the atomistic

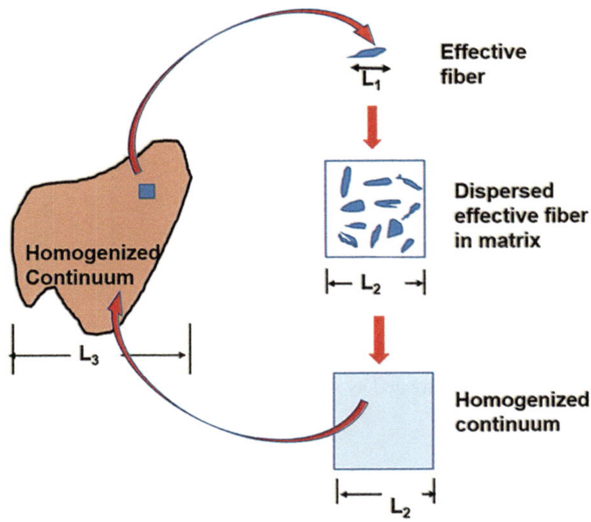

Figure 10.6 Separation of different length scales depicting $L_1 \ll L_2 \ll L_3$ that comprise the composite and the necessary homogenization process to develop an effective fiber for reinforcement of the epoxy matrix with a specific volume fractions

constituents, known as a "Representative Volume Element (RVE)." A number of techniques exist for the selection of a RVE. In this case study, it is decided to use the separation of length scales between atomistic, micro and macro ranges to develop the appropriate RVE. Accordingly, we must identify a characteristic length in the composite solid. Figure 10.6 shows the three different length scales identified: atomistic constituent scale (L_1), followed by a representative volume element (RVE) scale (L_2) and finally a homogenized continuum scale (L_3). These characteristic length scales must satisfy the condition that $L_1 \ll L_2 \ll L_3$, as illustrated in figure. The size differences between the characteristic lengths would facilitate the identification of the appropriate constitutive laws that govern the material behavior at every scale. In turn, the stress-strain response for a given RVE must be independent of the atomistic and microstructural variations within a devised scale. This will ensure that the RVE is statistically representative in the structural scale.

10.3.3 Development of an Effective Fiber

In this section, a two-step multiscale model is used to develop an effective fiber from a CNT surrounded by epoxy molecules of a given volume fraction using an appropriate homogenization technique. In the first step, we use MD simulations to: (i) determine the isotropic stiffness tensor of a representative volume element (RVE) surrounded by epoxy molecules, (ii) determine the transversely isotropic stiffness tensor of the RVEs representing an individual CNT embedded in an epoxy matrix, and (iii) determine the transversely isotropic stiffness of an agglomerated bundle of CNTs embedded in an epoxy matrix. In the second, the RVE is considered as an effective reinforcement and the epoxy polymer as the matrix phase. In this case, the

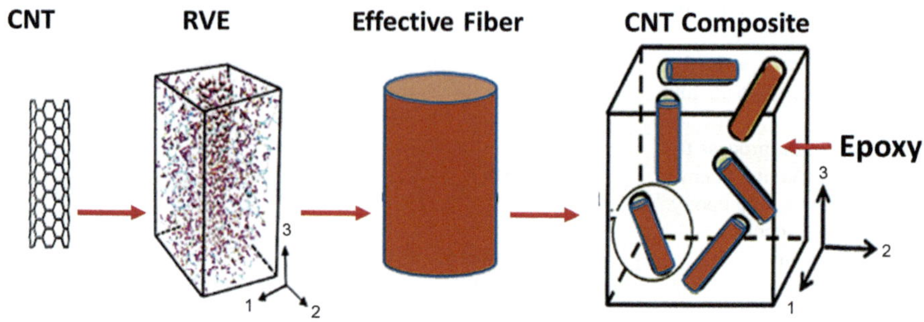

Figure 10.7 Multiscale modeling of CNT surrounded by a molecular epoxy that is homogenized to form RVE that is treated as an effective fiber. Effective fiber is then dispersed in the epoxy polymer matrix

effective elastic moduli of the nanocomposite are scaled up from the nanoscale to the microscale using Mori-Tanaka method. Figure 10.7 shows the steps involved in the hierarchical multiscale model examined in this case study.

All MD simulation runs were conducted using large-scale atomic/molecular massively parallel simulator (LAMMPS) using the consistent valence force field (CVFF). This force field has been used by other researchers to predict the mechanical properties of CNTs and epoxy polymers. A conjugate gradient algorithm was used to minimize the total potential energy of the initial configurations, while the velocity Verlet algorithm was used to integrate the equations of motion in all conducted MD simulations [10.5]. Periodic boundary conditions were imposed on all directions of the MD RVEs. Furthermore, only non-bonded interactions between the atoms of the embedded CNTs and the surrounding epoxy are considered (Figure 10.7). The non-bonded interactions between the atoms are represented by van der Waals (vdW) interactions and Coulombic forces. The cut-off distance for these non-bonded interactions was set to 14.0 Å.

Determination of the atomic-level elastic properties of the pure epoxy and the nanocomposite RVE was accomplished by straining the MD RVE which was then followed by constant-strain energy minimization. The averaged stress tensor of the MD RVE is defined in terms of virial stress definition; as follows,

$$\sigma = \frac{1}{V} \sum_{i=1}^{N} \left(\frac{m_i}{2} v_i^2 + F_i \, r_i \right) \tag{10.2}$$

where V is the volume of the RVE; v_i, m_i, r_i and F_i are the velocity, mass, position and force acting on the ith atom, respectively.

10.3.4 Interphase Layer and CNT Agglomeration

The structure of the epoxy matrix in the vicinity of the CNT surface differs from the bulk epoxy due to the formation of an ultra-thin interphase region [10.5]. This ultra-thin interphase layer consists of a highly packed crystalline polymer, which has higher elastic properties than the amorphous bulk polymer; illustrated in Figure 10.8.

In order to determine the thickness of the interphase layer, we performed MD simulations for a system consisting of a CNT of length 76.0 Å surrounded by 4 cross-linked 80:40 epoxy structures. The initial size of the periodic RVE was 150 Å × 150 Å × 150 Å. Subsequently, the volume of the simulation box was gradually reduced to 60 Å × 60 Å × 76 Å and equilibrated using the same steps as adopted in the pure epoxy case. The obtained values of the interphase thickness and the equilibrium separation distance were used to select the appropriate RVE sizes and in calculating the actual volume fraction of the embedded CNTs in the RVE [10.5].

Figure 10.9 shows two examples of dispersed CNTs in a polymer matrix, with the first depicting a random distribution, while the second shows agglomerated CNTs. The presence of agglomerated CNTs in the polymer matrix is due to vdW attractive forces at the atomistic scale. It is important to study the effect of these agglomerates upon the bulk properties of the nanoreinforced composite. This is carried out under the following subheadings.

10.3.5 MD of CNT-Reinforced Composite

Let us now investigate the influence of agglomerated CNTs on the bulk elastic properties of their nanocomposites (Figure 10.10). For such an investigation, three

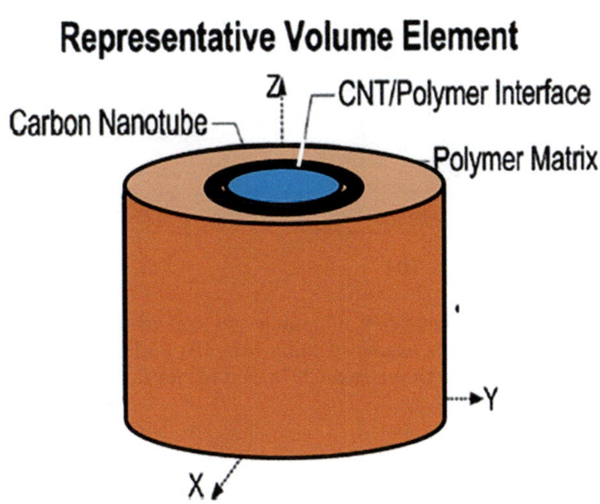

Figure 10.8 A schematic of CNT-Polymer depicting interphase/interface regions

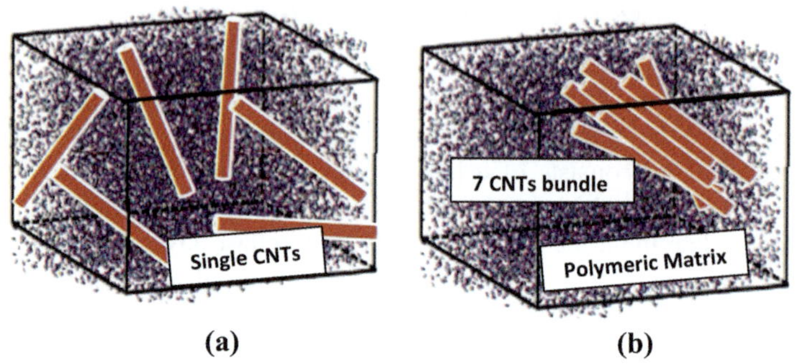

Figure 10.9 Illustrative examples of (**a**) randomly dispersed and (**b**) agglomerated CNTs in a polymeric matrix

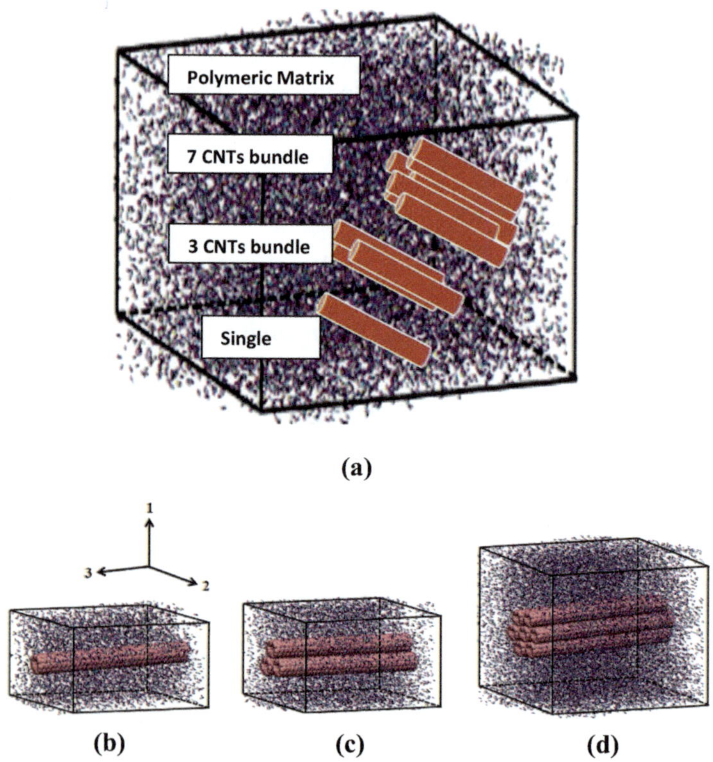

Figure 10.10 Illustrations of a single and agglomerated CNTs in a polymeric material: (**a**) general decription of cases examined using MD, (**b**) RVEs investigated of a single CNT, (**c**) RVE of a bundle of three agglomerated CNTs, and (**d**) RVE a bundle of seven agglomerated CNTs

RVEs were constructed to represent an epoxy matrix containing: (i) a single CNT (Figure 10.10b, (ii) a bundle of three CNTs (Figure 10.10c, and (iii) a bundle of seven CNTs (Figure 10.10d). Such RVEs are shown in Figure 10.10. The initial distance between the adjacent CNTs in the bundle was taken to be 3.4 Å, which is the assumed inter-tube separation distance in multi-walled CNTs.

The RVEs are assumed to be transversely isotropic with the third axis being the axis of symmetry. Therefore, only five independent material constants are required to fully define the elastic stiffness matrix. The constitutive relationship of the transversely isotropic RVE is given by the following constitutive law:

$$\begin{bmatrix} \sigma_{11} \\ \sigma_{22} \\ \sigma_{33} \\ \sigma_{23} \\ \sigma_{13} \\ \sigma_{12} \end{bmatrix} = \begin{bmatrix} C_{11} & C_{12} & C_{13} & 0 & 0 & 0 \\ C_{12} & C_{11} & C_{13} & 0 & 0 & 0 \\ C_{13} & C_{13} & C_{33} & 0 & 0 & 0 \\ 0 & 0 & 0 & C_{44} & 0 & 0 \\ 0 & 0 & 0 & 0 & C_{44} & 0 \\ 0 & 0 & 0 & 0 & 0 & (C_{11}-C_{12})/2 \end{bmatrix} \begin{bmatrix} \varepsilon_{11} \\ \varepsilon_{22} \\ \varepsilon_{33} \\ \varepsilon_{23} \\ \varepsilon_{13} \\ \varepsilon_{12} \end{bmatrix} \tag{10.3}$$

where σ_{ij} and ε_{ij} are the respective stress and strain components; with C_{ij} representing the elastic coefficients of the RVEs.

A series of MD simulations were conducted to determine the elastic moduli of the RVEs. The simulation box in each case was constructed by randomly placing the cross-linked epoxy structures around the CNT bundle. The size of the RVE in each case was adjusted to ensure that the CNT volume fraction in it remains the same at 6.5%, irrespective of the number of CNTs, as shown in Figure 10.10. The details of the three RVEs are summarized in Table 10.2.

In this case, it is necessary to determine five elastic constants. These constants can be obtained by subjecting the RVEs to five different loadings: simple longitudinal tension, transverse tension, in-plane tension, in-plane shear and out of-plane shear. The boundary and loading conditions applied to the RVE are listed in Table 10.3. A series of MD simulations were then carried out to determine the elastic properties of transversely isotropic RVEs reinforced with either an individual CNT or bundles of CNTs. Figure 10.10 demonstrates close views of the equilibrated RVEs. Table 10.4 summarizes the outcome of the MD simulations.

Table 10.2 Design parameters used in RVEs

Parameter	Single CNT	Bundle of three CNTs	Bundle of seven CNTs
CNT type	(5, 5)	(5, 5)	(5, 5)
Number of CNTs	1	3	7
Length of a CNT (Å)	73	73	76
CNT volume fraction	6.5%	6.5%	6.5%
RVE dimensions (Å^3)	31 × 31 × 76	55 × 55 × 76	88 × 88 × 76
Total number of DEGBA molecules	120	320	960
Total number of TETA molecules	60	160	480
Total number of atoms	7836	21,116	62,028

Table 10.3 Effective elastic coefficients of the RVEs and corresponding displacement fields

Elastic coefficients	Applied strains	Applied displacement
C_{11}	$\varepsilon_{11} = e$	$u_1 = ex_1$
C_{33}	$\varepsilon_{33} = e$	$u_3 = ex_3$
C_{44}	$\varepsilon_{23} = e/2$	$u_2 = \frac{e}{2}x_3, u_3 = \frac{e}{2}x_2$
C_{66}	$\varepsilon_{12} = e/2$	$u_1 = \frac{e}{2}x_2, u_2 = \frac{e}{2}x_1$
$K_{12} = \frac{C_{11}+C_{12}}{2}$	$\varepsilon_{11} = \varepsilon_{22}$	$u_1 = ex_1, u_2 = ex_2$

Table 10.4 Material properties of RVE containing either a single CNT or a bundle CNTs

RVE	RVE (V_f)	C_{11} (GPa)	C_{12} (GPa)	C_{13} (GPa)	C_{33} (GPa)	C_{44} (GPa)	C_{66} (GPa)
Single CNT	6.5%	**11.8**	0.71	**5.6**	**47.35**	3.16	2.35
Bundle of three CNTs		**10.5**	6.6	4.5	**37.4**	1.89	1.95
Bundle of seven CNTs		9.1	5.9	3.5	**29.2**	1.48	1.6

It may be observed from the results that the effective elastic properties of the RVEs fiber are significantly higher than those of the neat epoxy. It is also clear from the results that the CNT agglomeration reduce the reinforcing effect of the embedded CNTs, which eventually degrades the bulk elastic moduli of the nanocomposites. The axial elastic coefficients (C_{33}) of the RVEs containing bundles of three and seven CNTs decreased by some 21% and 38.5%, respectively, as compared with the RVE containing an individual CNT. The CNT agglomerates are also found to affect the transverse elastic coefficients of the RVEs. For example, the respective transverse elastic coefficient (C_{11}) of the RVEs containing bundles of three and seven CNTs decreased by some 11% and 23% compared with the RVE containing an individual CNT. Each RVE will be used as an effective fiber in the micromechanical model to calculate the effective elastic moduli of the nanocomposite at the microscale level.

10.3.6 Scaling Up Using Micromechanics

In this section, the elastic moduli of the pure epoxy and the nanocomposite RVE obtained from the MD simulations are used as an input to the micromechanics model in order to determine the elastic properties of the nanocomposite. With this process firmly established in several studies, the Mori-Tanaka method can be developed by utilizing the transversely isotropic elastic properties of the nanocomposite's RVE and the isotropic elastic properties of the pure epoxy matrix. In the case of a two-phase composite, where the inhomogeneity is randomly orientated in the three-dimensional space, the following relation can be used to determine the effective stiffness tensor [C] of the nanocomposite:

$$[C] = [C^m] + v_{RVE}\langle([C^{RVE}] - [C^m])\rangle\left([A][v_m[I] + v_{RVE}\langle[A]\rangle]^{-1}\right) \tag{10.4}$$

in which the mechanical strain concentration tensor [A] is given by

$$[A] = \left[[I] + [S^{RVE}]([C^m])^{-1}([C^{RVE}] - [C^m])\right]^{-1} \tag{10.5}$$

where $[C^m]$ and $[C^{RVE}]$ are the stiffness tensors of the epoxy matrix and the RVE, respectively; [I] is an identity matrix, v_m and v_{RVE} represent the volume fractions of the epoxy matrix and the RVE, respectively; and $[S^{RVE}]$ indicate the Eshelby tensor. It may be noted that the elastic coefficient matrix [C] directly provides the values of the effective elastic properties of the nanocomposite, where the RVE is aligned with the third axis.

In case of random orientations of CNTs, the terms enclosed with angle brackets in Equation (10.4) represent the average value of the term over all orientations defined by transformation from the local coordinate system of the RVE to the global coordinate system. The transformed mechanical strain concentration tensor for the RVEs with respect to the global coordinates is given by,

$$[\tilde{A}_{ijkl}] = t_{ip}t_{jq}t_{kr}t_{ls}[A_{pqrs}] \tag{10.6}$$

where t_{ij} are the direction cosines for the transformation and are given by,

$t_{11} = cos\phi\ cos\psi - sin\phi\ cos\gamma\ sin\psi,$
$t_{12} = sin\phi\ cos\psi + cos\phi\ cos\gamma\ sin\psi,$
$t_{13} = sin\psi\ sin\gamma,\ t_{21} = -cos\phi\ sin\psi - sin\phi\ cos\gamma\ cos\psi$
$t_{22} = -sin\phi\ sin\psi + cos\phi\ cos\gamma\ cos\psi,\ t_{23} = sin\gamma\ cos\psi,\ t_{31} = sin\phi\ sin\gamma,\ t_{32} = -cos\phi\ sin\gamma,$ and $t_{33} = cos\gamma$

Consequently, the random orientation average of the dilute mechanical strain concentration tensor $\langle[A]\rangle$ can be determined by using the following equation:

$$\langle[A]\rangle = \frac{\int_{-\pi}^{\pi}\int_{0}^{\pi}\int_{0}^{\pi/2}[\tilde{A}](\phi,\gamma,\psi)\ sin\gamma\ d\phi d\gamma d\psi}{\int_{-\pi}^{\pi}\int_{0}^{\pi}\int_{0}^{\pi/2} sin\gamma\ d\phi d\gamma d\psi} \tag{10.7}$$

where φ, γ, and ψ are the Euler angles. It may be noted that the respective averaged mechanical strain concentration tensors given by Equations (10.5) and (10.7) are used for the cases of aligned and random orientations of CNTs in Equation (10.4).

10.3.7 Effective Properties of Nanocomposite

Figure 10.11 shows the variations of the effective axial Young's modulus (E_{33}) of the nanocomposite against the CNT volume fraction. It can be observed that the effect of agglomeration of CNTs on E_{33} becomes more pronounced at higher CNT volume fraction (>2%). At 5 vol% of CNTs, the nanocomposite containing bundles of three and seven CNTs, the values of E_{33} are decreased by approximately 20% and 37% as compared with the nanocomposite reinforced with individual CNTs, respectively. Further details about analysis of results of this effort can be viewed in [10.5].

10.4 Case Study #II: Atomistic Based Continuum

Traditional continuum mechanics models are incapable of accurately characterizing the mechanical properties of CNT structures, because they lack the atomistic representation and appropriate constitutive relations that govern material behavior at that nanoscale. At that scale, traditional continuum mechanical concepts do not maintain their validity [10.7] and gross oversimplifications can arise from the use of a purely continuum model. Fully atomistic descriptions are also unrealistic. They are computationally intensive and are limited by the system sizes that they represent, because of the enormous number of degrees of freedom required to represent these systems. As stated earlier, even the use of state-of-the-art parallel supercomputers can only handle a limited number of atoms ($\sim 10^9$), corresponding to less than one cubic micron [10.7, 10.8]. Since various scales in the system depend on each other, it is necessary to formulate that system in terms of multiscale modeling.

In this case study, the atomistic-based continuum (ABC) multiscale modeling technique is used to predict the mechanical behavior of CNT-reinforced composites.

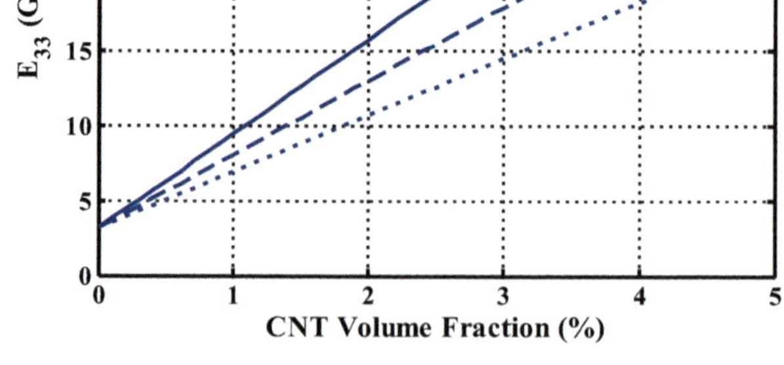

Figure 10.11 Variation of effective axial Young's modulus (E_{33}) of epoxy nanocomposite containing single and bundled volume fraction

It has the unique advantage of describing the atomic-property relations of the structure in a continuum framework. In this case, the continuum constitutive behavior is derived purely from atomistic principles. The deformation measures introduced in the model capture the underlying atomistic structure of the different phases considered. As a result, the significance of these measures is fundamentally different from those in the classical continuum theory. For example, in the case of CNT-reinforced systems, it is important to consider the atomic bonding and interaction between the carbon atoms in the CNT lattice. This is typically carried out at the atomistic scale using appropriate interatomic potentials. This leads to an atomistic constitutive law between the force and the displacement.

The strength of atomistic-based continuum techniques lies in their ability to avoid the large number of degrees of freedom typically encountered in coupled atomistic discrete modeling techniques. In the development of a computational multiscale model for the study of CNT-reinforced structures, it is necessary to consider the integration of atomistic and continuum scales, and the nonlinearity of the atomic interactions. In this case, we must include an atomistic description in the form of interatomic potentials. The selected interatomic potential should be capable of describing how the atomic interactions behave under different loading conditions and atomic environments. The general form of an interatomic potential, or equivalently, the energy of a molecule is given by:

$$E_{tot} = E_{bonded} + E_{non\text{-}bonded} \tag{10.8}$$

The bonded interactions can further be subdivided into,

$$E_{bonded} = E_s + E_b + E_t + E_I \tag{10.9}$$

where E_s, E_b, E_t *and* E_I correspond to bond stretching, bond angle-bending, torsion and inversion energies, respectively. The non-bonded interactions can also be subdivided into,

$$E_{non\text{-}bonded} = E_{vdW} + E_{es} + E_{hb} \tag{10.10}$$

where E_{vdW}, E_{es} and E_{hb} correspond to van der Waals forces, electrostatic forces and hydrogen bonding, respectively. The atomic bond deformation mechanisms are depicted in Figure 10.12; excluding the electrostatic and hydrogen bonding forces. Interatomic potentials are nonlinear in nature and this should be accounted for in the simulations.

It is with the above in mind that an atomistic-based continuum model is developed to investigate the nonlinear mechanical behavior of CNTs. The mechanical properties of the CNTs' structures are explored under tensile loading.

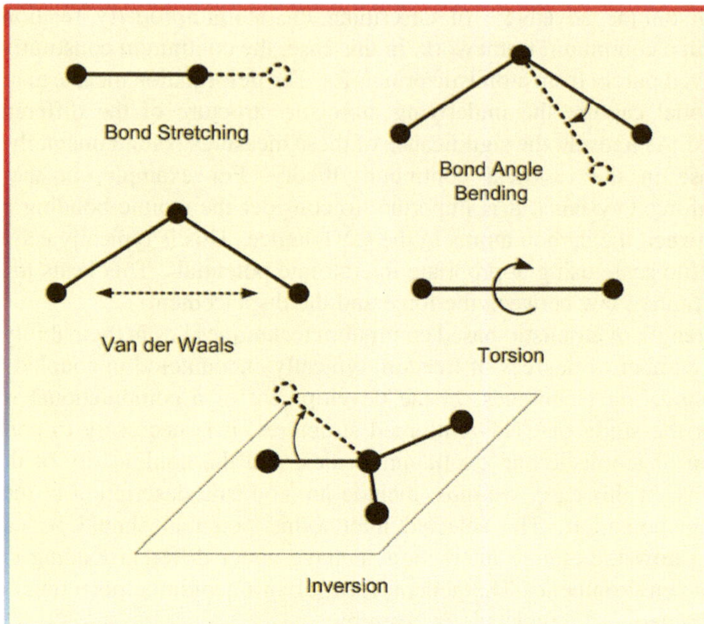

Figure 10.12 Kinematics of atomic bond deformation

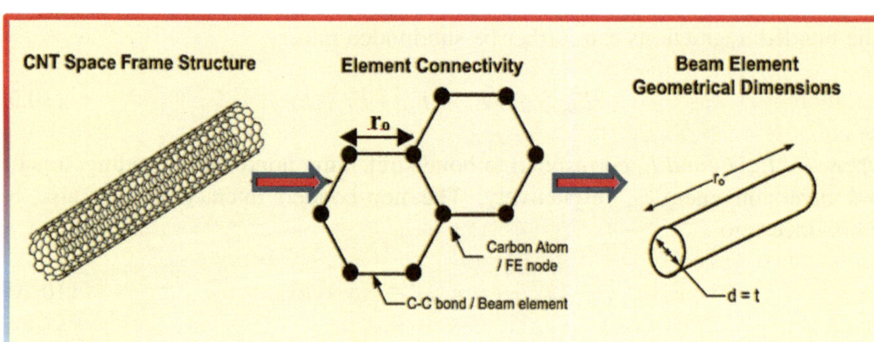

Figure 10.13 Carbon nanotube space frame structure

10.4.1 Modified Morse Interatomic Potential

In this second case study, two distinct CNT structures were investigated: (16,0) zigzag nanotube and (9,9) armchair nanotube, each having an approximate diameter of 1.2 nm. CNT was modeled as a space-frame structure, as depicted in Figure 10.13 The figure also illustrates a portion of the hexagonal lattice to demonstrate that beam elements were used in the representation of the individual C–C covalent bonds in the FE framework. As in traditional FE models, nodes were used to connect the beam elements to form the CNT structure. In this case the nodes represent the carbon atoms and are defined by the same atomic coordinates. The Modified Morse interatomic potential was used to describe the behavior of the atomic interactions in the CNT. This potential is well suited for characterizing the behavior of C–C covalent bonding, and has been used by several authors in both purely atomistic and atomistic-based continuum simulations of CNT fracture. This potential consists of two parts: a bond stretching component and an angle-bending component, as given by

$$E = E_s + E_b \tag{10.11}$$

$$E_s = D_e\left(\left[1 - e^{-\beta(r - r_0)}\right]^2 - 1\right) \tag{10.12}$$

$$E_b = \frac{1}{2}(\theta - \theta_0)^2\left[1 + k_{sextic}(\theta - \theta_0)^4\right] \tag{10.13}$$

where r is the current bond length, θ is the current angle of the adjacent bond, D_e is the dissociation energy, and β is a constant which controls the 'width' of the potential. This is the usual form of the Morse potential except that the angle-bending component has been added and the parameters are slightly modified to correspond with the Brenner potential for strains below 10% [10.8]. This is because the Brenner potential is considered more accurate and versatile due to its ability to account for bond hybridization and bonds with atoms other than carbon. It has been suggested that the angle-bending component can be neglected. However, this component has been found to contribute to the initial stability of the CNT. Without the addition of the angle-bending component, the CNT will become unstable and tends to collapse on itself. In this case study, nonlinear rotational spring elements were used to introduce the angle-bending component, while beam elements were used to represent the stretching component of the potential.

In order to ensure an appropriate connectivity and transmission of moments among adjacent beam elements, rotational springs were introduced between coincident nodes with coupled translational and rotational degrees of freedom, as depicted in Figure 10.14. To describe the behavior of the beam and rotational spring elements, we first derive material models for each that accurately represents the characteristics of the Modified Morse potential. First, by deriving the stretching potential (Equation 10.12) with respect to the change in bond length and by using the following relationship,

$$\varepsilon = (r - r_o)/r \tag{10.14}$$

we can arrive at the following expression:

$$F = 2\beta D_e\left(1 - e^{-\beta\varepsilon r}\right)e^{-\beta\varepsilon r} \tag{10.15}$$

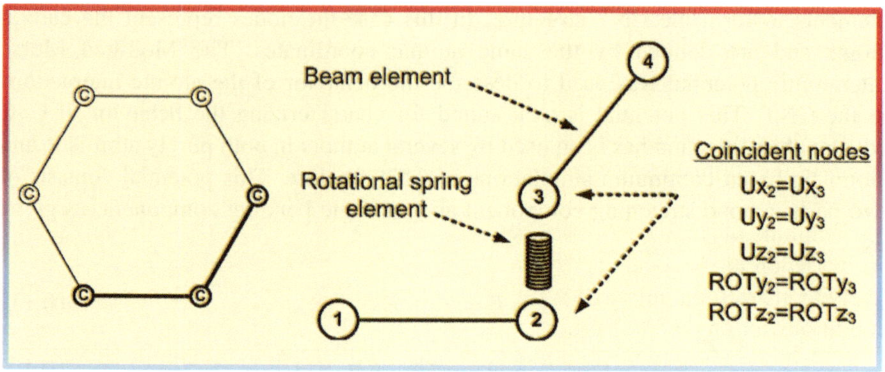

Figure 10.14 Coincident node locations in defining the rotational spring element

which represents the force required to stretch a C–C bond. The interlayer spacing of CNT walls was assumed to be 0.34 nm, which is widely regarded as the CNT's wall thickness. This implies that the diameter of each bond or beam element is also 0.34 nm. The corresponding cross-sectional area is determined to be 0.09079 nm^2. Utilizing this area and the above expression in Equation 10.14, we were be able to derive the stress–strain relationship for beam elements. This relationship is used to describe the behavior of the beam elements up to a critical strain of around 19% at which point the stiffness of the beams approaches zero and the bond will cease to exist. The initial stiffness of the beam elements was calculated to be 1.3 TPa. Similarly, recognizing that the derivative of the angle-bending component of the potential with respect to the change in rotation is equivalent to the moment required to bend neighboring bonds, we can arrive at the following expression:

$$M = k_\theta \Delta\theta \left[1 + 3k_{sextic}(\Delta\theta)^4\right] \tag{10.16}$$

Given the appropriate set of interatomic potentials and corresponding parameters, the other bond deformation mechanisms can also be included, i.e. dihedral angle torsion, out-of-plane torsion, van der Waals interactions and electrostatic interactions. However, several authors have indicated that these effects are negligible and play an insignificant role in the simulation of CNT structures.

10.4.2 Fracture Behavior of CNTs

It should be noted that the Modified Morse potential is not appropriate for describing the behavior of CNTs after the atomic bonds are broken. This is because it does not consider the formation of new bonds and the possibility that rehybridization and structural transformations may occur atomistically. However, it has been shown that

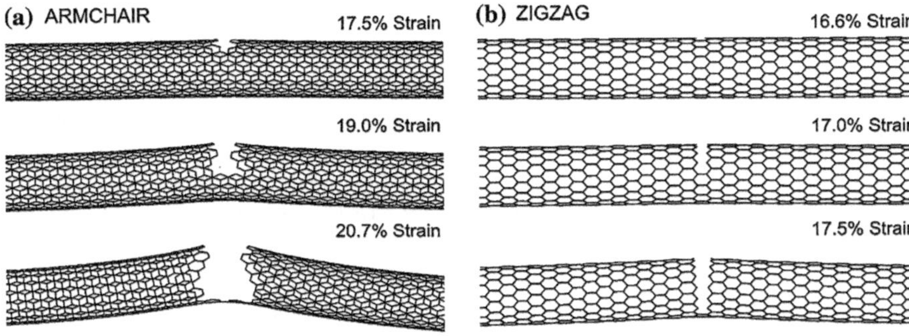

Figure 10.15 Nanotube fracture processes under tensile loading for (**a**) armchair and (**b**) zigzag configurations

the fracture strength of CNTs depends primarily on the inflection point of the potential. Therefore, since the inflection strain occurs well before the strain required to break the atomic bonds and form new ones, we are able to use the Modified Morse potential to obtain the tensile properties of CNTs up to their ultimate fracture.

Of interest to the present case study is the deformation mechanisms of CNTs under tensile loading. Figure 10.15a, b show the tensile fracture process for both armchair and zigzag nanotubes, respectively. For both cases the nanotubes simply elongate with little reduction in their cross-sectional area until they fracture. Fracture initiates at the weakened bond. For the case of the armchair nanotube, fracture occurs gradually. At 17.4% strain the a number of bonds are broken; leading to the successive elongation and failure of neighboring bonds. At the critical strain of 20.7%, all the bonds along the circumference of the nanotube break. In this case, the bonds fail in an irregular pattern and give rise to a large hole on the surface of the CNT. For the case of the zigzag nanotube, fracture occurs almost instantaneously. At a strain of 16.6% the weakened bond reaches its critical strain and break causing the adjacent bonds along the circumference of the nanotube to share the load. At a strain of 17.5%, nearly all the adjacent bonds break and the zigzag nanotube fractures. From this we can conclude that the zigzag nanotube exhibits brittle tendencies at fracture. Furthermore, the fracture process resembles the propagation of an ideal crack through the nanotube in a direction orthogonal to the nanotube longitudinal axis. For a detailed account of this work, the reader is encouraged to consult Ref. [10.8].

10.5 Case Study #III: Multiscale Modeling of Fracture Behavior of Polycrystalline Graphene

10.5.1 Polycrystalline CNTs

The ability to produce homogeneous defect-free CNTs represents a major challenge in terms of fabrication and cost. Specifically, the existence of grain boundaries (GBs) are ubiquitous in large diameter CNTs obtained by various large-scale growth methods and their effect on the mechanical properties of CNTs is not fully explored. Understanding the influence of the grain size and the GB orientation upon the mechanical behavior of CNTs is crucial for their functional and structural applications.

The significance and novelty of the current work lies in its ability to establish the mechanical properties and the fracture behavior of CNTs containing grain boundaries of varied size, orientation and structure to an applied uniaxial load using comprehensive molecular dynamics simulations. A specially developed program, using Voronoi tessellation method and Delaunay triangulation, was implemented to generate the atomistic descriptions of the considered polycrystalline CNTs with a variety of GB morphology [10.9].

10.5.2 Polycrystalline CNTs Using Voronoi Tessellation

In this third case study, we created polycrystalline CNTs by rolling graphene sheets containing GBs. The atomic structures of the graphene sheets with randomly distributed grain sizes and shapes were created by using Voronoi tessellation technique. First, based on the required average grain size, a specific number of points were randomly placed inside a square. Second, Delaunay triangulation was performed to obtain a set of spatial coordinates that forms a minimum distance with each nucleus from the randomly created grain nuclei. The circumcenter of neighboring triangles that was generated in the last step was connected to create GBs and form a particular grain for each nucleus in our simulation cell. Although the grain nuclei were placed randomly, the distribution of the grain sizes (grain area) can be fitted following a gamma distribution. This method has been widely used to model the grain structures in polycrystalline graphene sheets. Figure 10.16 demonstrates a typical grain size distribution in polycrystalline graphene sheet generated from the Voronoi construction. Figure 10.17 shows the systematic steps involved in creating a graphene sheet with randomly distributed GBs, which was then rolled to create a polycrystalline CNT. We generated a collection of polygons separated by planar cell walls perpendicular to lines connecting neighboring nucleation sites, as shown in Figure 10.17a. A polycrystalline graphene sheet was constructed by filling each cell with randomly oriented graphene domains, where the atoms adjacent to the planar cell walls represent the GBs. The initial C–C bond length in the graphene sheets was set to 1.41 Å, which has been confirmed experimentally. Several polycrystalline graphene sheets with dimensions of 30 nm × 30 nm and an average grain size ranging from 5 to 10 nm were constructed (see Figure 10.17b). Several regions at different locations inside the polycrystalline graphene with dimensions of 80 Å × 80 Å were selected and extracted to create the basic structure that is rolled to form CNTs, as shown in Figure 10.17c. Pentagons and heptagons were created at the boundaries of the grains in each extracted region and then the energy of a graphene system was minimized using the conjugate gradient method to obtain its

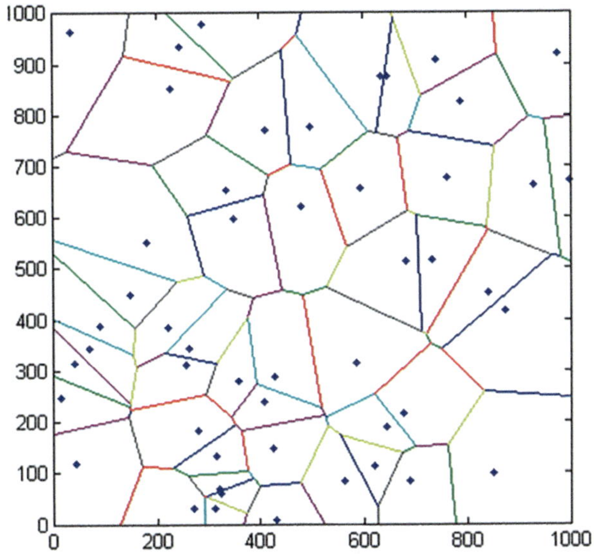

Figure 10.16 An example of randomly distributed grains with different sizes and shapes created using Delaunay triangulation and Voronoi tessellation techniques

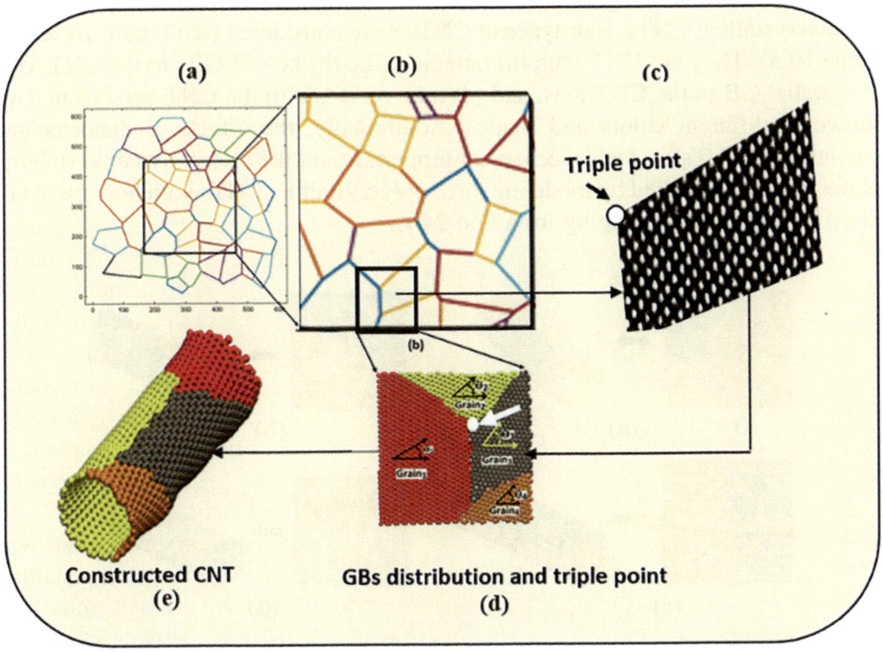

Figure 10.17 Development of a polycrystalline CNT: (**a**) A square graphene sheet filled with randomly distributed grains/grain boundaries with different sizes and shapes created using Voronoi tessellation technique. (**b**) An enlarged view of the graphene sheet with selected 4 grains. (**c**) Shows a sector of crystalline group of GBs. (**d**) Depicts crystallographic directions of the grains and the corresponding angles that are determined with respect to the horizontal axis. (**e**) Depicts the wrapping of the sheet to create a polycrystalline CNT with varied grain boundaries orientations and triple point

optimized structure. The obtained structure was then annealed by equilibrating it at room temperature in the constant temperature and volume canonical (NVT) ensemble over 100 ps with 0.5 fs time step using the velocity Verlet algorithm. The annealing process allows the rearrangement of C–C atomic positions at the GBs. The annealed structures were further equilibrated using the isothermal–isobaric (NPT) ensemble over another 100 ps at 300 K and 1 atm. to reach the atmospheric pressure.

In polycrystalline graphene, the GBs have 1D interface between two domains of a material with different crystallographic orientations. The structure of the grain is defined by its orientation angle Θ_i, where θ_i is the angle between the zigzag direction (n,0) and the horizontal axis We can also define the tilt angle φ_i of a GB as the joint between two mismatched angles (Θ_L and Θ_R) of adjacent grains.

Accordingly, the chiral vectors for the left (n_L, m_L) and the right (n_R, m_R) grains along the GB can be easily obtained.

In principle, different arrangements can develop along the GB. Formation of pentagon and heptagon defects along the GB can develop and defects such as quadrilaterals, octagons, and nonagons can also form due to higher formation energy, and are rarely observed in experiments. Therefore, we consider pentagon and heptagon defects in this work and ruled out all other types of defects.

To study the effect of grain shape, size and structure on the mechanical properties of polycrystalline CNTs, four types of CNTs were considered (see Figure 10.18 and Table 10.5). They are CNT with (a) random GBs, (b) helical GBs to the CNT axis, (c) parallel GB to the CNT axis, and (d) transverse GB to the CNT axis. Grains are shown in different colors and shapes. Additionally, the effects of diameter and curvature of CNTs on their mechanical properties and the thermodynamic stability of the GBs were studied by modeling three polycrystalline CNTs with horizontal GB of different diameters ranging from 9 to 24 Å.

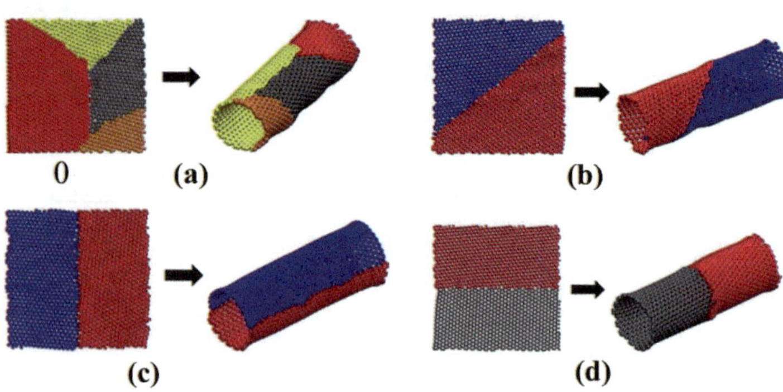

Figure 10.18 Different cases of polycrystalline CNTs: (**a**) CNT with random GBs. (**b**) CNT with helical GBs. (**c**) Parallel GB to the CNT axis. (**d**) Transverse GB to the CNT axis

Table 10.5 Orientation of grains in polycrystalline CNTs

Type of GBs	Grain number	Sample 1, θ_i (°)	Sample 2, θ_i (°)	Sample 3, θ_i (°)
Random GBs	1	5.0	1.5	2.0
	2	26.6	10.5	3.2
	3	18.7	23.0	0
	4	24.5	42.7	45.6
Helical GB	1	6.4	1.8	9.4
	2	7.4	22.2	9.6
Parallel GB	1	5.6	17.3	0.4
	2	27.6	14.7	12.9
Transverse GB	1	5.6	17.3	30.3
	2	27.6	14.7	12.9

10.5.3 Molecular Dynamics Simulations

All MD simulation runs were conducted using LAMMPS and the molecular interactions in CNTs are described in terms of Adaptive Intermolecular Reactive Empirical Bond Order (AIREBO) force fields. This interatomic potential accurately captures the interaction between carbon atoms as well as bond breaking and reforming. The cutoff parameter describing the short-range C–C interaction was selected to be 2.0 Å in order to avoid spuriously high bond forces and nonphysical results at large deformation. The interlayer separation distance between graphite layers, 3.4 Å, was taken as the effective thickness of the CNT.

The sequence of the conducted MD simulations is as follows. First, the initial structures of the CNTs were created by rolling the graphene sheets. The grains of the rolled CNT were then heated up to 300 K and equilibrated for 100 ps in the NVT ensemble to enable the formation of the pentagon-heptagon rings along the GBs. The obtained structure was then analyzed and the regions with high atomic density that failed to yield a perfect geometry were reconstructed by removing few carbon atoms to reduce the atomic density to the normal levels. The previous steps were repeated until perfect boundaries were obtained. This was then followed by energy minimization simulation using the conjugate gradient algorithm to obtain CNTs with optimized configurations. The minimized structure of a CNT was considered to be optimized once the change in the total potential energy (PE) of the system between subsequent steps is less than 1.0×10^{-10} kcal/mol. Subsequently, MD simulations were performed in the NVT ensemble at an interval of 0.5 fs for 50 ps to equilibrate the CNT structures, where the velocity Verlet algorithm was used to integrate the equations of motion. Finally, a uniaxial tension simulations were conducted by fixing one end of the CNT and applying a constant speed of 0.1 Å/ps at the other end. This corresponds to a nominal strain rate in the conducted MD simulations of 1.25×10^9 s^{-1}. The atoms of the CNT between the two ends were equilibrated during the simulation in the NVT ensemble at 300 K. The reaction forces on the fixed end atoms were calculated throughout the simulation and then used to obtain

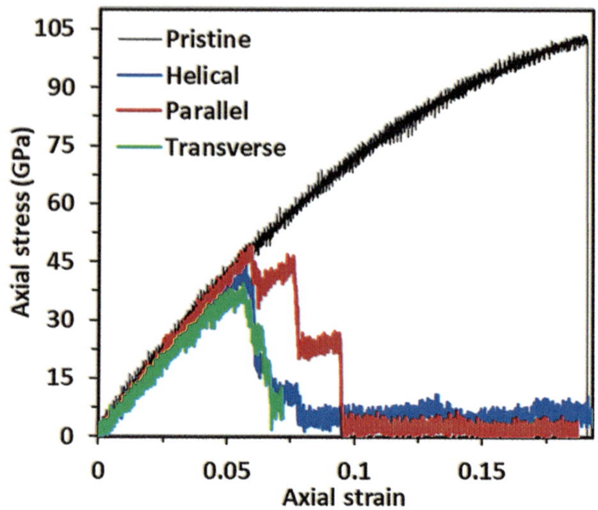

Figure 10.19 Averaged stress-strain curves of pristine and polycrystalline CNTs with helical, parallel and transverse grain boundaries (see Figure 10.18 for details)

the stress-strain curve. The tensile simulation was stopped when the CNT was completely fractured. Young's modulus of the CNT was calculated from the initial slope of the stress-strain curve and the averaged stresses were calculated assuming a uniform tensile stress distribution along the cross-sectional area of the CNT.

10.5.4 Sample Results

Figure 10.19 shows the constitutive behavior of the polycrystalline CNTs with helical GB, parallel GB with respect to CNT axis and transverse GB relative to CNT axis. The averaged Young's moduli (slope of stress strain in the elastic range) for CNTs is shown in Figure 10.19. Table 10.6 summarizes the obtained results. This table also lists the fracture strength, failure strain and PE for all cases.

The results of Table 10.6 show a noticeable reduction in Young's modulus of the polycrystalline CNTs in comparison with the pristine single-crystal CNT for which Young's modulus (E) is approximately 936 GPa. This reduction results from the presence of GBs which manifest themselves in highly localized defects and fracture initiation sites.

The fracture of polycrystalline CNTs was explored by performing MD simulations using realistic finite-grain-size models, emphasizing the role of their orientations. GBs may act as initiation sites and the potential precursor of their ultimate fractures. Therefore, we hypothesize that C-C bond along the GBs would ultimately lead to the fracture of CNTs, serving as accumulators of prestrain that consequently lead to a reduction in the CNT strength.

Figure 10.20 shows snapshots of the polycrystalline CNT with random GBs at different stages of the tensile test simulation illustrating the initiation and propagation of cracks along GBs that lead eventually to the complete fracture of the CNT.

Table 10.6 Comparisons between the mechanical properties and potential energy (PE) of different polycrystalline CNTs

CNT Type		Young's Modulus (GPa)	Fracture strength (GPa)	Fracture strain (%)	PE/atom (eV/atom)
Pristine (19,19)		936.2	103.2	18.9	−7.310
Random GBs	Sample #1	739.4	50.1	−7.24	−7.240
	Sample #2	719.3	54.2	−7.207	−7.207
	Sample #3	657.3	41.6	−7.222	−7.222
	Average	**705.3**	**48.6**	−7.223	**−7.223**
Parallel GBs	Sample #1	868.2	49.9	−7.236	−7.236
	Sample #2	832.5	68.6	−7.261	−7.261
	Sample #3	819.9	52.8	−7.261	−7.261
	Average	**840.2**	**57.1**	−7.253	**−7.253**
Helical GB	Sample #1	802.1	44.1	−7.264	−7.264
	Sample #2	785.8	44.8	−7.276	−7.276
	Sample #3	868.9	45	−7.278	−7.278
	Average	**818.9**	**44.6**	−7.273	**−7.273**
Transverse GB	Sample #1	712.5	39.3	−7.219	−7.219
	Sample #2	837.6	44.3	−7.245	−7.245
	Sample #3	718.2	39.5	−7.267	−7.267
	Average	**756.1**	**41**	−7.244	**−7.244**

Figure 10.20a shows a snapshot of the CNT prior to applying a uniaxial deformation of 7%, while Figure 10.20b shows a localized deformation at the triple junction at which three grain boundaries meet just prior to crack initiation. The out-of-plane ripples at the different grain boundaries undergo a straightening process during subsequent tension as shown in Figure 10.20b. Figure 10.20c shows that the fracture initiates at the triple junction of a grain boundary due to the associated high stress concentration at that junction. Once a crack initiation site is established, the crack extends in an intergrannular manner depending on the orientations of the GBs. Figure 10.20d shows a completely fractured CNT with some dangling carbon atoms.

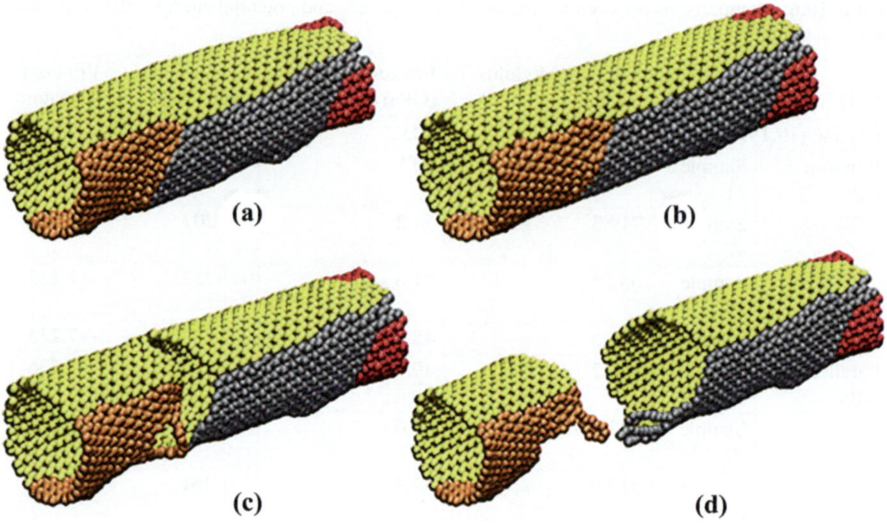

Figure 10.20 Failure progression of a CNT with crystalline grain boundaries under tensile deformation: (**a**) Undeformed CNT showing a tripple junction, (**b**) CNT showing localized deformation at the tripple junction where three orientations of GBs meet. (**c**) Crack nucleates at the tripple point and propagates directionally in least resistance path (**d**) CNT ultimate fracture with increased loading with carbon atoms dangling

References

10.1 S.A. Meguid (ed.), *Advances in Nanocomposites: Modeling* (Characterization and Applications, Springer, 2016). https://doi.org/10.1007/978-3-319-31662-8
10.2 S.A. Meguid, G.J. Weng, *Micromechanics and Nanomechanics of Composite Solids* (Springer, 2018). https://doi.org/10.1007/978-3-319-52794-9
10.3 Y.W. Kwon, D.H. Allen, R. Talreja, *Multiscale Modeling and Simulation of Composite Materials and Structures* (Springer, 2008). https://doi.org/10.1007/978-0-387-68556-4
10.4 E.B. Tadmor, M. Ortiz, R. Phillips, Quasicontinuum analysis of defects in solids. Philosophical Magazine A **71**, 1529–1563 (1996). https://doi.org/10.1080/01418619608243000
10.5 A.R. Alian, S.I. Kundalwal, S.A. Meguid, Multiscale modelling of carbon nanotube epoxy composites. Polymer **70**, 149–160 (2015). https://doi.org/10.1016/j.polymer.2015.06.004
10.6 J.D. Littell, C.R. Ruggeri, R.K. Goldberg, G.D. Roberts, W.A. Arnold, W.K. Binienda, Measurement of epoxy resin tension, compression, and shear stress–strain curves over a wide range of strain rates using small test specimens. J. Aerosp. Eng. **21**, 162–173 (2008). https://doi.org/10.1061/(ASCE)0893-1321(2008)21:3(162)
10.7 J.M. Wernik, S.A. Meguid, Recent developments in multifunctional nanocomposites using carbon nanotubes, Appl. Mech. Rev. **63**, 050801 (40 pages) (2010). https://doi.org/10.1115/1.4003503
10.8 J.M. Wernik, S.A. Meguid, Atomistic-based continuum modeling of the nonlinear behavior of carbon nanotubes. Acta Mech. **212**, 167–179 (2010). https://doi.org/10.1007/s00707-009-0246-4
10.9 A.R. Alian, S.A. Meguid, S.I. Kundalwal, Unraveling the influence of grain boundaries on the mechanical properties of polycrystalline carbon nanotubes. Carbon **125**, 180–188 (2017). https://doi.org/10.1016/j.carbon.2017.09.056

Index

S. A. Meguid, *Atomistic and Continuum Fracture Mechanics of Solids*,
https://doi.org/10.1007/978-3-031-56085-9

GPSR Compliance

The European Union's (EU) General Product Safety Regulation (GPSR) is a set of rules that requires consumer products to be safe and our obligations to ensure this.

If you have any concerns about our products, you can contact us on ProductSafety@springernature.com

In case Publisher is established outside the EU, the EU authorized representative is:

Springer Nature Customer Service Center GmbH
Europaplatz 3
69115 Heidelberg, Germany

Batch number: 10682290

Printed by Printforce, the Netherlands